BEHAVIOURAL ECOLOGY
OF FISHES

Ettore Majorana International Life Sciences Series
Edited by A. Zichichi, Director, Ettore Majorana Centre for Scientific Culture.
A series of books from courses and workshops held at the Ettore Majorana Centre.

This book is part of a series. The publisher will accept continuation orders which may be cancelled at any time and which provide for automatic billing and shipping of each title in the series upon publication. Please write for details.

BEHAVIOURAL ECOLOGY OF FISHES

Edited by

Felicity A. Huntingford
Department of Zoology, University of Glasgow, Great Britain

and

Patrizia Torricelli
Dipartmento di Biologia e Fisiologia,
Università di Parma, Italy

**Proceedings of a conference held at
the Ettore Majorana Centre for Scientific Culture,
Erice, Italy 30 September–5 October 1991**

CRC Press
Taylor & Francis Group
Boca Raton London New York

CRC Press is an imprint of the
Taylor & Francis Group, an **informa** business

First published 1993 by Harwood Academic Publishers

Published 2019 by CRC Press
Taylor & Francis Group
6000 Broken Sound Parkway NW, Suite 300
Boca Raton, FL 33487-2742

© 1993 by Taylor & Francis Group, LLC
CRC Press is an imprint of Taylor & Francis Group, an Informa business

First issued in paperback 2019

No claim to original U.S. Government works

ISBN 13: 978-0-367-44960-5 (pbk)
ISBN 13: 978-3-7186-5346-1 (hbk)

Visit the Taylor & Francis Web site at
http://www.taylorandfrancis.com

and the CRC Press Web site at
http://www.crcpress.com

Library of Congress Cataloging-in-Publication Data

Behavioural ecology of fishes / edited by Felicity A. Huntingford and
    Patrizia Torricelli.
        p. cm. -- (Ettore Majorana international life sciences
    series, ISSN 0896-4343 ; v. 11)
        "Proceedings of a conference held at the Ettore Majorana Centre
    for Scientific Culture, Erice, Italy, 30 September - 5 October
    1991."
    Includes bibliographical references and index.
    ISBN 3-7186-5346-X.
    1. Fishes--Behavior--Congresses. 2. Fishes--Ecology--Congresses.
I. Huntingford, Felicity. II. Torricelli, Patrizia, 1949- .
III. Title: Behavioral ecology of fishes. IV. Series.
QL639.3.B45 1992
597'.051--dc20
                                                92-44345
                                                CIP

CONTENTS

INTRODUCTION TO THE SERIES

The purpose of this series of Life Sciences is to provide the latest information on the research and development taking place at institutions throughout the world.

The courses, held at the Ettore Majorana Centre for Scientific Culture in Erice, Sicily, are invaluable forms for direct communication between specialists.

It is hoped that the dissemination of information via this series will provide international reference works for researchers, practitioners and students.

Antonino Zichichi

PREFACE

This volume is based on the proceedings of a conference of the same name held at the Ettore Majorana Centre, Erice, Sicily, in October 1991. The conference was organised because it was felt that the discipline of behavioural ecology is at something of a turning point, with the classic explanatory theories being reformulated in the light of new empirical evidence and with new techniques and areas of interest becoming apparent. For this reason, an overview of the subject was felt to be timely. Teleost fishes are, in many ways, extremely suitable subjects for research into behavioural ecology, and many important studies have used this group. Therefore, we felt that the current state of behavioural ecology and the prospects for the future could be well documented by considering work on fishes, allowing a reasonably concise but complete treatment of the topic. Four main areas of interest were identified, and key researchers invited to give talks in which they first reviewed the current status of their research field and then illustrated important points with reference to their own work.

The meeting duly went ahead and was generally agreed to be informative, stimulating and productive. Good conferences do not necessarily make good books, but it was felt that the talks made a sufficiently coherent whole to justify producing a publication along the same lines. The presentations at the meeting have therefore been altered and refereed to form the nineteen articles of this volume. The organisers and participants of the conference would like to thank the Ettore Majorana Foundation and the staff of Ettore Majorana Centre for both financial and logistic support for the meeting and Professor Danilo Mainardi, director of the School of Ethology at the Ettore Majorana Centre, for his support.

Felicity A. Huntingford
Patrizia Torricelli

BEHAVIOUR, ECOLOGY AND TELEOST FISHES

FELICITY A. HUNTINGFORD

Department of Zoology, University of Glasgow, Glasgow G12 8QQ, UK

During the 1960s, a number of separate fields of research into the general question of the adaptive significance of behaviour could be identified. Within the traditions of classical ethology, people were conducting simple, direct experimental studies of the survival value of behaviour (as pioneered by Tinbergen, 1951). For example, Krebs *et al.,* 1972 demonstrated experimentally that flocking in birds confers foraging advantages, and Neill and Cullen (1974) did the same for the anti-predator advantages of schooling in fishes. At the same time, in a rather different context, an increasing number of comparative studies, especially in birds (Lack, 1968; Crook, 1965) and primates (Crook and Gartlan, 1966), were beginning to show how social organisation can be related to ecological factors such as predation risk and food distribution. Also, people were seeking explanations for apparently altruistic behaviour, such as sacrificing personal reproductive output in favour of another individual or using ritualised aggression when fierce fighting would carry the day. Dissatisfaction with explanations couched in terms of group selection (Wynne-Edwards, 1962) led to recognition of the importance of inclusive fitness, and the development of the theory of kin selection (Williams, 1966; Hamilton, 1964; Maynard Smith, 1964, and see Dawkins, 1976). In the context of aggression, games theory analyses were used by Maynard Smith and collaborators (Maynard Smith and Price, 1973; Maynard Smith and Parker, 1976) to show how frequency-dependent costs and benefits are sufficient to explain the existence of ritualised fighting. This was a special case of the economic approach pioneered by MacArthur (e.g. MacArthur and Pianka, 1966), whereby models incorporating the costs and benefits of behaviour were used to identify behaviour patterns that maximise fitness and so to elucidate the ways in which natural selection acts on behavioural traits.

These various strands came together in the early 1970s as a separately-identifiable discipline that was given the name of "behavioural ecology" (Krebs and Davies, 1978). Like any successful branch of science, behavioural ecology has gone through a process of development and change. Initially, a number of new and relatively simple ideas and concepts were developed that offered exciting answers to important and difficult questions. These, inevitably, fell victim to their own success as the research that they stimulated (both theoretical and empirical) revealed their weaknesses. The next stage was one of seeking more satisfactory (if less charismatic) answers.

The development of techniques and ideas in behavioural ecology can be traced by looking at the three major multi-author books on this subject edited by John Krebs and Nick Davies in 1978, 1984 and 1991. The 1978 edition defined the main areas of interest at the time (space use and territoriality, foraging, predator avoidance, mating systems, mate choice and sexual selection, the evolution of sociality and cooperative breeding, life history strategies) and the main analytical techniques (direct experimentation, comparative studies, optimality modelling and games theoretical analyses). The 1984 edition covered the same broad areas of research, using the same techniques, but marked the end of what Krebs and Davies called the 'romantic era' of behavioural ecology. On more careful scrutiny, and in particular, in the light of hard empirical data, theories that seemed to provide satisfactory answers were found to be oversimple (optimal foraging theory is a case in point, Krebs and McCleery, 1984), and techniques that seemed simple (like the comparative approach, Clutton-Brock and Harvey, 1984), turned out to be fraught with problems. The third edition (1991) looks very different, reflecting in part a number of changes in emphasis within behavioural ecology, some of which are spelled out below:

1. As Krebs and Davies (1991) note, there has been a major rise of interest in parasitism and behaviour, particularly in the idea of sexual displays as indicators of disease resistance (initially proposed by Hamilton and Zuk, 1982).

2. Optimality models have been extended to take into account the existence of conflicting demands that have to be traded-off against each other. In optimal foraging theory, for example, the need to reduce predation risk means that simply maximising rates of food intake (or any other food-related currency) will not necessarily maximise fitness (e.g. Lima and Dill, 1990).

3. Classical optimality models have also been extended and greatly strengthened by introduction of stochastic dynamic modelling (Houston *et al.,* 1988; Mangel and Clarke, 1988; Krebs and Kacelnik, 1991). This helps to bridge the gap between behaviour now and fitness later (dealt with in classical optimality modelling by simply assuming that the currency being considered ultimately relates to fitness). It does so by incorporating the animal's physiological state into the model and allowing this to change continually as a result of prior decisions, taken in an unpredictable environment.

4. Although by definition, behavioural ecologists are concerned with the consequences of behaviour, it is becoming increasingly clear that to understand how selection acts on behaviour it is often necessary to consider mechanisms as well. In the field of foraging theory, animals use rules of thumb to generate responses that approximate to, but may not attain, theoretically optimal behavioural solutions; recognition of this fact has led to increasing interest in the proximate cues that elicit feeding and in the role of learning in foraging. It is also becoming clear that structural and physiological traits may constrain the range of options open to an animal and, where this is so, need to be incorporated into the models of behavioural ecologists.

5. The development of DNA fingerprinting means that genetic relatedness can be precisely defined; this has allowed direct tests of a number of theories in behavioural ecology about the role of relatedness of companions in behavioural decision making and has revised previous ideas about the evolution of mating systems.

6. As the name of their discipline promises, behavioural ecologists are beginning to bridge the gap between the analysis of behavioural adaptations at the individual level and processes in population and community ecology.

Thus the field of behavioural ecology is at something of a turning point, with new or more sophisticated ideas and techniques offering the potential of much more powerful analyses of the way selection act on behaviour. The aim of this volume is to illustrate these developments, but why fishes?

The comparative approach lies at the roots of behavioural ecology, and one of the strengths of this discipline is the fact that it draws on examples from across the Animal Kingdom. Therefore it might seem artificial to try to demonstrate the state of the art using just one animal group. On the other hand, the developments outlined above mean that the generality of earlier models in behavioural ecology has largely been lost. We may be able to find rules that can be applied in a more restricted area by concentrating on a single major group of animals with the same broad body design. In addition, teleost fishes have a number of convenient features that make them particularly suitable subjects for behavioural ecology; these are discussed in the following article, by Barlow. So many key studies have been done on this group.

To illustrate the importance of studies using fishes in the development of behavioural ecology, fishes were the subjects of several early demonstrations of the feeding (Robertson et al., 1976) and anti-predator (Neill and Cullen, 1974) advantages of group life. They also provided tests of classical optimality theory (Werner and Hall, 1974; Kislaliaglou and Gibson, 1976) and exemplified the use of rules of thumb during foraging (O'Brien et al., 1976) and of physiological constraints on diet selection (Werner, 1977). Studies using fish as subjects have tested and expanded Fretwell and Lucas's Ideal Free Distribution (Fretwell and Lucas, 1970; Milinski, 1979) and have elucidated the trade off between habitat choice, feeding and predator avoidance (Werner et al., 1983; Milinski and Heller, 1978). Fishes have also been used to demonstrate the way in which parasites manipulate the behaviour of their intermediate host in such a way as to promote transmission (Milinski, 1985). In the field of aggression research, cichlid fish were used to provide an important test of the prediction that animals should not signal intentions during fights (Jakobsson et al., 1979); this prediction arose from the War of Attrition model (Maynard Smith and Price, 1973), and was in contrast to the expectations of classical ethological theory. Many other experimental tests of the predictions of games theory have used fishes (e.g. Turner and Huntingford, 1986). A recent example is the use of cichlid fish again to test the predictions of one of the latest generation of games theory models, namely Enquist et al. (1990). A number of elegant studies of the costs and benefits of territoriality have used fish as subjects (Ebersole, 1980; McNicol and Noakes, 1984). In the context of reproduction, fish have been used in many studies of mate choice (for example, Downhower et al., 1983) and in investigations of the evolution of different patterns of parental care (Sargent and Gross, 1986). Indeed, sticklebacks gave their name to one of the strategies in Maynard Smith's model (1977) of the evolution of parental care. Finally, some of the best known and most intensively studied examples of alternative reproductive strategies come from the literature on fishes (Jones, 1959; Gross and Charnov, 1980).

The articles in this volume reflect the many contributions that students of the behaviour of fishes have made to behavioural ecology, and also illustrate the new developments in the discipline that were outlined above. Thus Bakker and Milinski (this volume) describe experimental studies of the role of parasites in the assessment of mate quality. An application of stochastic dynamic programming is presented by Hart and Gill. Examples of the increasing interest in the behavioural mechanisms that underlie strategic behavioural decisions (including learning and other developmental processes) are provided by Baerends, Losey, Csanyi and Doka, and Huntingford. The importance of morphological and physiological constraints in foraging and diet choice is illustrated by Gallis and also by Kaiser and Hughes. Among other things, Magurran and Seghers illustrate how an understanding of genetic relatedness (derived in this case from the allozyme analysis) can elucidate the evolution of behavioural traits. Links between behaviour and more ecological issues such as reproductive investment and life history strategies are considered by Hutchings, Magnhagen, and Metcalfe. Questions of resource defence and territory economics are addressed by Grant and by Barlow, while issues concerning reproductive behaviour, such as competition for mates, the evolution of size dimorphism and patterns of parental care are covered by Toricelli *et al.*, Bisazza, and Rosenqvist.

References

Clutton-Brock, T. H. and Harvey, P. H. (1984). Comparative approaches to investigating adaptation. In: *Behavioural Ecology: An Evolutionary Approach*, 2nd edn. (Ed. by J. R. Krebs and N. B. Davies), pp. 7–29. Oxford: Blackwell Scientific.

Crook, J. H. (1965). The adaptive significance of avain social organisation. *Symp. Zool. Soc. Lond.*, **14**, 181–218.

Crook, J. H. and Gartlan, J. S. (1966). Evolution of primate societies. *Nature, Lond.*, **210**, 1200–1203.

Dawkins, R. (1976). *The Selfish Gene*. Oxford: Oxford Univ. Press.

Downhower, J. F., Brown, L., Pedersen, R. and Staples, G. (1983). Sexual selection and sexual dimorphism in mottled sculpins. *Evolution*, **37**, 96–103.

Ebersole, J. P. (1980). Food density and territory size: an alternative model and a test on the reef fish *Eupomacentrus leucostrictus*. *Amer. Natl*, **115**, 492–505.

Enquist, M., Leimar, O., Ljungberg, T., Mallner, T. and Segerdahl, N. (1990). A test of the sequential assessment game: fighting in the cichlid fish *Nannocara anomala*. *Animal Behaviour*, **40**, 1–15.

Fretwell, S. D. and Lucas, J. R. (1970). On territorial behaviour and other factors influencing habitat distribution in birds. I. Theoretical development. *Acta Biotheoretica*, **19**, 16–36.

Gross, M. R. and Charnov, E. L. (1980). Alternative male life histories in bluegill sunfish. *Proc. Natl. Acad. Sci. USA*, **77**, 6937–6940.

Hamilton, W. D. 1964. The genetical theory of social behaviour. I, II. *J. Theoretical Biology*, **7**, 1–52.

Hamilton, W. D. and Zuk, M. (1982). Heritable true fitness and bright birds: a role for parasites. *Science*, **218**, 384–387.

Houston, A. I., Clarke, C. W., McNamara, J. M. and Mangel, M. (1988). Dynamic models in behavioural and evolutionary biology. *Nature*, **332**, 29–34.

Jakobsson, S., Radesater, T. and Jarvi, T. (1979). On the fighting behaviour of *Nannacara anomala* (Pisces: Cichlidae). *Z. Tierpsychologie*, **49**, 210–220.

Jones, J. W. (1959). *The Salmon*. London: Collins New Naturalist.

Kislalioglou, M. and Gibson, R. (1976). Prey 'handling time' and its importance in food selection by the 15-spined stickleback, *Spinachia spinachia*. *J. Exp. Mar. Biol. Ecol.*, **25**, 151–158.

Krebs, J. R. and Kacelnic, A. (1991). Decision-making. In: *Behavioural Ecology: An Evolutionary Approach*, 3rd edn (Ed. by J. R. Krebs and N. B. Davies), pp. 105–135. Oxford: Blackwell Scientific.

Krebs, J. R. and Davies, N. B. (1978). *Behavioural Ecology: An Evolutionary Approach*. Oxford: Blackwell Scientific.

Krebs, J. R. and Davies, N. B. (1984). *Behavioural Ecology: An Evolutionary Approach*, 2nd edn. Oxford: Blackwell Scientific.

Krebs, J. R. and Davies, N. B. (1991). *Behavioural Ecology: An Evolutionary Approach*, 3rd. edn. Oxford: Blackwell Scientific.

Krebs, J. R. and McLeerly, R. H. (1984). Optimization in behavioural ecology. In: *Behavioural Ecology: An Evolutionary Approach*, 2nd edn. (Ed. by J. R. Krebs and N. B. Davies), pp. 91–121. Oxford: Blackwell Scientific.

Krebs, J. R., MacRoberts, M. H. and Cullen, J. M. (1972). Flocking and feeding in the Great Tit *Parus major*—an experimental study. *Ibis*, **114**, 507–530.

Lack, D. (1968). *Ecological Adaptations for Breeding in Birds*. London: Methuen.

Lima, S. L. and Dill, L. M. (1990). Behavioural decisions made under the risk of predation: a review and prospectus. *Canadian J. Zoology*, **68**, 619–640.

MacArthur, R. H. and Pianka, E. R. (1966). On the optimal use of a patchy environment. *Amer. Natl.*, **100**, 603–609.

Mangel, M. and Clarke, C. W. (1988). *Dynamic Programming in Behavioural Ecology*. Princeton: Princeton University Press.

Maynard Smith, J. (1964). Group selection and kin selection. *Nature, Lond.*, **201**, 1145–1147.

Maynard Smith, J. (1977). Parental investment—a prospective analysis. *Animal Behaviour*, **25**, 1–9.

Maynard Smith, J. and Parker, G. A. (1976). The logic of asymmetric contests. *Animal Behaviour*, **24**, 159–175.

Maynard Smith, J. and Price, G. R. (1973). The logic of animal conflict. *Nature, Lond.*, **246**, 15–18.

McNicol, R. G. and Noakes, D. L. G. (1984). Environmental influences in territoriality of juvenile brook char *Salvelinus fontinalis* in a stream environment. *Env. Biol. Fishes*, **10**, 29–42.

Milinski, M. (1979). An evolutionarily stable feeding strategy in sticklebacks. *Z. Tierpsychologie*, **51**, 36–40.

Milinski, M. (1985). Risk of predation taken by parasitised sticklebacks foraging under competition for food. *Behaviour*, **93**, 203–216.

Milinski, M. and Heller, R. (1978). Influence of a predator on the optimal foraging behaviour of stickleback *Gasterosteus aculeatus*. *Nature, Lond.*, **275**, 642–644.

Neill, S. R. St. J. and Cullen, J. M. (1974). Experiments on whether schooling by their prey affects the hunting behaviour of cephalopods and fish predators. *J. Zoology, Lond.*, **172**, 549–569.

O'Brien, W. J., Slade, N. A. and Vinyard, G. L. (1976). Apparent size as the determinant of prey selection by Bluegill sunfish (*Lepomis macrochirus*). *Ecology*, **57**, 1304–1311.

Robertson, D. R., Sweatman, H. P. A., Fletcher, E. A. and Cleland, M. G. (1976). Schooling as a mechanism for circumventing the territoriality of competitors. *Ecology*, **57**, 1208–1220.

Sargent, R. C. and Gross, M. R. (1986). William's principle: an explanation of parental care in teleost fishes. In: *Behaviour of Teleost Fishes* (Ed. by T. J. Pitcher), pp. 275–293. London: Croom-Helm.

Tinbergen, N. (1951). *The Study of Instinct*. Oxford: Oxford University Press.

Turner, G. F. and Huntingford, F. A. (1986). A problem for games theory analysis: assessment and intention in male mouthbrooder contests. *Animal Behaviour*, **34**, 1961–1970.

Werner, E. E. (1977). Species packing and niche complementarity in three sunfishes. *Amer. Natl.*, **111**, 553–578.

Werner, E. E. and Hall, D. J. (1974). Optimal foraging and the size selection of prey by the Bluegill sunfish (*Lepomis macrochirus*). *Ecology*, **55**, 1042–1052.

Werner, E. E., Gilliam, J. F., Hall, D. J. and Mittelbach, G. G. (1983). An experimental test of the effects of predation risk on habitat use in fish. *Ecology*, **64**, 1540–1548.

Williams, G. C. (1966). *Adaptation and Natural Selection*. Princeton: Princeton Univ. Press.

Wynne-Edwards, V. C. (1962). *Animal Dispersion in Relation to Social Behaviour*. Edinburgh and London: Oliver and Boyd.

FISH BEHAVIORAL ECOLOGY:
PROS, CONS AND OPPORTUNITIES

GEORGE W. BARLOW

*Department of Integrative Biology, and Museum of Vertebrate Zoology,
University of California, Berkeley, California 94720, USA*

INTRODUCTION

From time to time, my ethologically minded friends and I play an animal-chauvinist game. If this diversion had a name it would be something like, "Why my kind of animal is better than yours for studying behavioral ecology." Having fishes to champion, and by fishes I mean teleosts, is an almost unfair position *vis à vis* other vertebrate biologists, especially when contrasting diversity in modes of reproduction (Warner, 1978, 1984; Policansky, 1982). Of course, no one of us likes to play this game with those who study invertebrates such as insects and molluscs. But how much of this game playing is rationalization after the fact?

I imagine most of the contributors to this volume on fish behavioral ecology did not give serious thought to why they chose fishes as the objects of their investigations, though some may have. Rather, the typical ichthyophile probably gravitated toward fishes almost by accident. Perhaps pivotal was an early experience with angling, with aquarium fish, or with an inspirational teacher.

How each of us became involved with fishes is no longer an important question. It remains a useful exercise, nonetheless, to reflect on the pros and cons of using fishes in the study of behavioral ecology (or in other areas of research: Powers, 1989). Such reflection may highlight propitious research problems, or may help some uncommitted reader make a more objective choice about studying fishes than the decisions we made that entangled our lives with those enchanting animals. At the least, contemplating why one should, or should not, study fishes may help others when playing the game.

The game takes various forms. At the outset, diversity scores heavily. (We conveniently ignore the insects here. Never mention that bees alone may have as many or more species than do the teleost fishes, even though bees are more uniform than teleosts.) The gamut of reproductive biology, encompassing internal and external fertilization, puts fishes out in front, even surpassing insects. As the game gets more subtle, however, points are lost. Nursing among mammals, for instance, is more highly evolved than is the eating of mucus by young fish from the skin of their parents.

But I don't want to get ahead of my thesis. To be fair I begin with behavior for which fish are unsuited and progress to behavior that falls into a gray area between unsuitable and attractive. Then I turn to aspects of fish biology and behavior that make them splendid subjects for behavioral ecology. In this short introduction to the symposium, I cannot begin to cover all the possibilities; some will be developed, others just mentioned, and many ignored.

BEHAVIOR BEST OR EQUALLY WELL STUDIED IN OTHER KINDS OF ANIMALS

I start this section with behavior whose study offers little hope of payoff among fishes. I progress to behavior that might be profitably pursued. Indeed, some recent examples are promising. Nonetheless, the behavior is of a type for which fish offer no special opportunities, compared with other kinds of animals.

The first kind of behavior, play, is indeed fruitless. Play among fishes is rarely reported, and is usually easily discounted (Hubbs, 1948). One intriguing though implausible case has been described for a weakly electric fish (Meyer-Holzapfel, 1960). A study of tool-using among fishes would also be frustrating. Cases are either nonexistent or are open to different interpretations, such as the archer fish shooting drops of water at aerial insects (Dill, 1977). Some might argue that the building of complex spawning nests demonstrates the use of tools, but that stretches the concept so much as to make it uninteresting.

Other nonprofitable behavior to study among fishes is that which occurs sporadically, or is poorly developed. Some might reason, on the other hand, that if the behavior is weakly expressed that is all the more reason to study it because it serves as a model for how such behavior started. Fair enough. Nevertheless, if one wishes to explore such behavior in all its richness and complexity, perhaps to serve as a model of parallel or convergent evolution with more advanced vertebrates, then the following aspects of fish behavior might best be avoided.

Behavior that calls for advanced mental ability has not been demonstrated for fishes. This may be a matter of our inability to recognize it, however, or of not having devoted enough effort to exploring more sophisticated facets of their behavior. Fishes learn about their environment in complex ways that can only be revealed through experimentation. A recent study of daily feeding migrations in a grunt, for instance, demonstrated what amounts to the "cultural" perpetuation of the migratory route (Helfman and Schultz, 1984). And female blueheaded wrasses develop traditional preferences for spawning sites (Warner, 1988). Still, refined mental performance among fishes will never compare to that regularly shown by mammals.

Studies on the effect of early experience are becoming more common. Some of these have paid off, such as habitat imprinting in salmonids (Hara and Macdonald, 1975; Cooper and Hasler, 1976). And this volume contains articles by Huntingford, and by Magurran and Seghers, demonstrating the impact of early experience on antipredator behavior.

Most such research, nevertheless, has so far revealed a remarkable lack of effect of early development on the behavior of adults, in contrast to the profound effects of the environment on morphology (e.g. Meyer, 1987). Sexual imprinting, for instance, has been claimed in a small number of reports, but none of these is persuasive (Barlow, 1986; Noakes and Godin, 1988). A recent detailed study of the effects of early experience failed to find evidence for sexual imprinting in a cichlid fish (Barlow *et al.*, 1990)

At the individual level, fishes show little long-term bonding. Reese (1973) did establish that pairs of butterflyfish (Chaetodontidae) in some species may remain together several years. Persistent bonding may be typical for several monogamous species of marine fishes (Barlow, 1984, 1986; A. Vincent, pers. comm.).

However, the examples of enduring bonds among fishes that are so often mentioned by aquarists come from the monogamous cichlids. Those claims are suspect because too often a given pair of cichlids is kept together as the lone twosome in an aquarium, or with some conspecific fish that are subordinates. Thus the breeding pair is dominant and the opportunity to change partners is nil. I have found no evidence that cichlids are monogamous in nature for longer than one breeding cycle; proper studies to determine this, however, have never been done. This general lack of enduring bonds among fishes may help explain why fishes do not engage in polyandry at the level shown by some mammals and birds in which a female bonds with a small number of males (e.g. Jenni and Collier, 1972).

The paucity of enduring relationships may also contribute to the rarity of cooperative behavior. Cooperative foraging seems almost incidental rather than organized (Major, 1978; Partridge et al., 1983). Pairs of a small serranine fish, however, seem to feed cooperatively (Pressley, 1980). Another kind of cooperation is mobbing, so common among birds. It has been described for a few species of fishes (Motta, 1983; Dominey, 1983; Donaldson, 1984). But such behavior might instead be inspecting a predator (Dugatkin, 1988).

Mate provisioning apparently does not exist. Behavior such as parents alone, or jointly, assisting their young is merely parental care and is not newsworthy, although provisioning of young by means of trophic eggs in a catfish merits passing notice (McKaye, 1986). Differential provisioning of young, say according to sex or size, is unlikely because distinguishing sex is probably not possible and because the young are so small, numerous and probably sexually plastic (see further on).

Rarely, young fish help at the nest, as reported for the cichlid *Lamprologus brichardi* (Taborsky, 1984). Similar helping has been observed in related species in Lake Tanganyika (Taborsky and Limberger, 1981) and may occur in an unrelated cichlid fish in Sri Lanka (Ward and Wyman, 1977). In a particularly bizarre twist, fish of one species sometimes help another species. Thus unmated males of *Cichlasoma nicaraguense* assist females of *C. dovii* guard her young; as a result, more young are reared to independence (McKaye, 1977, 1981). Even more startling is the shared brood care of catfish, *Bagrus meridionalis*, and cichlids in Lake Malawi, Africa (McKaye, 1985).

Related to shared care-taking is the adopting or even kidnapping young by one fish or pair to add to their own brood. Once thought to be exceptional, and certainly controversial, examples continue to grow (Sjölander, 1972; McKaye, 1981). McKaye and McKaye (1977) suggested that the adopting pair benefits from adding unrelated young, sometimes even of a different species, to their school of young because it dilutes the effect of predation. Such adoption, however, raises questions about the possibility of differential distribution of parental care between own and adopted young, and how they might know which is which.

A few years ago nothing was known about kin recognition among fishes and it seemed an unpromising area of research. Kin recognition might be expected, however, when fish live in groups, as when schooling as young or adults, following the parallel among amphibian tadpoles (Blaustein and O'Hara, 1982; Waldman et al., 1984). And one could anticipate kin recognition in the face of cannibalism. While scarce, such evidence for kin recognition now exists (salmonids: Quinn and Busack, 1985; Quinn and Hara, 1986; Olsen, 1989; sticklebacks: Havre and FitzGerald, 1987;

poeciliids: Loekle *et al.,* 1982). Infrequent, but perhaps often overlooked, is visual alarm signalling despite early accounts that should have alerted investigators (Keenleyside, 1955; Verheijen, 1956), and despite the possibility that alarm signalling is common among schooling fishes. Recently, Smith and Smith (1989) documented the use of visual signals by nonschooling gobies. Acoustic alarm signals have not been reported, but I have heard triggerfish grunt loudly as they darted into their hiding places on coral reefs; an ornithologists would almost automatically label such behavior alarm calling and look for reciprocal altruism.

Other interspecific behavior might be labeled reciprocal altruism, a point Trivers (1971) made with regard to a cleaner wrasse removing parasites from large host species of fishes. Losey (this volume), however, discounts cleaning behavior as reciprocal altruism. Various other kinds of mutualism could fit here, such as anemonefishes and their anemones (Fautin, 1991), or gobies and the shrimp with whom they share burrows (Karplus, 1987); that is an area of study rich in opportunities. Predator inspection has become the focus of experiments on reciprocal altruism (Milinski *et al.,* 1990; see also Magurran and Seghers, this volume), generating a lively and provocative literature. The controversy probably has its roots in the difficulty of finding an example among animals that satisfies all the criteria for reciprocal altruism (Emlen, 1991).

PRACTICAL DIFFICULTIES IN STUDYING FISH BEHAVIORAL ECOLOGY

I mention here only three kinds of problems. These arise typically, but not exclusively, in studies done in the field. The first obstacle stems from the slippery, slimy nature of fishes. The second results from the fact that we live in air and fishes live in water. The third turns on the dispersal of numerous tiny offspring, each of which has an exceedingly low probability of survival.

A serious hindrance to progress in fish behavioral ecology is the lack of a good method of marking fish individually. We need techniques that make individuals easily recognized at a modest distance and over long periods of time but that do not harm the fish. The traditional tags used in fisheries biology are difficult to read unless the fish is in hand, usually become covered with algae, often become sources of infection, are frequently shed, may increase susceptibility to predation and are impractical for small species. In some instances, more subtle types of marking have worked well (see minireviews in Chapman and Bevan, 1990; Moring, 1990). Marking fish, moreover, depends on the characteristics of the species. The same technique that succeeds on one species too often gives disappointing results on others. The literature is geared to reporting only positive results, so the high frequency of unsuccessful attempts is not generally known.

The challenge lies in the nature of the skin of fishes. In many species the skin is too densely pigmented for injected, colored, substances to show through. Even when the skin is reasonably transparent, vital stains are rapidly removed by the actively metabolizing cells. Seemingly inert material such as acrylic paint may disappear after a while. Likewise, hot or cold brands (e.g. Knight, 1990), while sometimes enduring, often vanish within a short period.

The skin is covered with mucus that is constantly renewed. Thus any pigments placed on the surface of the fish are soon flushed away. For the same reason, attaching objects to the surface of a fish using, say, surgical glue is frustrating. This is unfortunate because it limits field experiments on fish behavioral ecology.

Ornithologists can alter the color of bird plumage (Marler, 1955), or the length of ornamental feathers (Andersson, 1982), powerful tools in experiments. Ichthyologists have been stymied when attempting similar manipulations on fishes (but see Noble and Curtis, 1939). Now Basolo (1990) has shown the way to manipulate sword length (i.e. length of fin) in a species of swordtail (*Xiphophorus*). Her technique might be used with profit on other fishes with elaborate, sexually selected ornaments such as occur in the Glandulocaudinae. On the other hand, some hues in fish are eliminated by bathing them in filtered light (Wunder, 1934; Bakker and Milinski, this volume). And, if the fish is white, its hue can be manipulated with monochromatic illumination.

Practical difficulties are also encountered when observing fish in their medium. To be sure, modern diving equipment makes this possibility a reality, but compared to observing birds on land, severe obstacles exist (Helfman, 1983). If the observations are done at depths where the pressure exceeds two atmospheres (10 m) the observer's time is limited by nitrogen saturation, and the more so the deeper the water. This commonly means a maximum of two to four hours of observing per day. Furthermore, weather conditions are a more severe and limiting factor than when observing on land. For safety's sake, the observer must be physically fit.

By selecting the right species, much observing can be done from the surface in shallow protected water. But that constrains the investigator to fewer species. The overall result is that remarkably few papers have been published, based on field observations of fish in the marine environment, given the vast array of easily observed fishes there and the sophisticated equipment available for diving.

Much the same can be said for freshwater fishes, though the reasons are different. One can indeed dive in freshwater lakes and large slow rivers (e.g. Barlow, 1976, McKaye and Barlow, 1976; Perrone, 1978; Helfman, 1979; Kuwamura, 1986; Yanagisawa, 1987), but so far little of that has been done. More surprising is that fishes have been so seldom observed from the banks of streams or ponds. One reason is that fresh waters are often turbid or clogged with vegetation or debris, removing the fish from view. Nonetheless, excellent field studies have been done by observers looking down into streams and ponds where conditions permit (Reighard, 1910; Keenleyside and Yamamota, 1962; van den Berghe and Gross, 1984; Blumer, 1985).

The third drawback to studying fish behavioral ecology in the field relates to the lack of continuity between offspring and parents. The offspring of individuals or pairs are dispersed over great distances, particularly among marine species (Barlow, 1981a). The smaller the offspring, the more numerous they are and the higher the expected mortality among them. Even when fewer, better developed young are produced, as among viviparous embiotocids or clinids, it is virtually impossible to assess Darwinian fitness excepting as an index, say the number of eggs fertilized or hatched, or young born.

The scattering and mixing of offspring is less pronounced among freshwater species for the obvious reason that the population is confined to a relatively small body of water, or because well developed (though few) young are produced (Barlow, 1981a). Rarely, family care is protracted, as in the Asian cichlid *Etroplus maculatus* (Ward and Wyman, 1977), which protects its young until they are nearly adult. Nevertheless, scattering of offspring remains the general situation. Thus not only is fitness virtually impossible to measure, the effects of early experience cannot be assessed in nature.

That the study of fishes presents problems, particularly in the field, should not detract from research on their behavior there. Other kinds of animals have difficulties inherent in their biology. Fishes, moreover, offer an impressive array of practical reasons for studying them.

PRACTICAL ADVANTAGES TO STUDYING FISH BEHAVIORAL ECOLOGY

One of the great features of fishes is that they are so abundant in so many kinds of environments. This is most obvious on a coral reef where fishes dazzle the viewer with their profusion, beauty, and endless variety. A diver can approach and watch most species without disturbing them. Likewise, in clear freshwater environments, many kinds of fishes tolerate observers (nonbreeding salmonids and some others, however, are legendary for their shyness). And freshwater fishes can be impressively diverse; just think of groups such as characins and catfishes in South America and cichlids in the rift lakes of Africa.

Also consequential is that so many fishes are packaged in ideal sizes for aquarium studies. Because most people are mainly familiar with the larger species that are eaten, it comes as a surprise even to many biologists that teleost fishes are mostly, about 100 mm long as adults (Marshall, 1971; Barlow, 1981a; Kerr, 1989). Being so small, they adapt well to aquaria and yet are large enough to observe accurately. Many freshwater species live in small streams and ponds and regularly thrive in the narrow confines of captivity, as do several small marine fishes. The swim bladders of midwater species make them neutrally buoyant so they readily negotiate the confined setting. Bottom-dwelling fishes adjust even more easily.

Their small size also means they mature rapidly. This can be important to a graduate student or starting faculty who is anxious to get his or her fish up to a breeding size to analyze the effects of some prior manipulation or genetic difference. Further, the size at maturity can vary enormously. Breeding Arctic charr range in mass from around 25 g to 1000 g (Johnson, 1980). Some *Tilapia* have a comparable spread of size among breeding adults (Fryer and Iles, 1968). While extreme, such variation is common among fishes and provides a convenient tool for experimentation that is not available among other vertebrates.

Another practical benefit to working with fishes is that they are so fecund. Some species propagate only a few offspring at a time, such as *Starksia* (Greenfield, 1979; see also Noakes and Balon, 1982). In general, however, hundreds to thousands or more gametes are produced by a single female. The researcher consequently has a large cohort of siblings available for parallel experiments in which mean genetic differences are ruled out (Danzmann *et al.*, 1993).

If genetic identity is called for, certain remarkable species are available. A number of gynogenetic species are found in the genus *Poeciliopsis* (Schultz, 1961, 1977). Another atherinomorph, *Rivulus marmoratus* is self-fertilizing and is the ultimate easy species to maintain (Harrington, 1968; Davis, 1988).

In general, fishes are less trouble to maintain than are other vertebrates in most cases, and particularly for freshwater species. Keeping marine fish in aquaria, admittedly, is more expensive and demanding, though not unmanageable. The smaller the species, the easier its husbandry. Freshwater species such as guppies flourish in aquaria as small as 20 1. A *Betta splendens* or a *Rivulus marmoratus* thrives in a container not much larger than a coffee cup. For most freshwater species, the dimensions of the system can be scaled to the size of the species. Food is inexpensive, and water temperature is easily controlled for tropical species. Filtration requires only a little more investment, and none at all if the fresh water is of sufficient quality to have an open system.

In the absence of predators in captivity, nearly complete cohorts of many freshwater species can be reared with ease from the earliest stage of development to adulthood. As a consequence, comparative studies of ontogeny are seldom hindered by special requirements of the developing offspring.

Rearing the young marine fishes, however, is much more difficult than for freshwater species, although procedures for doing this have become routine (Hunter, 1984). Special care must be given to water quality and other physical conditions. Being themselves tiny, the early stages require exceedingly small plankton. Microscopic food organisms can be cultured away from marine environments. However, marine fish larvae survive and grow best on "wild" plankton (Hunter, 1984).

OPPORTUNITIES PRESENTED BY FISHES

The vogue is to analyze behavior in terms of adaptiveness within a given species (e.g. Krebs and Davies, 1991; Lott, 1991). That approach is exciting and illuminates evolutionary processes. Nonetheless, the success of that strategy has distracted from the more traditional comparative method, which is a more direct attack on the evolution of behavior, as compared with its current utility. Surprisingly, few programmatic comparative studies have ever been done on the behavior of taxonomically cohesive groups, despite the often-cited pioneering analysis of Konrad Lorenz (1941) on the behavior of anatid ducks (see also Selander, 1964; Nelson, 1978; Zimmermann and Zimmermann, 1988).

Teleost fishes offer extraordinary opportunities for comparative studies. Genera and families frequently comprise large numbers of species with the promise of a spectrum of varying degrees of specialization in any given behavioral system. The spectacular radiations of fishes that illustrate this point are well known, such as the cichlids of the rift lakes of Africa (Fryer and Isles, 1972) or the cottoid fishes of Lake Baikal (Nikol'skii, 1961). But a great many such opportunities exist all around us. They have attracted little attention, perhaps because they are so standard among fishes. If instead the species were birds, they would be hailed as impressive radiations.

In Central America the poeciliid fishes, which are so effortlessly cultured, and their allies have undergone an incredible radiation (Rosen and Gordon, 1953). Their relatives, the goodeids of Mexico, are even more remarkable though less speciose (Hubbs and Turner, 1939). In North America, Africa and Asia cyprinids have speciated with abandon. One could add the percids and centrarchids of North America, or the radiation among three-spine sticklebacks in the Northern Hemisphere (Bell and Foster, unpublished data). I have already mentioned the characins and catfishes of South America, and I could list many more "mini-radiations."

The marine environment is equally rich. The cottids and scorpaenids of the North Pacific have evolved large numbers of species with profound differentiations. Gobies and flatfishes, especially in tropical waters, are so diverse as to boggle the mind, as are syngnathids. The coral reefs are a story unto themselves, but especially notable are the labrids, scarids, pomacentrids and gobiids. The opportunities are so endless that they are daunting.

This is especially true of comparative analyses of behavioral ontogeny, which are almost nonexistent but are sorely needed and hold great promise. Such studies could be done easily on many families of freshwater fishes, such as the Cichlidae. Describing the trajectories of modal action patterns and social organization across a coherent taxon such as a genus or subfamily, would be enormously rewarding. However, such studies would be more credible if the ontogeny of at least some of the species could be tracked in the field to verify interpretations of adaptive significance and evolutionary pathways.

I digress here to mention a singular opportunity that exists in the study of behavioral development. Despite my discouraging comments earlier about the difficulty of detecting the effects of early experience, the unique way the nervous system of fishes develops calls for closer examination of its behavioral consequences. In reptiles, birds and mammals, the generally accepted view (except for the acquisition of bird song) is that the nervous system, and especially the brain, stops developing early in ontogeny; further change is through sculpturing the existing nervous system, called regressive development. Fritzsch and Crapon de Caprona (1990), however, reminded us that fishes are just the opposite: neurogenesis occurs throughout life. We need to explore the meaning of this progressive development of the nervous system in fishes. The nature of the difference may be informative about the neural substrates of behavior of other vertebrates as well as of fishes. The difference also raises a host of interesting questions about the evolution of regressive from progressive development.

The opportunities to study co-evolution, while not so numerous, are equally alluring. Some examples have already received attention, though major questions remain unresolved. Most written about is the association between the damselfishes of the genus *Amphiprion* and a number of species of anemones (Fautin, 1991). Many species of shrimp and gobies form mutualistic pairs with varying degrees of specificity (Karplus, 1987). Trees in the Amazon Basin have evolved fruit to be dispersed by characins (Goulding, 1980).

Other evolved relationships between species are poorly understood, the most striking of these being mimicry. Little has been done beyond description since the seminal paper by Randall and Randall (1960). But enough examples have been reported now to establish that remarkable and complex connections exist (Wickler, 1968; Springer and Smith-Vaniz, 1972; Russell *et al.,* 1976; Ormond, 1980; Kuwamura, 1981).

Examples of convergence and parallel evolution arise, it seems, wherever one looks. Internal fertilization with varying degrees of internal development has evolved independently many times among the toothcarps. Apparently, this has been facilitated by the juxtaposition of vents of the two sexes at the time of fertilization, together with the male embracing the female with his dorsal and anal fins (Rosen and Gordon, 1953).

Fertilization among fishes is commonly done by the male releasing milt in close conjunction with the eggs. But in the males of some bony fishes elements of the pelvic fins (Phallostethidae) or anal fin (Poeciliidae, Cottidae) have been modified for the transfer of sperm. More spectacularly, in certain cichlids the eggs are fertilized in the mouths of the females (Wickler, 1962); how that is accomplished differs across species, apparently independently evolved, but in each case has been preceded evolutionarily by female mouthbrooding.

Taking a larger view, fishes have evolved stunningly different ways of bearing young (Breder and Rosen, 1966; Balon, 1984). In addition to the most frequent form, bearing the zygotes in the lumen of the ovary, some species carry embryos in their mouths, branchial chambers, a hook on the head, embedded in their skin, between their pelvic fins, and in specialized pouches (e.g. Rosenqvist, this volume). No other group of animals approaches this expression of fundamentally different ways of bearing young.

Almost as opportune is the variation in which sex provides parental care. The predominant mode among fishes is male care (Breder and Rosen, 1966; Blumer, 1979). This has sometimes led authors to talk of such species as sex-role reversed, which is a misrepresentation. In such species the female remains the limiting sex. The male characteristically shows the masculine syndrome (Williams, 1966) and indications of being sexually selected (Baylis, 1981). In some species, however, the male may be the limiting sex, with important consequences for the masculine and feminine syndromes (see Rosenqvist, this volume).

In many kinds of fishes both parents care for the eggs and fry (Barlow, 1981a). Less common is care by the female only, although this is prevalent among tilapiine and haplochromine cichlids (Fryer and Isles, 1972; Keenleyside, 1991). More interesting is that some species show both biparental and uniparental care, depending on circumstances (Barlow, 1991). This diversity of care-taking provides an exceptional opportunity to test the model of parental care formulated by Maynard Smith (1977) and Clutton-Brock and Godfray (1991).

The protracted care of free-swimming fry in a variety of fishes (Barlow, 1984, 1986) might be seen as fertile ground for examining parent-offspring conflict, especially since nothing has appeared on that subject. When the young derive benefits from schooling with the parents, should they be selected to continue to demand care when the parents would do so at a cost to their future reproduction? My experience with Central American cichlids indicates that no conflict exists, that the young simply drift away from the parents when they reach a certain stage of development (see also Townshend and Wootton, 1985).

Another aspect of parental care that has drawn attention lately is filial cannibalism (Rohwer, 1978; FitzGerald, 1991). In many species, the parenting male eats a few eggs. That can help sustain him through a parental cycle and leave him better prepared to enter into another cycle sooner, thus increasing his fitness at a small cost to current offspring. The female pays the cost of egg cannibalism so selection should work on females to spawn with the male that presents the lowest likelihood of eating her eggs (Pressley, 1981).

Cannibalism is also common among juveniles and nonbreeding adults (Dominey and Blumer, 1984; FitzGerald, 1991). Although little studied, it can affect the breeding structure and demography of local populations (FitzGerald, 1991). For too long, cannibalism among fishes was ignored, seemingly out of the belief that fishes are too dumb not to be cannibals and the issue is therefore of little interest. Much needs doing here.

Fishes are also especially suited to analyses of mating, hence of mate choice. The common guppy is becoming a paradigmatic species for the study of intersexual selection (Houde and Endler, 1990; Kodric-Brown, 1990). Its minuscule size and rapid generation time are conducive to the analysis of both ultimate and proximate mechanisms. For the first time, these researchers have demonstrated the trade-offs to males between risk of predation and success in mating in relation to conspicuous displays and ornamentation (Endler, 1987). They have also shown that male ornamentation and female choice are genetically correlated (Houde and Endler, 1990). Fortunately, the genetics of color patterns are known, so sperm competition can be brought in as well.

Other poeciliid fishes also offer opportunities to analyze the ramifications of mate choice, hence intersexual selection. Intersexual selection is still difficult to explain, as attested by the number of alternative hypotheses available (Kirkpatrick and Ryan, 1991). One hypothesis that has received little attention, but holds much promise, is that of sensory bias, nicely demonstrated in the Mexican green swordtail (Basolo, 1990).

A special twist on sexual selection in poeciliids is "reverse" size dimorphism. Why males should so often be smaller than females, even when the males area brilliantly colored, as in guppies, has long puzzled biologists. Several alternative hypotheses have been proposed and tested (see Bisazza, this volume).

Some 'species' depend on alien sperm in order to reproduce. This is the case in matroclinous poeciliids of the genus *Poeciliopsis*, in which the females count on male 'errors' in mate choice in order to parasitize sperm to trigger the development of their eggs (Schultz, 1961, 1977). The sperm, however, contribute no genes to the zygote. This is a most peculiar mating system, and one whose behavioral substrate has been little studied (McKay, 1971; Schlupp *et al.*, 1991) in relation to its interest.

Mate choice easily flows into the issue of species recognition and hybridization. It is the latter I wish to discuss here. Fishes stand out among vertebrates for the ease with which fertile hybrids are produced. Hubbs (1955) reported a naturally occurring interfamilial hybrid and several intergeneric hybrids. Some species, however, cross only with difficulty, or produce offspring with reduced or nonexistent fertility.

Hybrids open a window into the genetics of behavior (Heinrich, 1967; Franck, 1974; Danzmann *et al.*, 1993) and perhaps the plasticity of behavioral development (Barlow, 1981b). They also shed insights into species-isolating mechanisms (Falter and Charlier, 1989), and conceivably even into the process of rapid speciation (Crapon de Caprona and Friztsch, 1984).

The mating systems of fishes are as diverse as other groups, polyandry aside. They are predominantly polygynandrous. Polygyny and harems occur (Moyer, 1979; Warner and Hoffman, 1980; Hoffman, 1983; Clark *et al.*, 1991), and lekking has arisen in many species (Fryer and Isles, 1972; Loiselle and Barlow, 1978; McKaye, 1983). Sneaking and deception are practiced in several species (e.g. Barlow, 1967; Warner, 1984). Monogamy is widespread (Barlow, 1984, 1986). What makes monogamy different, however, is that all but one marine species lack biparental care of swimming young, but all monogamous freshwater species have biparental care of swimming young.

Given the opportunities for comparative studies that are so abundant among fishes, one type of study is striking by it near absence. Little attempt has been made to relate mating systems of fishes to ecology. A few beginning essays have appeared (Barlow, 1974, 1984, 1986, 1991; Reese, 1975; Thresher, 1977; Hourigan, 1989) but they have only scratched the surface. Nothing exists that compares with the methodical classical studies on weaver finches (Crook, 1964) or ungulates (Jarman, 1974). Fish biologists have been content instead to seek correlations between mating systems and other features of the fish, such as size, mobility and color (e.g. Colin and Bell, 1991). The lack of thorough studies may reflect the prodigious effort required to gather the ecological data needed for comparative analyses of mating systems.

Two features of mating systems of fishes call for mention. One relates to the great range in sizes of mature adults within a species. That facilitates alternative life styles (Bruton, 1989; see also Metcalfe, this volume). Thus under certain circumstances tilapiine cichlids breed at a size normally associated with juveniles, greatly shortening their life cycle and increasing their population much faster than normal (Fryer and Iles, 1968). Behavioral ecologists have not exploited this aspect of fish biology.

The other feature is sexual plasticity. The more species are examined the more hermaphrodites are discovered (Ross, 1990). Most are protogynous, but some are protandrous (Warner, 1984), and a few are simultaneous hermaphrodites (Harrington, 1968; Fischer, 1981). That affords a special opportunity to test sex-ratio theory (Fisher, 1930). Sexual plasticity results in a rich variety of mating games among fishes, including females that change sex to males but still resemble females, the better to deceive the dominant male (Warner, 1984).

Among the simultaneous hermaphrodites, the situation is more complex. Recently studied serranine fishes have combinations of types of sex, such as a dominant male whose harem consists of simultaneous hermaphrodites (Hastings and Peterson, 1986). Further study of these little marine groupers may reveal additional variations on this theme.

As our body of knowledge grows, a pattern emerges of labile sexuality among fishes (Francis, in press). For some time we have known that physical factors such as pH (Heiligenberg, 1965) and temperature (Harrington, 1968; Conover and Kynard, 1981) can deflect the sex to male or female during development. Now it is evident that even among presumed gonochoristic, monogamous cichlids (Francis and Barlow, unpublished data) sex may be regulated by social interactions during development.

This pattern of sexual lability may result from the general lack of sex-determining chromosomes among fishes (Francis, in press). Understanding how sex is resolved in such species should shed light on the origin of sex-determining chromosomes in higher vertebrates. Metcalfe's study (this volume) of alternative life-history strategies in Atlantic salmon may indicate where to look.

Metcalfe reports that among cohorts of salmon parr, some over-winter one year and others two years before migrating to sea. One-winter parr already have higher metabolic rates at hatching. These parr grow faster than the two-winter parr and thus achieve larger size faster. Size is positively correlated with dominance. This finding may relate to sex determination in unrelated fishes, such as the Midas cichlid.

Sex in Midas cichlids appears to be established through dominance relationships, the dominant, larger individuals becoming males (Francis and Barlow, unpublished data). But how does a fish become dominant or large? At hatching, the young of a cohort are remarkably uniform in size (Lagomarsino et al., 1988), and aggressiveness appears to have low heritability (Francis, in press). Extrapolating from Metcalfe's finding, the cichlids that grow faster and become dominant may have higher metabolism. Thus the genetic loci for higher metabolism might be the precursors of sex-determining genes, hence sex chromosomes, and in teleost fishes in general.

Another area that has received too little attention is coloration. Fishes are the only vertebrates that change colors rapidly and radically. In addition, the color patterns of fishes, especially on coral reefs, transcends the imagination of artists. Despite

Lorenz's (1962) provocative writings on the poster-color hypothesis and its relation to niche partitioning, analyses of fish color in a behavioral context remain relatively few. When one seeks the literature on fish coloration the result is an abundance of articles on physiological and morphological substrates and speculative papers on the functional significance of coloration. Particularly neglected are analyses of color change within a given species.

That is not to say that no such papers exist. Serviceable descriptions have been provided, for instance, of color changes in cichlids (Neil, 1964; Baldaccini, 1973; Nelissen, 1975) and in *Badis* (Barlow, 1963). A signal analysis of the dynamics of color pattern is that on courtship among guppies (Baerends *et al.,* 1955). Other investigators have employed dummies colored in different ways (Heiligenberg, 1972; Rowland, 1975). Nevertheless, the ability of fishes to change color quickly remains fragmentarily studied in the context of social interaction. The inhibition of closer study in the past may have resulted from the difficulty inherent in quantifying acute color changes. Modern computer technology, coupled with television, should get around that obstacle.

The final opportunity I want to take up is that of staged fights. Fishes are ideal subjects for the analysis of combat. Many species regularly kept in aquaria not only fight readily but are so pugnacious that they must be held under conditions that preclude fighting. Further, staged fights must be closely monitored to minimize injury to the subjects.

Fights start with emphatic displays. Typically, the fins are raised while their colors darken or brighten, increasing the apparent size of the fish. The opercles may or may not be spread, and spreading is more likely with frontal orientation. Arrayed lateral to one another, the fish often tail beat, dousing the opponent with turbulence. The displaying may escalate into actual combat, and then the fish bite one another on the fins or reciprocally on the mouth. Following a bout of active fighting, the fish typically pause, then resume after a while. If the fish are separated immediately when one gives up, injuries are minor and heal quickly.

The fight is thus structured and is ideal for analysis. I stress this because theory (e.g. Hammerstein and Parker, 1982) has outstripped empirical studies in this area. I also hasten to add that much of the theory was stimulated by the pioneering empirical study by Michael Simpson (1968) on combat among bettas and the near impossibility of predicting the winner. The rich literature on the theory of combat is ready to be exploited and fishes are excellent subjects for the experiments, as demonstrated, for example, by Enquist and Jacobsson (1986). In the past, fights were too data-rich to record, but now microcomputers and television are available to analyze combats at a high level of accuracy and relatively quickly.

CLOSING COMMENTS

In writing this essay I have not concerned myself with being comprehensive. I wanted to paint with a broad brush. In so doing I may have created a picture that reveals more about my views of the field of fish behavioral ecology than it does about the realities of what makes fishes attractive as subjects of study. I have neglected areas such as bioluminescence and annual fishes (cyprinodontids that live in ephemeral ponds and die even if kept in aquaria when the pond dries up). Sometimes the neglect was for good reason, other times just out of oversight. I did not emphasize research for which fish are practically well suited but offer no special opportunities by virtue of their natural history. Foraging is a case in point and, besides, it receives ample attention in this volume.

TOPICS COVERED IN THE SYMPOSIUM

As I wrote, I became curious about the match up between my portrait and that of the organizers of this symposium. I had no reason to believe they were trying to capture a representative view of the field. I doubt that was their goal. Rather, I suppose they got together and agreed on who might present a collection of articles on what is the vogue in fish behavioral ecology. To some degree, that had to reflect their own involvement. And to some degree the selection of authors must have been constrained by economic considerations.

I analyzed the topics of the lectures, nevertheless (Table 1). My groupings are broad and my conclusions general. Some papers appear in more than one category, and I omitted this article because it could fit into so many of them.

About half the lectures fall into the slot I call predator–prey, including feeding. The critical reader will have noticed I didn't even mention predator–prey relationships in the foregoing essay. Note well that my own article, on feeding territories, belongs in this category.

Slightly fewer than 1/3 of the articles treat mating or reproducing in one form or another. The next largest category is aggression, but comprising only three papers. Two papers take up the role of experience, two are on life history and one is on motivation. Thus, about half of the papers deal with aspects of fish behavior that present no special opportunity. Fishes can be wonderful subjects for such studies, but so can several kinds of animals. For some of those authors the motivating factor was most likely the ease of working with fish in aquaria, in miniaturized systems. That is important.

The other half of the articles are more in keeping the special opportunities I mentioned in the foregoing, although the fit varies from close to approximate. Five treat mating and reproducing, including parental care. Three at least touch on aggression, though they are not on the analysis of combat.

Table 1 An analysis of the general content of the Lectures on the Behavioural Ecology of Fishes, International School of Ethology, given in Erice

Category	Number	Percent[*]
Predator/prey	8	47
Mating/reproducing	5	29
Aggression	3	18
Experience	2	12
Life history	2	12
Motivation	1	6

* Sum >100% because of overlapping categories

Reading periodicals such as *Animal Behaviour, Behaviour* or *Ethology*, or any text book on animal behavior, underlines the lack of attention to fishes. Even though bony fishes are the most abundant and diverse vertebrates, and even though they are better suited to captive studies than are birds and mammals, the literature is dominated by investigations of birds and mammals, and to a lesser degree by insects. I appreciate the reasons for that. But that focus limits the horizons of theory. Fishes are the stem vertebrates and the champions of diversity. They are also a lot less expensive and easier to work with than are birds and mammals.

My apologies to those who study poikilothermic tetrapods. They are swell animals that do wonderful things, but they are less speciose and space was limiting.

ACKNOWLEDGMENTS

I am grateful to David L. G. Noakes and Felicity Huntingford for their constructive criticism, which greatly improved the manuscript. Participants in the course also gave me useful suggestions. The writing was supported by NSF Grant BNS 91-09852.

References

Andersson, M. (1982). Female choice selects for extreme tail length in a widowbird. *Nature*, **299**, 818–820.

Baerends, G. P., Brouwer R. and Waterbolk, H. Tj. (1955). Ethological studies on *Lebistes reticulatus* (Peters). I. An analysis of the male courtship pattern. *Behaviour*, **8**, 249–332.

Baldaccini, E. O. (1973). An ethological study of the reproductive behaviour including colour patterns of the cichlid fish *Tilapia mariae* (Boulenger). *Monit. Zool. Ital.*, **7**, 247–290.

Balon, E. K. (1984). Patterns in the evolution of reproductive styles in fishes. In: *Fish Reproduction: Strategies and Tactics* (Ed. by G. W. Potts and R. J. Wootton), pp. 35–53. New York: Academic Press.

Barlow, G. W. (1963). Ethology of the Asian teleost *Badis*. II. Motivation and signal value of the colour patterns. *Anim. Behav.*, **11**, 97–105.

Barlow, G. W. (1967). Social behavior of a South American leaf fish, *Polycentrus schomburgkii*, with an account of recurring pseudofemale behavior. *Amer. Midl. Natur.*, **78**, 215–234.

Barlow, G. W. (1967). The functional significance of the split-head color pattern as exemplified in a leaf fish, *Polycentrus schomburgkii*. *Ichthyologica*, **39** (2), 57–70.

Barlow, G. W. (1974). Contrasts in social behavior between Central American cichlid fishes and coral-reef surgeon fishes. *Amer. Zool.*, **14**, 9–34.

Barlow, G. W. (1976). The Midas cichlid in Nicaragua. In: *Investigations of the Ichthyofauna of Nicaraguan Lakes* (Ed. by T.B. Thorson), pp. 333–358. Lincoln: School of Life Sciences, University of Nebraska.

Barlow, G. W. (1981a). Patterns of parental investment, dispersal and size among coral-reef fishes. *Env. Biol. Fish.*, **6**, 65–85.

Barlow, G. W. (1981b). Genetics and the development of behavior, with special reference to patterned motor output. In: *Behavioral Development: The Bielefeld Interdisciplinary Project* (Ed. by K. Immelmann, G. W. Barlow, L. Petrinovich and M. Main), pp. 191–251. Cambridge: Cambridge University Press.

Barlow, G. W. (1984). Patterns of monogamy among teleost fishes. *Arch. FischWiss.*, **35** (Beiheft 1), 75–123.

Barlow, G. W. (1986). A comparison of monogamy among freshwater and coral-reef fishes. *Proc. Second Intern. Conf. Indo-Pac. Fish.*, pp. 767–775.

Barlow, G.W. (1991). Mating systems among cichlid fishes. In: *Cichlid Fishes. Behaviour, Ecology and Evolution* (Ed. by M. H. A. Keenleyside), pp. 173–190. New York: Chapman and Hall.

Barlow, G. W., Francis, R. C. and Baumgarter, J. V. (1990). Do the colours of parents, companions and self influence assortative mating in the polychromatic Midas cichlid? *Anim. Behav.*, **40**, 713–722.

Basolo, A. L. (1990). Female preference predates the evolution of the sword in swordtail fish. *Science*, **250**, 808–810.

Baylis, J. R. (1981). The evolution of parental care in fishes, with reference to Darwin's rule of male sexual selection. *Env. Biol. Fish.*, **6**, 223–251.

Berghe, E. P. van den and Gross, M. R. (1984). Female size and nest depth in coho salmon (*Oncorhynchus kisutch*). *Canad. J. Fish. Aquat. Sci.*, **41**, 204–206.

Blaustein, A. R. and O'Hara, R. K. (1982). Kin recognition in *Rana cascadae* tadpoles: Maternal and paternal effects. *Anim. Behav.*, **30**, 1151–1157.

Blumer, L. S. (1979). Male parental care in the bony fishes. *Quart. Rev. Biol.*, **54**, 149–161.

Blumer, L. S. (1985). Reproductive natural history of the brown bullhead *Ictalurus nebulosus* in Michigan. *Amer. Midl. Natur.*, **114**, 318–330.

Breder, C. M. and Rosen, D. E. (1966). *Modes of Reproduction in Fishes*. Garden City, New Jersey: Natural History Press.

Bruton, M. N. (ed.) (1989). *Alternative Life-History Styles In Animals*. Dordrecht: Kluwer.

Chapman, L. J. and Bevan, D. J. (1990). Development and field evaluation of a mini-spaghetti tag for individual identification of small fishes. *Amer. Fish. Soc. Symp.*, **7**, 101–108.

Clark, E., Pohle, M. and Rabin, J. (1991). Stability and flexibility through community dynamics of the spotted sandperch. *Nat. Geogr. Res. Explor.*, **7**, 138–155.

Clutton-Brock, T. and Godfray, C. (1991). Parental investment. In: *Behavioural Ecology, third ed.* (Ed. by J.R. Krebs and N.B. Davies), pp. 234–262. Boston: Blackwell.

Colin, P. L. and Bell, L. J. (1991). Aspects of the spawning of labrid and scarid fishes (Pisces:Labroidei) at Enewetak Atoll, Marshall Islands with notes on other families. *Env. Biol. Fish.*, **31**, 229–260.

Conover, D. O. and Kynard, B. E. (1981). Environmental sex determination: Interaction of temperature and genotype in a fish. *Science*, **213**, 577–579.

Cooper, J. C. and Hasler, A. D. (1976). Electrophysiological studies of morpholine imprinted coho salmon (*Oncorhynchus kisutch*) and rainbow trout (*Salmo gairdneri*). *J. Fish. Res. Bd. Can.*, **33**, 688–694.

Crapon de Caprona, M. D. and Fritzsch, B. (1984). Interspecific fertile hybrids of haplochromine Cichlidae (Teleostei) and their possible importance for speciation. *Neth. J. Zool.*, **34**, 503–538.

Crook, J. H. (1964). The evolution of social organisation and visual communication in the weaver birds (Ploceinae). *Behaviour Suppl.*, **10**, 1–178.

Danzmann, R. G., Ferguson, M. M. and Noakes, D. L. G. (1993). The genetic basis of fish behaviour. In: *Behaviour of Teleost Fishes, Second Edition* (Ed. by T.J. Pitcher), pp. 3–30. London: Chapman and Hall.

Davis, W. P. (1988). Reproductive and developmental responses in the self-fertilizing fish, *Rivulus marmoratus*, induced by the plasticizer, di-n-buyylphthalate. *Env. Biol. Fish.*, **21**, 81–90.

Dill, L. M. (1977). Refraction and the spitting behavior of the archerfish (*Toxotes chatareus*). *Beh. Ecol. Sociobiol.*, **2**, 169–184.

Dominey, W. (1983). Mobbing in colonially nesting fish, especially the bluegill. *Copeia*, 1086–1088.

Dominey, W. J. (1984). Effects of sexual selection and life history on speciation: Species flocks in African cichlids and Hawaiian *Drosophila*. In: *Evolution of Fish Species Flocks* (Ed. by A. A. Echelle and I. Kornfield), pp. 231–249. Orono: University of Main at Orono Press.

Dominey, W. J. and Blumer, L. (1984). Cannibalism of early life stages of fishes. In: *Infanticide: Comparative and Evolutionary Perspectives* (Ed. by G. Hausfater and S. Hardy Blaffer), pp. 43–64. Chicago: Aldine.

Donaldson, T. J. (1984). Mobbing behavior by *Stegastes albifasciatus* (Pomacentridae), a territorial mosaic damselfish. *Jap. J. Ichthyol.*, **31**, 345–348.

Dugatkin, L. A. (1988). Do guppies play TIT FOR TAT during predator inspection visits? *Beh. Ecol. Sociobiol.*, **23**, 395–399.

Emlen, S. T. (1991). Evolution of cooperative breeding in birds and mammals. In: *Behavioural Ecology: An Evolutionary Approach,* 3rd edn (Ed. by J. R. Krebs and N. B. Davies), pp. 301–337. Boston: Blackwell.

Endler, J. A. (1987). Predation, light intensity and courtship behaviour in *Poecilia reticulata* (Pisces: Poeciliidae). *Anim. Behav.*, **35**, 1376–1385.

Enquist, M. and Jakobsson, S. (1986). Decision making and assessment in the fighting behaviour of *Nannacara anomola* (Cichlidae, Pisces). *Ethology*, **72**, 143–153.

Falter, U. and Charlier, M. (1989). Mate choice in pure-bred and hybrid females of *Oreochromis niloticus* and *O. mossambicus* based upon visual stimuli (Pisces: Cichlidae). *Biol. Behav.*, **14**, 218–228.

Fautin, D. G. (1991). The anemonefish symbiosis: What is known and what is not. *Symbiosis*, **10**, 23–46.

Fischer, E. A. (1981). Sexual allocation in a simultaneously hermaphroditic coral reef fish. *Amer. Nat.*, **117**, 64–82.

Fisher, R. A. (1930). *The Genetical Theory of Natural Selection.* New York: Dover.

FitzGerald, G. J. (1991). The role of cannibalism in the reproductive ecology of the threespine stickleback. *Ethology*, **89**, 177–194.

FitzGerald, G. J. and Havre, N. van (1987). The adaptive significance of cannibalism in sticklebacks (Gasterosteidae: Pisces). *Beh. Ecol. Sociobiol.*, **20**, 125–128.

Francis, R. C. (In press). Sexual lability in teleosts: Developmental factors. *Quart. Rev. Biol.*

Franck, D. (1974). The genetic basis of evolutionary changes in behaviour patterns. In: *The Genetics of Behaviour* (Ed. by J. H. F. van Abeelen), pp. 119–140. Amsterdam: North Holland Publ. Co.

Fritzsch, B. and Crapon de Caprona, M.-D. (1990). Neurogenesis and learning. *Trends Neurosci.*, **13**, 328.

Fryer, G. (1969). Speciation and adaptive radiation in African lakes. *Verh. Internat. Verein. Limnol.*, **17**, 303–322.

Fryer, G. and Iles, T. D. (1968). Alternative routes to evolutionary success as exhibited by African cichlid fishes of the genus *Tilapia* and the species flocks of the Great Lakes. *Evolution*, **23**, 359–369.

Fryer, G. and Iles, T. D. (1972). *The Cichlid Fishes of the Great Lakes of Africa: Their Biology and Evolution.* Edinburgh: Oliver and Boyd.

Goulding, M. (1980). *The Fishes and the Forest: Explorations in Amazonian Ecology.* Berkeley: University of California Press.

Greenfield, D. W. (1979). A review of the western Atlantic *Starksia ocellata-complex* (Pisces: Clinidae) with the description of two new species and proposal of superspecies status. *Fieldiana Zool.*, **73**, 9–48.

Hammerstein, P. and Parker, G. A. (1982). The asymmetric war of attrition. *J. Theor. Biol.*, **96**, 647–682.

Hara, T. J. and Macdonald, S. (1975). Morpholine as olfactory stimulus in fish. *Science*, **187**, 81–82.

Harrington, R. W. Jr. (1968). Delimination of the thermolabile phenocritical period of sex determination and differentiation in the ontogeny of the normally hermaphroditic fish *Rivulus marmoratus* Poey. *Physiol. Zool.*, **41**, 447–460.

Hastings, P. A. and Petersen, C. W. (1986). A novel sexual pattern in serranid fishes: Simultaneous hermaphrodites and secondary males in *Serranus fasciatus*. *Env. Biol. Fish.*, **15**, 59–68.

Havre, N. van and FitzGerald, G. J. (1988). Shoaling and kin recognition in threespine sticklebacks (*Gasterosteus aculeatus* L.). *Biol. Behav.*, **13**, 190–201.

Heiligenberg, W. (1965). Color polymorphism in the males of an African cichlid fish. *J. Zool.*, **146**, 95–97.

Heiligenberg, W., Kramer, U. and Schulz, V. (1972). The angular orientation of the black eye-bar in *Haplochromis burtoni* (Cichlidae, Pisces) and its relevance to aggressivity. *Z. Vergl. Physiol.*, **76**, 168–176.

Heinrich, W. (1967). Untersuchungen zum sexualverhalten in der Gattung *Tilapia* (Cichlidae, Teleostei) und bei Artbastarden. *Z. Tierpsychol.*, **24**, 684–754.

Helfman, G. S. (1979). Twilight activities of yellow perch, *Perca flavescens*. *J. Fish. Res. Bd. Canad.*, **36**, 173–179.

Helfman, G. S. (1983). Underwater methods. In: *Fisheries Techniques* (Ed. by L. A. Nielsen and D. L. Johnson), pp. 249–369. Bethesda: American Fisheries Society.

Helfman, G. S. and Schultz, E. T. (1984). Social transmission of behavioural traditions in a coral reef fish. *Anim. Behav.*, **32**, 379–384.

Hoffman, S. G. (1983). Sex-related foraging behavior in sequentially hermaphroditic hogfishes (*Bodianus* spp.). *Ecology*, **64**, 798–808.

Houde, A. E. and Endler, J. A. (1990). Correlated evolution of female mating preferences and male color patterns in the guppy *Poecilia reticulata*. *Science*, **248**, 1405–1408.

Hourigan, T. F. (1989). Environmental determinants of butterflyfish social systems. *Env. Biol. Fish.*, **25**, 61–78.

Hubbs, C. L. (1948). "Leapfrogging" by topsmelt and shark. *Copeia*, 298.

Hubbs, C. L. (1955). Hybridization between fish species in nature. *Syst. Zool.*, **4**, 1–20.

Hubbs, C. L. and Turner, C. L. (1939). Studies of the fishes of the order Cyprinodontes. XVI. A revision of the Goodeidae. *Mich. Univ. Mus. Zool. Misc. Publ.*, **42**, 1–80.

Hunter, J. R. (1984). Synopsis of culture methods for marine fish larvae. In: *Ontogeny and Systematics of Fishes* (Ed. by H. G. Moser, W. J. Richards, D. M. Cohen, M. P. Fahay, A. W. Kendall, Jr. and S. L. Richardson), pp. 24–27. Special Publication Number 1, American Society of Ichthyologists and Herpetologists.

Jarman, P. J. (1974). The social organisation of antelope in relation to their ecology. *Behaviour*, **48**, 215–267.

Jenni, D. A. and Collier, G. (1972). Polyandry in the American jacana *Jacana spinosa*. *Auk*, **88**, 743–765.

Johnson, L. (1980). The Arctic charr, *Salvelinus alpinus*. In: *Charrs, Salmonid Fishes of the Genus Salvelinus* (Ed. by E. K. Balon), pp. 15–98. The Hague: Dr. W. Junk Publishers.

Karplus, I. (1987). The association between gobiid fishes and burrowing alpheid shrimps. *Oceanogr. Mar. Biol. Ann. Rev.*, **25**, 507–562.

Keenleyside, M. H. A. (1955). Some aspects of the schooling behaviour of fish. *Behaviour*, **8**, 183–248.

Keenleyside, M. H. A. (1991). Parental care. In: *Cichlid Fishes: Behaviour, Ecology and Evolution* (Ed. by M. H. A. Keenleyside), pp. 191–208. New York: Chapman and Hall.

Keenleyside, M. H. A. and Yamamoto, F. T. (1962). Territorial behavior of juvenile Atlantic salmon (*Salmo salar* L.). *Behaviour*, **19**, 138–169.

Kerr, S. R. (1989). The switch to size. D. Pauly and G. R. Morgan (ed.) (1987), Length-based Method in Fisheries Research. *Env. Biol. Fish.*, **24**, 157–159.

Kirkpatrick, M. and Ryan, M. J. (1991). The evolution of mating preferences and the paradox of the lek. *Nature*, **350**, 33–38.

Knight, A. E. (1990). Cold-branding techniques for estimating Atlantic salmon parr densities. In: *Fish-marking Techniques* (Ed. by N. C. Parker, A. E. Giorgi, R. C. Heidinger, D. B. Jester, Jr., E. D. Prince and G. A. Winans), pp. 36–37. Bethesda: American Fisheries Society.

Kodric-Brown, A. (1990). Mechanisms of sexual selection: Insights from fishes. *Ann. Zool. Fennici*, **27**, 87–100.

Krebs, J. R. and Davies, N. B. (1991). *Behavioural Ecology: An Evolutionary Approach.* 3rd Edn. Oxford: Blackwell Scientific.

Kuwamura, T. (1981). Mimicry of the cleaner wrasse *Labroides dimidiatus* by the blennies *Aspisdontus taeniatus* and *Plagiotremus rhinorhynchos. Nankiseibutu: Nanki Biol. Soc.,* **23**, 61–70.

Kuwamura, T. (1986). Parental care and mating systems of cichlid fishes in Lake Tanganyika: A preliminary field survey. *J. Ethol.,* **4**, 129–146.

Lagomarsino, I. V., Francis, R. C. and Barlow, G. W. (1988). The lack of correlation between size of egg and size of hatchling in the Midas cichlid, *Cichlasoma citrinellum. Copeia,* 1086–1089.

Loekle, D. M., Madison, D. M. and Christian, J. J. (1982). Time dependency and kin recognition of cannibalistic behavior among poeciliid fishes. *Behav. Neural Biol.,* **35**, 315–318.

Loiselle, P. V. and Barlow, G. W. (1978). Do fishes lek like birds? In: *Contrasts in Behavior* (Ed. by E. S. Reese and F. J. Lighter), pp. 33–75. New York: Wiley.

Lorenz, K. (1962). The function of colour in coral reef fishes. *Proc. Roy. Inst. Great Brit.,* **39**, 282–296.

Lorenz, K. Z. (1941). Vergleichende Bewegungsstudien an Anatinen. *J. Ornithol., Suppl.,* **3**, 194–293.

Lott, D. F. (1991). *Intraspecific Variation in the Social Systems of Wild Vertebrates.* Cambridge: Cambridge University Press.

Major, P. F. (1978). Predator–prey interactions in two schooling fishes, *Caranx ignobilis* and *Stolephorus purpureus. Anim. Behav.,* **26**, 760–777.

Marler, P. (1955). Studies of fighting in chaffinches. (2) The effect on dominance relations of disguising females as males. *Brit. J. Anim. Behav.,* **3**, 137–146.

Marshall, N. B. (1971). *Explorations in the Life of Fishes.* Cambridge: Harvard University Press.

Maynard Smith, J. (1977). Parental investment: A prospective analysis. *Anim. Behav.,* **25**, 1–9.

McKay, F. (1971). Behavioral aspects of population dynamics in unisexual–bisexual *Poeciliopsis* (Pisces: Poeciliidae). *Ecology,* **52**, 778–790.

McKaye, K. R. (1977). Defense of a predator's young by a herbivorous fish: An unusual strategy. *Amer. Nat.,* **111**, 301–315.

McKaye, K. R. (1981). Natural selection and the evolution of interspecific brood care in fishes. In: *Natural Selection of Social Behavior* (Ed. by R. Alexander and D. Tinkle), pp. 173–183. New York: Chiron.

McKaye, K. R. (1983). Ecology and breeding behavior of a cichlid fish, *Cyrtocara eucinostomus,* on a large lek in Lake Malawi. *Env. Biol. Fish.,* **8**, 81–96.

McKaye, K. R. (1985). Cichlid-catfish mutualistic defense of young in Lake Malawi, Africa. *Oecologia (Berlin),* **66**, 358–364.

McKaye, K. R. (1986). Trophic eggs and parental foraging for young by the catfish *Bagrus meriodionalis* of Lake Malawi, Africa. *Oecologia (Berlin),* **69**, 367–369.

McKaye, K. R. (1986). A unique form of biparental care by an African catfish: An evolutionary puzzle. *ANIMA Mag. Nat. Hist.,* **6**, 96–97.

McKaye, K. R. (1986). Trophic eggs and parental foraging for young by the catfish *Bagrus meridionalis* (Pisces: Bagridae) of Lake Malawi, Africa. *Oecologia (Berlin),* **69**, 367–369.

McKaye, K. R. and Barlow, G. W. (1976). Competition between color morphs of the Midas cichlid, *Cichlasoma citrinellum,* in Lake Jiloa, Nicaragua. In: *Investigations of the Ichthyofauna of Nicaraguan Lakes* (Ed. by T.B. Thorson), pp. 465–475. Lincoln: School of Life Sciences, University of Nebraska.

McKaye, K. R. and McKaye, N. M. (1977). Communal care and kidnapping of young by parental cichlids. *Evolution,* **3**, 674–681.

Meyer, A. (1987). Phenotypic plasticity and heterochrony in *Cichlasoma managuense* (Pisces, Cichlidae) and their implications for speciation in cichlid fishes. *Evolution,* **41**, 1357–1369.

Meyer-Holzapfel, M. (1960). Über das Spiel bei Fischen, insbesondere beim Tapirrusselfish (*Mormyrus kannume* Forskal). *Zool. Garten*, **25**, 189–202.

Milinski, M., Kulling, D. and Kettler, R. (1990). Tit for tat: Sticklebacks (*Gasterosteus aculeatus*) "trusting" a cooperative partner. *Behav. Ecol.*, **1**, 7–11.

Moring, J. R. (1990). Marking and tagging intertidal fishes: Review of techniques. *Amer. Fish. Soc. Symp.*, **7**, 109–116.

Motta, P. J. (1983). Response by potential prey to coral reef fish predators. *Anim. Behav.*, **31**, 1257–1259.

Moyer, J. T. (1979). Mating strategies and reproductive behavior of ostraciid fishes at Miyake-jima, Japan. *Jap. J. Ichthyol.*, **26**, 148–160.

Neil, E. H. (1964). An analysis of color changes and social behavior of *Tilapia mossambica*. *Univ. Calif. Publ. Zool.*, **75**, 1–58.

Nelissen, M. (1975). Contribution to the ethology of *Simochromis diagramma* (Gunther) (Pisces, Cichlidae). *Acta Zool. Path. Antv.*, **61**, 31–46.

Nelson, J. B. (1978). *The Sulidae: Gannets and Boobies. Aberdeen University Studies (154)*. Oxford: Oxford University Press.

Nikol'skii, G. V. (1961). *Special Ichthyology*. 2nd Edn (Translated from Russian). Washington, D.C.: Office of Technical Services, U.S. Department of Commerce.

Noakes, D. L. G. and Balon, E. K. (1982). Life histories of tilapias: An evolutionary perspective. In: *The Biology and Culture of Tilapias* (Ed. by R. S. V. Pullin and R. H. Lowe-McConnell), pp. 61–82. Manila: ICLARM Conference Proceedings 7, International Center for Living Aquatic Resources Management.

Noakes, D. L. G. and Godin, J.-G. J. (1988). Ontogeny of behavior and concurrent developmental changes in sensory systems in teleost fishes. In: *Fish Physiology, Volume XI The Physiology of Developing Fish, Part B: Viviparity and Posthatching Juveniles* (Ed. by W. S. Hoar and D. J. Randall), pp. 345–395. New York: Academic Press.

Noble, G. K. and Curtis, B. (1939). The social behavior of the jewel fish *Hemichromis bimaculatus* Gill. *Bull Amer. Mus. Nat. Hist.*, **75**, 1–46.

Olsen, K. H. (1989). Sibling recognition in juvenile Arctic charr, *Salvelinus alpinus* (L.). *J. Fish. Biol.*, **34**, 571–581.

Ormond, R. F. G. (1980). Aggressive mimicry and other interspecific feeding associations among Red Sea coral reef predators. *J. Zool. Lond.*, **191**, 247–262.

Partridge, B. L., Johansson, J. and Kalish, J. (1983). The structure of schools of giant tuna in Cape Cod Bay. *Env. Biol. Fish.*, **9**, 253–262.

Perrone, M. (1978). Mate size and breeding success in a monogamous cichlid fish. *Env. Biol. Fish.*, **3**, 193–201.

Policansky, D. (1982). Sex change in plants and animals. *Ann. Rev. Ecol. Syst.*, **13**, 471–495.

Powers, D. A. (1989). Fish as model systems. *Science*, **246**, 352–358.

Pressley, P. H. (1980). Pair formation and joint territoriality in a simultaneous hermaphrodite: The coral reef fish *Serranus tigrinus*. *Z. Tierpsychol.*, **56**, 33–46.

Pressley, P. H. (1981). Parental effort and the evolution of nest-guarding tactics in the threespine stickleback, *Gasterosteus aculeatus* L. *Evolution*, **35**, 282–295.

Quinn, T. P. and Busack, C. A. (1985). Chemosensory recognition of sibling in juvenile coho salmon (*Oncorhynchus kisutch*). *Anim. Behav.*, **33**, 51–56.

Quinn, T. P. and Hara, T. J. (1986). Sibling recognition and olfactory sensitivity in juvenile coho salmon. *Canad. J. Zool.*, **64**, 921–925.

Randall, J. E. and Randall, H. A. (1960). Examples of mimicry and protective resemblance in tropical marine fishes. *Bull. Mar. Sci. Gulf Carib.*, **10**, 440–480.

Reese, E. S. (1973). Duration of residence by coral reef fishes on "home" reefs. *Copeia*, 145–149.

Reighard, J. (1910). Methods of studying the habits of fishes, with an account of the breeding habits of the horned dace. *Bull. Bur. Fisher.*, **28**, 1111–1136.

Rohwer, S. (1978). Parent cannibalism of offspring and egg raiding as a courtship strategy. *Amer. Nat.*, **112**, 429–440.

Rosen, D. E. and Gordon, M. (1953). Functional anatomy and evolution of male genitalia in poeciliid fishes. *Zoologica*, **38**, 1–47.

Ross, R. M. (1990). The evolution of sex-change mechanisms in fishes. *Env. Biol. Fish.*, **29**, 81–93.

Rowland, W. J. (1975). The effects of dummy size and color on behavioral interaction in the jewel cichlid, *Hemichromis bimaculatus* Gill. *Behaviour*, **53**, 109–125.

Russell, B. C., Allen, G. R. and Lubbock, H. R. (1976). New cases of mimicry in marine fishes. *J. Zool. Lond.*, **180**, 407–423.

Schlupp, I., Parzefall, J. and Schartl, M. (1991). Male mate choice in mixed bisexual/unisexual breeding complexes of *Poecilia* (Teleostei: Poeciliidae). *Ethology*, **88**, 215–222.

Schultz, R. J. (1961). Reproductive mechanisms on unisexual and bisexual strains of the viviparous fish *Poeciliopsis*. *Evolution*, **15**, 302–325.

Schultz, R. J. (1977). Evolution and ecology of unisexual fishes. In: *Evolutionary Biology*. Volume 10 (Ed. by M. K. Hecht, W. C Steere and B. Wallace), pp. 277–331. New York: Plenum.

Selander, R. B. (1964). Sexual behavior in blister beetles (Coleoptera: Meloidae), I: The genus *Pyrota*. *Canad. Entomol.*, **96**, 1037–1082.

Simpson, M. J. A. (1968). The display of the Siamese fighting fish, *Betta splendens*. *Anim. Behav. Monogr.*, **1**, 1–73.

Sjölander, S. (1972). Feldbeogachtungen an einigen westafrikanischen Cichliden. *Aquar. Terr.*, **19** (2), 42–45; **19** (3), 86–88; **19** (4), 116–118.

Smith, R. J. F. and Smith, M. J. (1989). Predator-recognition behaviour in two species of gobiid fishes, *Asterropteryx semipunctatus* and *Gnatholepis anjerensis*. *Ethology*, **83**, 19–30.

Springer, V. G. and Gold, J. P. (1989). *Sharks in Question*. Washington: Smithsonian Institution Press.

Taborsky, M. (1984). Broodcare helpers in the cichlid fish *Lamprologus brichardi*: Their costs and benefits. *Anim. Behav.*, **32**, 1236–1252.

Taborsky, M. and Limberger, D. (1981). Helpers in fish. *Beh. Ecol. Sociobiol.*, **8**, 143–145.

Thresher, R. E. (1977). Ecological determinants of social organization of reef fishes. *Proc. Third Intern. Coral Reef Symp.*, pp. 551–559.

Townshend, T. J. and Wootton, R. J. (1985). Variation in the mating system of a biparental cichlid fish, *Cichlasoma panamense*. *Behaviour*, **95**, 181–197.

Trivers, R. L. (1971). The evolution of reciprocal altruism. *Quart. Rev. Biol.*, **46**, 35–57.

Verheijen, F. J. (1956). Transmission of a flight reaction amongst a school of fish and the underlying sensory mechanisms. *Experientia*, **12**, 202–204.

Waldman, B. (1984). Kin recognition and sibling association among wood frog (*Rana sylvatica*) tadpoles. *Beh. Ecol. Sociobiol.*, **14**, 171–180.

Ward, J. A. and Wyman, R. L. (1977). Ethology and ecology of cichlid fishes of the genus *Etroplus* in Sri Lanka: Preliminary findings. *Env. Biol. Fish.*, **2**, 137–145.

Warner, R. R. (1978). The evolution of hermaphroditism and unisexuality in aquatic and terrestrial vertebrates. In: *Contrasts in Behavior* (Ed. by E. S. Reese and F. J. Lighter), pp. 77–101. New York: Wiley.

Warner, R. R. (1984). Mating behavior and hermaphroditism in coral reef fishes. *Amer. Sci.*, **72**, 128–136.

Warner, R. R. (1988). Traditionality of mating-site preferences in a coral reef fish. *Nature*, **335**, 719–721.

Warner, R. R. and Hoffman, S. G. (1980). Local population size as a determinant of mating system and sexual composition in two tropical marine fishes (*Thalassoma* spp.). *Evolution*, **34**, 508–518.

Warner, R. R. and Robertson, D. R. (1978). Sexual patterns in the labroid fishes of the Western Caribbean. I: The wrasses (Labridae). *Smithson. Contr. Zool.*, **254**, 1–27.

Wickler, W. (1962). Eiattrappen und Maulbruten bei afrikanischen Cichliden: Zur Stammes-geschichte funktionnell korrelierter Organ- und Verhaltensmerkmale. *Z. Tierpsychol.*, **19**, 129–164.

Wickler, W. (1968). *Mimicry in Plants and Animals.* New York: McGraw-Hill.

Williams, G. C. (1966). *Adaptation and Natural Selection: A Critique of Some Current Evolutionary Thought.* Princeton: Princeton University Press.

Wunder, W. (1934). Gattungswahlversuche bei Stichling und Bitterling. *Verhandl. Deutsch. Zool. Gesellsch.*, **36**, 152–158.

Yanagisawa, Y. (1987). Social organization of a polygynous cichlid *Lamprologus furcifer* in Lake Tanganyika. *Jap. J. Ichthyol.*, **34**, 82–90.

Zimmermann, H. and Zimmermann, E. (1988). Etho-Taxonomie und zoogeographische Artengruppenbildung bei Pfeilgiftfroschen (Anura: Dendrobatidae). *Salamandra*, **24**, 125–160.

EVOLUTION OF ADAPTIVE VARIATION IN ANTIPREDATOR BEHAVIOUR

ANNE E. MAGURRAN*, BENONI H. SEGHERS*, GARY R. CARVALHO+
and PAUL W. SHAW+

*Department of Zoology, University of Oxford, South Parks Road,
Oxford, OX1 3PS, UK
+School of Biological Sciences, University College of Swansea, Singleton Park,
Swansea SA2 8PP, UK

In many species of fish, behaviour varies adaptively amongst populations in response to predation risk. One of the best examples is provided by the guppy, *Poecilia reticulata*, in Trinidad. Although separated by distances of a few km, or less, guppy populations vary in terms of predator assessment and avoidance, schooling, foraging behaviour, resource defence, female choice and mating tactics. We show that there are behavioural costs (such as lower levels of individual aggression and reduced female choice) associated with selection for a heightened antipredator response. In the majority of cases population variation in guppy behaviour can be clearly linked to the predation regime. Nevertheless, we have begun to uncover situations where there is behavioural divergence amongst populations apparently experiencing equivalent risk. We consider explanations for these differences including the possibility that they may be related to high levels of genetic divergence.

INTRODUCTION

Intraspecific variation in behaviour provides a forum for investigating evolution in action. The traditional approach has been to focus on a selective force and ascertain the extent to which a change in selection regime correlates with behavioural modification. Since untimely death due to predation has profound fitness consequences, there is little intuitive difficulty in accepting predation as an important form of selection. In addition, the relative risk of predation (at least at the level of the species of predator present in particular communities) is comparatively easy to quantify. The behavioural and evolutionary responses to a change in predation can be assessed by experiment and observation of natural populations. Consequently, there is now a large literature demonstrating that an increase in predation pressure leads to an adaptive shift in behaviour. Table 1 lists examples drawn from three species of fish—guppies, *Poecilia reticulata*, three-spined sticklebacks, *Gasterosteus aculeatus*, and European minnows, *Phoxinus phoxinus*—and illustrates the wide range of behaviours influenced by a change in predation regime.

Table 1 Adaptive variation in behaviour as a consequence of increased predation risk

Behaviour effect of increased predation pressure	Species	Reference
Schooling		
larger and more cohesive schools	P. reticulata	Seghers, 1974a
		Magurran and Seghers, 1991
		Breden et al., 1987
	P. phoxinus	Magurran and Pitcher, 1987
Evasion tactics		
more effectively integrated in high-risk	P. reticulata	Seghers, 1973
populations	G. aculeatus	Giles and Huntingford, 1984
	P. phoxinus	Magurran and Pitcher, 1987
Inspection and predator assessment		
increase in inspection frequency	G. aculeatus	Huntingford, (this volume)
	P. phoxinus	Magurran, 1986
increase in inspection group size	P. phoxinus	Magurran, 1990a, b
predator-specific attack cone avoidance	P. reticulata	Magurran and Seghers, 1990a
more efficient assessment of predator motivation	P. reticulata	Licht, 1989
more likely to employ tit for tat	G. aculeatus	Huntingford, 1992
	P. reticulata	Dugatkin and Alfieri, 1992
Response to overhead threat		
greatest in high-risk population	P. reticulata	Seghers, 1974b
	G. aculeatus	Giles and Huntingford, 1984
Response to alarm pheromone		
more pronounced behavioural change	P. phoxinus	Magurran and Pitcher, 1987
following exposure to Schreckstoff		Levesley and Magurran, 1988
Acquisition of defence skills		
predisposition to respond to experience	G. aculeatus	Tulley and Huntingford, 1987
greater in high-risk populations		Huntingford and Wright, 1989
	P. phoxinus	Magurran, 1990b
Habituation of antipredator responses		
slower in high-risk populations	G. aculeatus	Huntingford and Coulter, 1989
Habitat selection		
remain near surface and seek cover at edge of river	P. reticulata	Seghers, 1973
Foraging		
increased feeding tenacity	P. reticulata	Fraser and Gilliam, 1987
Female choice		
preference for less brightly coloured males	P. reticulata	Breden and Stoner, 1987
		Stoner and Breden, 1988
		Houde and Endler, 1990
avoidance of sneaky mating attempts compromised by predator avoidance	P. reticulata	Magurran and Nowak, 1991
Male mating tactics		
increased use of sneaky mating tactics in high risk populations	P. reticulata	Luyten and Liley, 1985
		Luyten and Liley, 1991
		Magurran and Seghers, 1990b

Table 1   Adaptive variation in behaviour as a consequence of increased predation risk   (Cont.)

Behaviour effect of increased predation pressure	Species	Reference
Aggression individual aggression reduced	*G. aculeatus*	Huntingford, 1982 Bakker and Feuth-De Bruijn, 1988
	P. reticulata	Ballin, 1973 Magurran and Seghers, 1991

BEHAVIOURAL DIVERGENCE: TRINIDADIAN GUPPIES

Population variation amongst guppy populations in Trinidad (Figure 1) is especially well documented. Not only is the species a very tractable organism for laboratory study, but there is also a clear gradation of predation regime within, as well as across, many rivers. As Table 1 reveals, there are population differences in a wide range of behaviours. A further extensive literature details the way in which predation influences guppy sexual selection (Endler, 1983) and life history (Reznick and Endler, 1982). Additional evidence for the pivotal role of selection in inducing population differentiation has been provided by experiments in which guppies transplanted across predation regimes within the same drainage rapidly evolve colour patterns and life history attributes appropriate to their new location (Reznick *et al.*, 1990).

Despite the wealth of information on population variation in Trinidadian guppies two areas have been little explored. First, few studies have considered behavioural changes contingent upon selection for more efficient predator avoidance. For instance, the ability to find food or mates may be compromised by the need to avoid predators. Second, there has been little attempt to explain variation that does not fit the adaptive model. In this paper we begin to address these issues.

For convenience we divide the Trinidad predation regimes into three main categories: sites with a range of piscivores including the pike cichlid (*Crenicichla alta*), an important guppy predator (Haskins *et al.*, 1961); sites with *Rivulus hartii* only (a minor predator which primarily targets juvenile guppies (Seghers, 1978; Liley and Seghers, 1975); and sites with *Macrobrachium crenulatum*, a freshwater prawn (often found in association with *R. hartii*) which is known to predate guppies (Endler, 1991). See Figure 1 for details.

BEHAVIOURAL TRADE-OFFS IMPOSED BY EFFECTIVE PREDATOR AVOIDANCE

Competition for Limited Resources

One interesting outcome of the studies of adaptive variation is that an increase in risk may modify behaviours not directly related to predator avoidance. For instance, well developed antipredator behaviour may reduce an individual's options for securing limited resources. Although schooling is an effective defence against

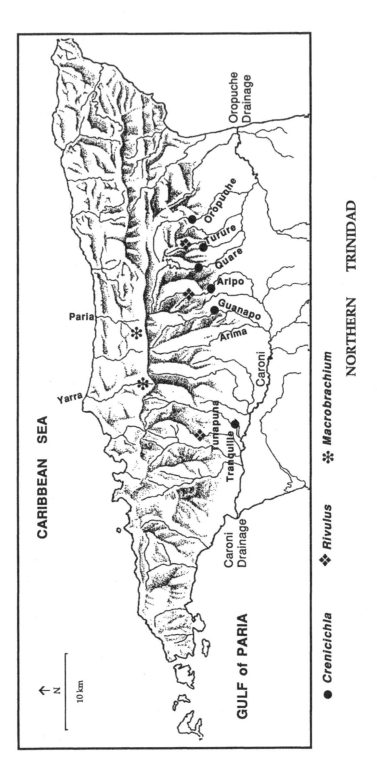

Figure 1 Location of study populations in Trinidad. The predator status of each site is indicated.

predation (Magurran, 1990a), it depends on coordination amongst school members and may restrict individual competition. We tested the hypothesis that levels of individual aggression will be reduced in fish with a high schooling tendency by comparing the behaviour of male guppies from 8 populations. Schooling behaviour was measured in the wild using the elective group size technique (Magurran and Pitcher, 1987). The observers watched quietly at the side of a stream and recorded the size and composition of adult guppy schools. Schools were defined as groups of fish in which individuals were within 5 body lengths of their nearest neighbour. As many as possible, but at least 30, different schools were recorded at each site. The median group size in the distribution of individuals against group size was taken as the measure of schooling tendency. EGS scores ranged from 1, for populations such as Paria (where there is only minimal risk from fish predators) to 21 for Tranquille (where guppies coexist with at least 12 predatory fish species (Magurran and Seghers, 1991)). Aggression, measured in the context of food patch defence, varied across populations with Paria and Tunapuna being significantly more aggressive and Tranquille guppies significantly less aggressive than the other populations (Figure 2).

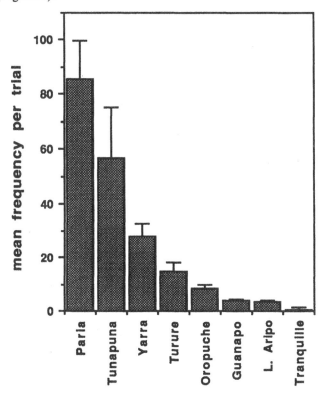

Figure 2 Mean (and SE) frequency of aggression in groups of 8 (wild) male guppies competing for a small food patch. Aggression was measured in the laboratory where the frequency of aggressive acts was recorded for a period of 20 minutes. Numbers of replicates (each with different fish) were as follows: Paria (n=5), Tunapuna (n=6), Yarra (n=6), Lower Turure (n=6), Oropuche (n=5), Guanapo (n=6), Lower Aripo (n=5), Tranquille (n=6). Aggression scores varied significantly (1 way ANOVA $F_{7,37}=49.0$, $P<0.001$). Redrawn from Magurran and Seghers (1991).

There was a significant inverse relationship between the ranked level of aggression (measured in the laboratory) and schooling tendency (measured in the wild) in males from the 8 populations (r_s=–0.90, P<0.01: Figure 3). These results, together with those obtained by Huntingford (1982) for sticklebacks, point towards a trade-off between antipredator behaviour and resource defence. A motivational link between antipredator behaviour and intraspecific aggression could provide an economical method of achieving this type of behavioural compromise (Huntingford, 1982; Magurran and Seghers, 1991). The behavioural consequences of trade-offs imposed by predation are explored in more detail by Magnhagen later in this volume.

Risk Sensitive Courtship

Tactics of courtship and mate choice can also be shown to vary adaptively in response to predation risk. Male guppies have essentially two ways of achieving a mating. They may either perform sigmoid displays in an attempt to persuade a receptive female to mate or opt for sneaky mating tactics. Females from low-risk populations prefer more brightly-coloured males (Stoner and Breden, 1988; Houde and Endler, 1990) while high-risk males show greater behavioural versatility under threat. One of our studies (Magurran and Seghers, 1990b) compared the courtship behaviour of males from two Trinidadian populations. Lower Aripo fish (see Figure 1) occur sympatrically with a range of piscivores including the pike cichlid, *C. alta*. Barrier waterfalls have prevented the upstream migration of these predators so that Upper Aripo guppies (which are genetically similar to their downstream

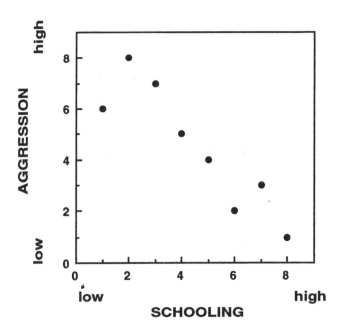

Figure 3 Relationship between (ranked) schooling tendency and aggression for male guppies from 8 Trinidadian populations. Redrawn from Magurran and Seghers (1991).

counterparts—see Figure 10) are subject to reduced risk. Upper Aripo males displayed risk-reckless courtship behaviour and did not reduce their sigmoid display rate or otherwise modify their courtship behaviour when threatened by two characins (*Astyanax bimaculatus*) which are potential predators (Liley and Seghers, 1975). The courtship behaviour of the Lower Aripo males was, by contrast, risk sensitive. These fish performed a lower proportion of sigmoid displays and increased their level of sneaky mating attempts in the presence of predators (Figure 4). Interestingly, males from upstream sites such as the Upper Aripo are more successful (in terms of actual inseminations) when mating in clear predator-free waters whereas downstream males do better in turbid (and potentially risky) conditions (Luyten and Liley, 1991).

Male behaviour under risk is not simply a direct reaction to danger but also an exploitation of female antipredator behaviour. In a subsequent study we focused on male/female interactions as a response to threat. Two high risk populations, Lower Aripo and Oropuche, (both from sites with high densities of the pike cichlid, *C. alta* (Douglas and Endler, 1982)) were investigated. Fish were tested in groups of 10 (5 adults of each sex). The behaviour of each individual was monitored for 5 min in a low risk (undisturbed) situation and in the presence of a threatening, but non-attacking, blue acara (*Aequidens pulcher*). Data were collected for 20 individuals of each sex and population. Males from both populations significantly reduced their display rate (Wilcoxon signed-ranks test, LA: z=3.18, P<0.01; Oro: z=2.02, P<0.05) and increased their sneaky mating attempts (LA: z=1.97, P<0.05; Oro z=3.25,

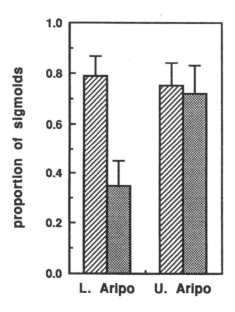

Figure 4 Courtship behaviour of Lower Aripo and Upper Aripo male guppies in the absence (stripes) and presence (stipples) of predators. The mean (and 95% confidence limits) proportion of mating attempts that involved sigmoids is shown for both experimental conditions. Each male was observed for 5 min and could be recognised by his unique colour pattern. There were n=30 males per population. Redrawn from Magurran and Seghers (1990b).

$P<0.01$—Figure 5—in the presence of the predator. Females made fewer attempts to avoid the advances of males when under threat (LA: $z=3.41$, $P<0.001$; Oro: $z=3.60$, $P<0.001$) and received more gonopodial thrusts as a consequence (LA: $z=3.12$, $P<0.01$; Oro $z=3.26$, $P<0.01$)—Figure 6. Most inspections of the predator are initiated and led by female guppies. This provides males with an excellent opportunity for attempting a sneaky mating while the females are otherwise preoccupied (Magurran and Nowak, 1991).

Females in the experiments would have been, like the majority of females in the wild, in a sexually non-receptive phase. Such fish continuously try to avoid the unceasing mating activity of males. This male avoidance behaviour is prejudiced by the presence of predators.

NON-ADAPTIVE VARIATION?

Schooling

The schooling tendency and inspection tactics of most populations of guppies can be predicted from prevailing predation risk. We have, however, discovered a number of cases where there are substantial levels of behavioural divergence within given predation regimes. Two comparisons underline this point. The upstream sections of the Paria and Yarra rivers contain high densities of the freshwater prawn, *M. crenulatum* (Douglas and Endler, 1982). Laboratory studies show that Yarra fish have substantially higher levels of schooling than Paria fish (Figure 7).

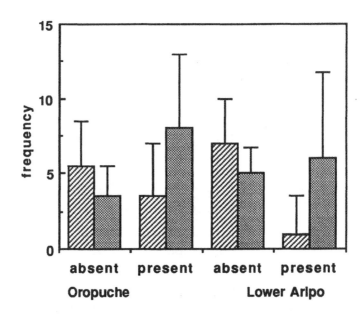

Figure 5 Male courtship behaviour as a response to risk. Median frequencies (and upper quartile) of sigmoid displays (stripes) and gonopodial thrusts (stipples) are shown.

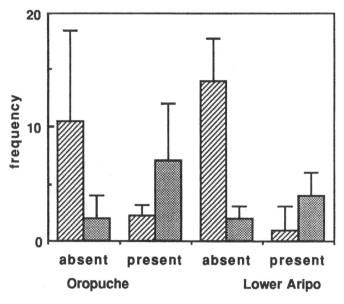

Figure 6   Female behaviour as a consequence of risk. Median number (and upper quartile range) of times that females avoided males (stripes) and received thrusts (stipples) per 5 minutes is indicated.

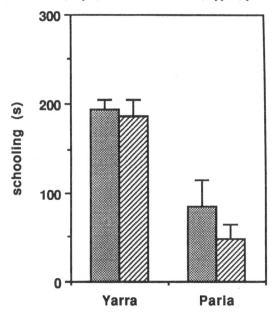

Figure 7   Mean (and SE) schooling tendency of Paria and Yarra guppies. Females are denoted by stipples, males by stripes. There were n=6 replicates of Paria fish per sex and n=8 replicates of Yarra fish. All fish were bred and raised in the laboratory. In this test the focal fish was allowed to associate freely with 4 individuals of the same sex and population. Schooling behaviour was measured by recording the number of seconds out of 5 minutes that the focal fish spent within 5 body lengths of its nearest neighbour. Males and females were tested separately. The portion of the Yarra River where the guppies originated is also known as the Limon River.

Likewise, Oropuche and Lower Aripo guppies (Figure 8) have equivalent levels of predation but quite different schooling tendencies. The low schooling scores of the Oropuche fish cannot be attributed to reduced numbers of predators since our own observations, and those of other workers (for example Douglas and Endler (1982)) point to high densities of *C. alta* in this river. The elevated schooling of female guppies is probably related to an increased perception of risk—see Magurran and Nowak (1991) for a discussion.

A comparison of four populations, Paria and Yarra, Guanapo and Oropuche also showed that inspection tactics are determined by schooling tendency rather than predation regime. Guanapo guppies (from the Caroni drainage) resemble Lower Aripo fish in their behaviour.

	High Fish Predation	*Prawn Predation*
High Schooling	Guanapo	Yarra
Low Schooling	Oropuche	Paria

In females, for example, there is a significant negative relationship (Figure 9) between schooling tendency and the proportion of solo inspections (data with arcsin transformation: $F_{1,28}=658.67$, $P<0.001$)

Since schooling behaviour offers important protection against predators such as the pike cichlid (Magurran, 1990a) it is not clear why levels of schooling are reduced in Oropuche fish. Similarly, it is difficult to explain why Paria and Yarra guppies

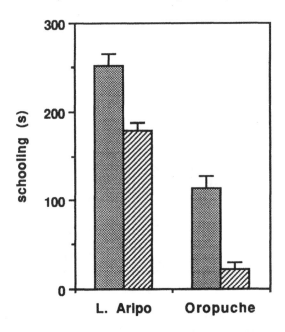

Figure 8 Mean (and SE) schooling tendency of guppies from the Lower Aripo and Oropuche Rivers. Females are denoted by stipples, males by stripes. All fish were bred and raised in the laboratory. There were 24 replicates (i.e. tests using separate focal fish) for each sex and population. In this test schooling behaviour was recorded as the number of seconds that a focal fish spent within 5 body lengths of a clear plastic bottle containing 6 guppies of the same sex.

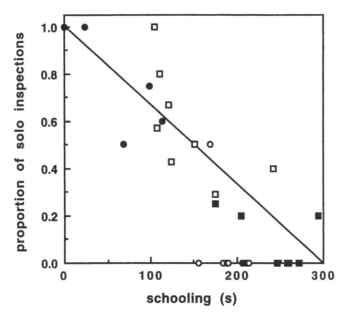

Figure 9 Proportion of solo inspections by female guppies in relation to schooling tendency (defined as seconds spent within 5 body lengths of conspecifics during a 5 min trial). The four populations are denoted as follows: Paria (solid circle); Yarra (open circle); Oropuche (open square); Guanapo (closed square).

behave so differently. Intriguingly, this behavioural divergence is paralleled by considerable genetic divergence.

GENETIC DIVERGENCE

Analysis of biochemical genetic variation among guppy populations in N. Trinidad using allozyme electrophoresis (Carvalho *et al.*, 1991) demonstrates high levels of genetic differentiation amongst populations from different rivers. Significant differences in allele frequencies (see Carvalho *et al.* (1991)) exist between the Oropuche and Lower Aripo, and Paria and Yarra populations. Figure 10 illustrates this extreme divergence between populations in terms of an overall similarity statistic, Nei's (1972) Genetic Similarity I (Oropuche – Lower Aripo I = 0.926; Paria – Yarra I = 0.886). The values of I shown are very low compared to most inter-population values within other fish (for example Buth *et al.* (1991) observed that I > 0.94 in 78% of comparisons). For comparison of levels of differentiation found between guppy populations in the same river a sample from the Upper Aripo (data from Carvalho *et al.* (1991)) is also included in Figure 10. The genetic similarity of the Guanapo and Lower Aripo population is I = 0.956.

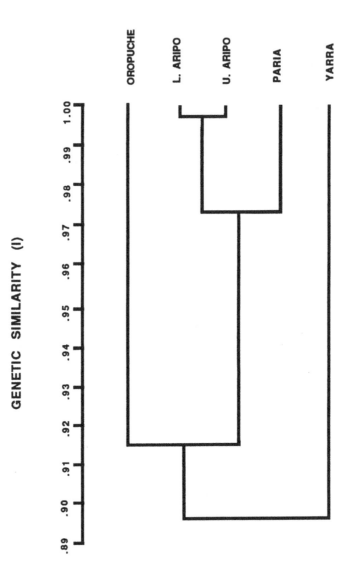

Figure 10 Dendrogram of mean genetic identities between five populations of *Poecilia reticulata* in N. Trinidad. Calculations (UPGMA–cophenetic correlation = 0.954) are based on 35 alleles at 23 loci. Redrawn from Carvalho *et al.* (1991).

DISCUSSION

Two previous studies, one on life-history tactics, the other on male coloration, have rejected the notion that either historical factors, or watershed differences contribute to observed patterns of variation.

Douglas and Endler (1982) performed a multivariate comparison (using the Mantel procedure) of 41 guppy sites in Trinidad. They assessed male colour patterns in the context of predation regime, altitude and watershed. Four alternative evolutionary models were put forward. Models 1 and 2 proposed that colour patterns vary clinally and patchily and could not be rejected. Model 3 considered the effect of distance while Model 4 explicitly tested historical factors by comparing sites within the same watersheds. Neither was supported by their analysis. Thus, there appears to be no evidence that colour patterns cluster within watershed independently of predation regime.

Strauss (1990) reanalysed Reznick and Endler's (1982) data using principal components analysis. He found a large amount (17%) of residual variation after adjusting for female body size and predation effects. This residual variation could not be attributed to drainage system or to geographical proximity. Again, the phylogenetic history of a population seems to play little part in shaping its life history tactics. Interestingly, however, a recent assessment of the life history tactics of Paria and Yarra guppies (Rodd and Reznick, 1991) points towards substantial differences in fecundity.

Why do we observe watershed differences in some aspects of behaviour when they appear to be absent from investigations of other traits? One hypothesis which could explain the surprising difference between the Yarra and Paria populations is the relative geographical isolation of each population from areas of high predation from other fish species. Although the lower portions of both rivers contain a wide variety of predators, notably *Eleotris pisonis* (Endler, 1991a), a high waterfall on the lower section of the Paria serves as a complete barrier to these predators thereby protecting the entire guppy population in this river.

By contrast, the situation in the Yarra River is quite different: the lower Yarra guppies are exposed to predators and have well-developed antipredator behaviour. Although the predators do not occur in the upper portions of the Yarra (where the present study is focused), gene flow between adjacent demes presumably links sub-populations along the entire length of the river. Consequently, behavioural adaptations may not be as finely tuned to local conditions as they are in the Paria, where gene flow from high predation populations is virtually impossible.

The difference between the Oropuche and Lower Aripo also eludes simple answers. Since the predisposition to respond to experience varies across populations of other fish species (for example minnows, *P. phoxinus* (Magurran, 1990b)) it could be that Oropuche and Lower Aripo fish react differentially to early behavioural cues. Experiments are required in order to disentangle the effects of inheritance and early experience.

Another possibility is that different behaviour patterns could be equally successful in terms of long term reproductive fitness. Thus, Oropuche guppies could be trading-off schooling against some other (as yet unknown) behaviour or life-history tactic. Finally, behaviour could genuinely be less malleable than colour or life-history traits. It may, for example, be that genetic constraints, arising as a legacy from founding events or past histories, limit the extent to which natural selection can induce change.

CONCLUSIONS

Intraspecific variation in morphology, behaviour and life-history tactics provides some of the most compelling evidence for natural selection in the wild (Endler, 1985). The work we report here does not undermine the importance of selection in inducing adaptive variation in behaviour. Indeed, most of the behaviour patterns described can be unequivocally linked to the prevailing predation regime. We have, however, uncovered a number of examples where population differences cannot be explained purely in terms of current selection pressures (at least as they have been hitherto understood). In particular we have shown that there can be considerable variation in schooling and associated behaviours of fish occurring in similar ecological communities. At present we do not know whether the different behaviour patterns are equally successful (in terms of reproductive fitness) or whether sub-optimal strategies are retained in some populations and represent a genetic legacy from the founding fish. Nevertheless, our investigation does underline the importance of considering stochastic as well as deterministic factors when attempting to explain observed behavioural variation.

ACKNOWLEDGMENTS

Financial support from the Royal Society and the Natural Environment Research Council (UK) is gratefully acknowledged.

References

Bakker, T. C. M. and Feuth-de Bruijn, E. (1988). Juvenile territoriality in stickleback *Gasterosteus aculeatus* L. *Anim. Behav.*, **36**, 1556–1558.

Ballin, P. J. (1973). Geographic variation in courtship behaviour of the guppy, *Poecilia reticulata*. M.Sc. thesis. British Columbia.

Breden, F., Scott, M. and Michel, E. (1987). Genetic differentiation for anti-predator behaviour in the Trinidad guppy, *Poecilia reticulata. Anim. Behav.*, **35**, 618–620.

Breden, F. and Stoner, G. (1987). Male predation risk determines female preference in the Trinidad guppy. *Nature*, **329**, 831–833.

Buth, D. G., Dowling, T. E. and Gold, J. R. (1991). Molecular and cytological investigations. In: *Cyprinid Fishes: Systematics, Biology and Exploitation* (Ed. by I. J. Winfield and J. S. Nelson), pp. 83–126. London: Chapman and Hall.

Carvalho, G. R., Shaw, P. W., Magurran, A. E. and Seghers, B. H. (1991). Marked genetic divergence revealed by allozymes among populations of the guppy, *Poecilia reticulata* (Poeciliidae) in Trinidad. *Biol. J. Linn. Soc.*, **42**, 389–405.

Douglas, M. E. and Endler, J. A. (1982). Quantitative matrix comparisons in ecological and evolutionary investigations. *J. Theor. Biol.*, **99**, 777–795.

Dugatkin, L. A. and Alfieri, M. (1991). Guppies and the Tit for Tat strategy: preference based on past interaction. *Behav. Ecol. Sociobiol.*, **28**, 243–246.

Endler, J. A. (1983). Natural and sexual selection on color patterns in poeciliid fishes. *Environ. Biol. Fish.*, **9**, 173–190.

Endler, J. A. (1985). *Natural Selection in the Wild.* Princeton: Princeton University Press.

Endler, J. A. (1991). Interactions between predators and prey. In: *Behavioural Ecology: an Evolutionary Approach* (Ed. by J. R. Krebs and N. B. Davies), pp. 169–196. Oxford: Blackwell.

Endler, J. A. (1991). Variation in the appearance of guppy color patterns to guppies and their predators under different visual conditions. *Vision Res.*, **31**, 587–608.

Fraser, D. F. and Gilliam, J. F. (1987). Feeding under predation hazard: response of the guppy and Hart's rivulus from sites with contrasting predation hazard. *Behav. Ecol. Sociobiol.*, **21**, 203–209.

Giles, N. and Huntingford, F. A. (1984). Predation risk and interpopulation variation in anti-predator behaviour in the three-spined stickleback, *Gasterosteus aculeatus* L. *Anim. Behav.*, **32**, 264–275.

Haskins, C. P., Haskins, E. F., McLaughlin, J. J. A. and Hewitt, R. E. (1961). Polymorphism and population structure in *Lebistes reticulatus*, a population study. In: *Vertebrate Speciation* (Ed. by W. F. Blair), pp. 320–395. Austin: University of Texas Press.

Houde, A. E. (1988). Genetic difference in female choice between two guppy populations. *Anim. Behav.*, **36**, 510–516.

Houde, A. E. and Endler, J. A. (1990). Correlated evolution of female mating preferences and male color pattern in the guppy, *Poecilia reticulata*. *Science*, **248**, 1405–1408.

Huntingford, F. A. (1982). Do inter- and intra-specific aggression vary in relation to predation pressure in sticklebacks? *Anim. Behav.*, **30**, 909–916.

Huntingford, F. A. and Coulter, R. M. (1989). Habituation of predator inspection in the three-spined stickleback, *Gasterosteus aculeatus* L. *J. Fish Biol.*, **35**, 153–154.

Huntingford, F. A. and Wright, P. J. (1989). How sticklebacks learn to avoid dangerous feeding patches. *Behav. Processes*, **19**, 181–189.

Levesley, P. B. and Magurran, A. E. (1988). Population differences in the reaction of minnows to alarm substance. *J. Fish Biol.*, **32**, 699–706.

Licht, T. (1989). Discriminating between hungry and satiated predators: the response of guppies (*Poecilia reticulata*) from high and low predation sites. *Ethology*, **82**, 238–242.

Liley, N. R. and Seghers, B. H. (1975). Factors affecting the morphology and behaviour of guppies in Trinidad. In: *Function and Evolution in Behaviour* (Ed. by G. P. Baerends, C. Beer and A. Manning), pp. 92–118. Oxford: Clarendon Press.

Luyten, P. H. and Liley, N. R. (1985). Geographic variation in the sexual behaviour of the guppy, *Poecilia reticulata* (Peters). *Behaviour*, **95**, 164–179.

Luyten, P. H. and Liley, N. R. (1991). Sexual selection and competitive mating success of male guppies (*Poecilia reticulata*) from four Trinidad populations. *Behav. Ecol. Sociobiol.*, **28**, 329–336.

Magurran, A. E. (1986). Predator inspection behaviour in minnow shoals: differences between population and individuals. *Behav. Ecol. Sociobiol.*, **19**, 267–273.

Magurran, A. E. (1990a). The adaptive significance of schooling as an anti-predator defence in fish. *Ann. Zool. Fennici*, **27**, 51–66.

Magurran, A. E. (1990b). The inheritance and development of minnow anti-predator behaviour. *Anim. Behav.*, **39**, 834–842.

Magurran, A. E. and Nowak, M. N. (1991). Another battle of the sexes: the consequences of sexual asymmetry in mating costs and predation risk in the guppy, *Poecilia reticulata*. *Proc. R. Soc. Lond. B.*, **246**, 31–38.

Magurran, A. E. and Pitcher, T. J. (1987). Provenance, shoal size and the sociobiology of predator evasion behaviour in minnow shoals. *Proc. R. Soc. Lond. B.*, **229**, 439–465.

Magurran, A. E. and Seghers, B. H. (1990a). Population differences in predator recognition and attack cone avoidance in the guppy, *Poecilia reticulata*. *Anim. Behav.*, **40**, 443–452.

Magurran, A. E. and Seghers, B. H. (1990b). Risk sensitive courtship in the guppy (*Poecilia reticulata*). *Behaviour*, **112**, 194–201.

Magurran, A. E. and Seghers, B. H. (1991). Variation in schooling and aggression amongst guppy (*Poecilia reticulata*) populations in Trinidad. *Behaviour*, **118**, 214–234.

Nei, M. (1972). Genetic distance between populations. *Am. Nat.*, **106**, 283–292.

Reznick, D. A., Bryga, H. and Endler, J. A. (1990). Experimentally induced life-history evolution in a natural population. *Nature*, **346**, 357–359.

Reznick, D. N. and Endler, J. A. (1982). The impact of predation on life history evolution in Trinidadian guppies (*Poecilia reticulata*). *Evolution*, **36**, 160–177.

Rodd, F. H. and Reznick, D. N. (1991). Life history evolution in guppies: III. The impact of prawn predation on guppy life histories. *Oikos*, **62**, 13–19.

Seghers, B. H. (1973). An analysis of geographic variation in the antipredator adaptations of the guppy, *Poecilia reticulata*. Ph.D. thesis. University of British Columbia.

Seghers, B. H. (1974a). Schooling behavior in the guppy (*Poecilia reticulata*): an evolutionary response to predation. *Evolution*, **28**, 486–489.

Seghers, B. H. (1974b). Geographic variation in the responses of guppies (*Poecilia reticulata*) to aerial predators. *Oecologia*, **14**, 93–98.

Seghers, B. H. (1978). Feeding behavior and terrestrial locomotion in the cyprinodontid fish, *Rivulus hartii* (Boulenger). *Verh. Internat. Verein. Limnol.*, **20**, 2055–2059.

Stoner, G. and Breden, F. (1988). Phenotypic differentiation in female preference related to geographic variation in predation risk in the Trinidad guppy (*Poecilia reticulata*). *Behav. Ecol. Sociobiol.*, **22**, 285–291.

Strauss, R. E. (1990). Predation and life-history variation in *Poecilia reticulata* (Cyprinodontiformes: Poeciliidae). *Env. Biol. Fishes*, **27**, 121–130.

Tulley, J. J. and Huntingford, F. A. (1987). Parental care and the development of adaptive variation in anti-predator responses in sticklebacks. *Anim. Behav.*, **35**, 1570–1572.

THE DEVELOPMENT OF ADAPTIVE VARIATION IN PREDATOR AVOIDANCE IN FRESHWATER FISHES

FELICITY A. HUNTINGFORD and PETER J. WRIGHT[†]

Department of Zoology, University of Glasgow, Glasgow G12 8QQ, UK

POPULATION DIFFERENCES IN PREDATOR AVOIDANCE

In many prey species, the nature and intensity of anti-predator responses vary in relation to local predation regimes. For example, in ground squirrels (Towers and Coss, 1990), prairie dogs (Loughry, 1989), garter snakes (Herzog and Schwartz, 1990), salamanders (Dowdey and Brodie, 1989) and spiders (Riechert and Hedrick, 1990) animals from populations that are naturally exposed to a high degree of predation risk show particularly well-developed anti-predator responses. Small freshwater fish have proved particularly suitable subjects for investigating this phenomenon. On the one hand, they often occur in isolated populations exposed to different suites of predators; on the other, they adapt well to laboratory aquaria and so behavioural differences can readily be identified and quantified under standardised conditions. A classic example is provided by Segher's work on guppies (*Poecilia reticulata*), in which the strength of schooling (a protective strategy, see Krebs and Davies, 1987) in fish from different sites is directly related to the level of risk from piscivorous fish that they experience in nature (Figure 1; Seghers, 1974). More recent work on the same system has demonstrated that other aspects of anti-predator responses such as predator inspection (see Pitcher *et al.*, 1986; Magurran, 1986) show a similar pattern of risk-related variation and that the suite of responses in any given population are those that offer greatest protection against indigenous predator species. For example, guppies from a site where piscivorous fish are abundant but freshwater prawns *Macrobrachium* spp. (which eat guppies) are rare kept a large distance from such predatory fish during staged confrontations and avoided the region of the predator's jaw during inspection visits; they failed to show such 'attack-cone avoidance' when confronted with a prawn. In contrast, guppies from a site with abundant prawns but few predatory fish avoided the prawn (especially the mouth and claws) but not the fish predator. Finally, guppies from a site where risk of predation from both types of predator was low responded weakly to both (Magurran and Seghers, 1990, and see the article by Magurran *et al.* in this volume). Similar results have been found for minnows (*Phoxinus phoxinus*; Magurran, 1986; Magurran and Pitcher, 1987) and for sticklebacks (*Gasterosteus aculeatus*; Huntingford, 1982; Giles and Huntingford, 1984, Tulley and Huntingford, 1987a; and see below). In a slightly

---

†. Present academic address: SOAFD Marine Laboratory, Victoria Rd, Torry, Aberdeen AB9 8DB, UK.

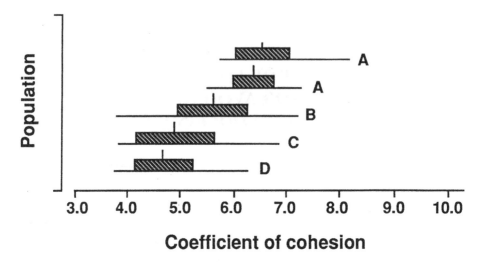

Figure 1 Strength of schooling (coefficient of cohesion) in laboratory-reared guppies from 5 populations in Trinidad with different regimes of predation by piscivorous fish. A = abundant carachins and cichlids, which eat adult guppies; B = high densities of *Rivulus* spp., which eat young guppies; C = medium densities of *Rivulus*; D = low densities of *Rivulus* (redrawn from Seghers, 1974).

different context, in sites where egg cannibalism is common (but not otherwise), when confronted with a cannibalistic raiding party, male sticklebacks with a brood of eggs in their nest show distraction displays (for example, making the rooting actions typical of a stickleback feeding on a nest at a site well away from his own brood; Foster, 1988).

Relating such between-population variation to local environmental conditions has been an important tool in behavioural ecology, since it can, potentially, show how phenotypes evolve in response to local selection regimes. For example, population comparisons of guppies have been used to test the hypothesis that reduced adult survival selects for an early age of maturation and increased reproductive effort (and the converse). At sites containing fish that feed selectively on adults, guppies mature earlier and invest more heavily in reproduction than is the case at sites containing fish that prey on younger age classes (Reznick, 1982). In the same way, a plausible explanation for the population differences described above is that in sites where a particular category of predator is abundant natural selection has favoured phenotypes with well developed protective responses. Protective fright responses interfere with other important activities such as feeding (Milinski and Heller, 1978; Ibrahim and Huntingford, 1989; Lima and Dill, 1990), fighting (Ukegbu and Huntingford, 1986; Magurran and Seghers, 1991) and reproduction (Magnhagen, 1990, and see the article by Magnhagen in this volume), so where predators are rare, prey individuals in which fright is hard to elicit have done better. Therefore, the argument would go, with time, protective behaviour has undergone local adaptive radiation.

THE DEVELOPMENT OF POPULATION DIFFERENCES

This scenario assumes that the behavioural differences are inherited, rather than being the result of individual experience acting in each generation; to establish if this is the case, it is necessary to investigate how these differences develop. As far anti-predator responses are concerned, there is good reason to question the assumption of inherited differences, since there is extensive evidence that small fish (and other prey species) readily alter their behaviour as a result of experience. For example, in coho salmon (*Onchorhynchus kisutch*; Olla *et al.*, 1992) experience of predatory attack results in improvement in anti-predator responses. Minnows with no previous experience of predation given a single 10min exposure to a model pike show more predator inspection than naive fish and are more likely to school rather than to take cover (Figure 2; Magurran, 1990 and see below). Conversely, paradise fish (*Macropodus opercularis*) given a single brief opportunity to inspect a large goldfish (which is potentially dangerous on account of its size, but actually does not attack) recognise that such a fish poses no threat and behave accordingly months after the initial exposure (Csanyi *et al.*, 1989, and in this volume).

Thus learning mechanisms certainly exist whereby differential experience of predatory attack might generate adaptive variation in anti-predator behaviour. However, where such effects have been demonstrated, this does not rule out the existence of inherited effects and, indeed, such effects have been identified for a number of species. For example, Figure 2 shows that, even when inexperienced, laboratory-reared minnows from a site with abundant pike school more strongly than do those from a site with no pike (Magurran, 1990). Population-specific suites of protective responses also develop in predator-naive guppies (the data in Figure 1 refer to guppies reared in the laboratory for five generations, Seghers, 1974; Breden *et al.*, 1987), and sticklebacks (Giles, 1983; Tulley and Huntingford, 1987a and see below). It therefore seems that in these species behavioural differences between

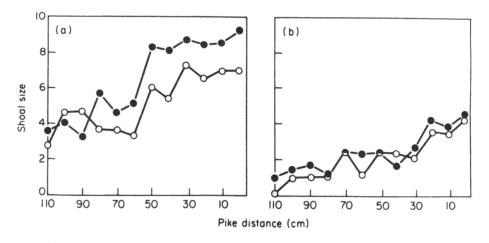

Figure 2   Shoaling tendency in minnows from (a) a high-risk site (Dorset) and (b) a low-risk site (Gwynedd) in relation to the distance of an approaching model pike. Open circles represents naive fish; closed circles represent fish with prior experience of a hunting pike (from Magurran,1990).

populations are at least partly the result of inherited behavioural traits, but that these can, potentially, be amplified by differential experience of direct attack. In this article, we consider the way in which inherited behavioural differences interact with experience to generate the marked differences in anti-predator behaviour that characterise sticklebacks from sites exposed to different levels of predation risk.

DEVELOPMENT OF ANTI-PREDATOR RESPONSES IN STICKLEBACKS

In spite of their protective spines, sticklebacks are vulnerable to a variety of predators and have a suite of behavioural adaptations that provide additional protection against attack (Hoogland *et al.,* 1957; Huntingford, 1976). These include: vigilance in potentially dangerous environments, inspection visits on siting a potential predator, a variety of evasive manoeuvres in response to attack and a subsequent period during which other responses are suppressed. All sticklebacks have the capacity to show these responses, but the extent to which they do so is extremely variable. Some of this variability can be related to site of origin; broad surveys of response to a standardised encounter with predators showed that sticklebacks from sites where predators (which might be fish or birds) are abundant (high-risk fish) show stronger protective responses when exposed to such predators than do those from safer sites (low-risk fish; Huntingford, 1982; Giles and Huntingford, 1984).

Following these initial broad surveys, more detailed studies were conducted of sticklebacks from a subset of these sites, to characterise precisely the nature of the behavioural differences between high-risk and low-risk fish. The three sites chosen were: Inverleith Pond, Edinburgh, a long-established urban pond where there are no piscivorous fish; the River Endrick, in the Loch Lomond catchment area where piscivorous birds and fish are relatively common and Strumore River on North Uist in the Outer Hebrides, where the sticklebacks have lost their armature (Campbell and Williamson, 1979; Giles, 1983; Bell, 1988) and, in terms of morphological protection, are especially vulnerable to the predatory fish and birds with which they coexist. When confronted in a laboratory aquarium with a realistic fibreglass model trout moved according to a standard protocol, sticklebacks from the two high-risk sites were particularly vigilant (responding more strongly to the initial movement of the predator; Figure 3), showed fast and well-oriented escape responses when attacked and had a long period of behavioural suppression following an attack; this is not the case for fish from the low risk site (Huntingford and Wright, unpublished data). These adaptive differences might be inherited responses to different selection regimes or they might be the result of differential experience of predatory attacks. Alternatively, and most probably, both might be involved.

To investigate the possibility of inherited differences, young sticklebacks from the three study sites were reared in the laboratory, with no experience of predatory attack. When they were 3–4 months old (*c.* 35 mm, by which time anti-predator responses are well established in wild-caught, high-risk fish, Giles, 1983; Tulley and Huntingford, 1987a) their responses to an appropriately-sized model predatory fish, moved according to a standard protocol, were compared. High-risk fish showed stronger responses both to the first movement of the predator (Figure 4a) and to a direct attack, escaped at a faster speed and in a more appropriate direction and took longer to return to normal activity afterwards; Huntingford and Wright, unpublished

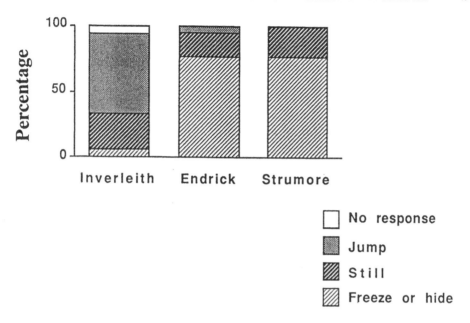

Figure 3 Responses of wild-caught stickleback from the three study sites to the first movement of a model piscivorous fish. Open bar = no response; stippled bar = jump; dark hatching = still; light hatching = freeze or hide. Freeze is distinguished from still by cessation of fin movements and suppression of respiratory movements. $Chi^2 = 21.9$, df = 2, $P \ll 0.001$; low-risk fish are distinct from the two high-risk groups (from Huntingford and Wright, unpublished data).

data). In other words, the special features of the behaviour of sticklebacks from dangerous sites (high vigilance, effective escape and long behavioural inhibition) are present in predator-naive fish, so direct experience of predatory attack is clearly not necessary for the development of these responses.

THE ROLE OF PATERNAL CARE

The subjects of this experiment were cared for by their father in the first few days after hatching. Previous studies suggest that this social experience might be important in the development of anti-predator responses, since sticklebacks denied contact with their father are less effective in later encounters with predatory fish (Benzie, 1965, Figure 5a; Tulley and Huntingford, 1987b). In a slightly different context Goodey and Liley (1986) have shown that encounters with older conspecifics promote the development of predator avoidance in guppies. Other groups of sticklebacks from the same sites were therefore reared as orphans and then tested with a model fish as before. None of the differences described for wild-caught and normally-reared fish from the two mainland sites were found in these orphaned fish; high-risk sticklebacks now showed the weak responses characteristic of their low-risk counterparts (Figure 4b). In marked contrast, orphaned fish from the North Uist site responded strongly and effectively to the model predator throughout the test (Huntingford and Wright, unpublished data).

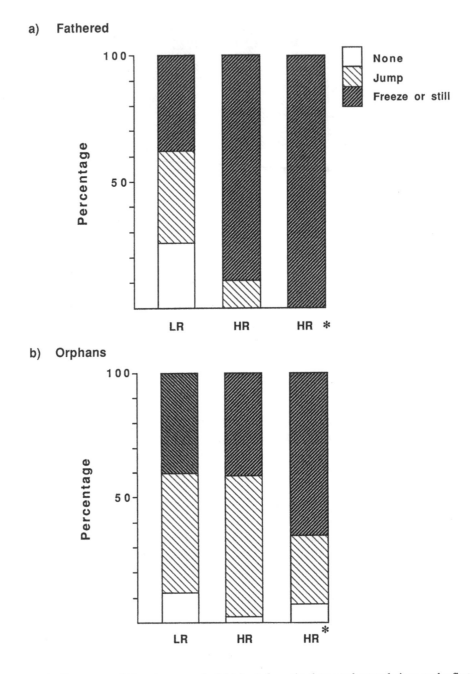

Figure 4 Responses of laboratory-reared stickleback from the three study populations to the first movement of a model piscivorous fish (a) Normally-reared, fathered fish; Chi^2 = 28.4, df = 4, $P \ll$ 0.001; low-risk fish are distinct from the two high-risk groups. (b) Orphans. Chi^2 = 9.3, df = 4, 0.10 > P > 0.05; high-risk fish from North Uist are (marginally) distinct from the other two groups (from Huntingford and Wright, unpublished data).

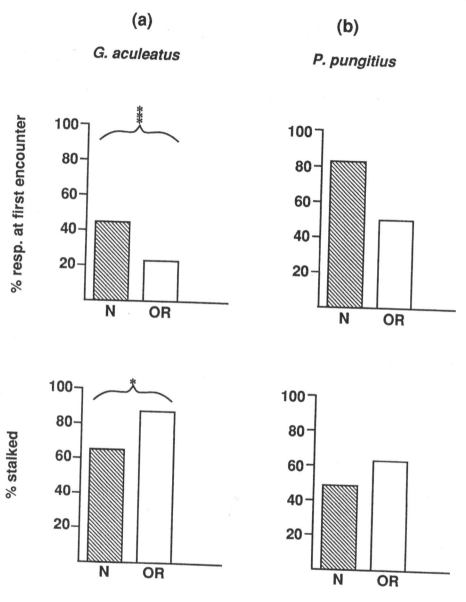

Figure 5 The proportion of normally-reared (N) and orphaned (OR) three-spined sticklebacks, *G. aculeatus* (a) and ten-spined sticklebacks, *Pungitius pungitius* (b) that took shelter in weed at the first encounter with a hunting pike and that were subsequently stalked by the predator (redrawn from Benzie, 1965).

Thus the well developed anti-predator responses typical of high risk fish develop regardless of rearing conditions in sticklebacks from North Uist, but for the mainland high risk fish, their full development depends on a period of parental care.

That the attentions of a father should have an effect is not so bizarre as it might seem, given that parental males chase and catch their fry (before spitting them back into the nest) when they first emerge from the nest (see Wootton, 1976). But why should this effect vary from population to population?

The obvious suggestion is that the quality or quantity of paternal care differs in sticklebacks from the three study sites. Analysis of filmed interactions between stickleback fathers and their fry in the laboratory and in the field showed no difference in the overall duration of paternal care in the two mainland sites (2–3 days post hatching in the laboratory and somewhat longer in the wild, where broods were larger), nor in the percentage of time during the parental period that fathers spent patrolling their territory looking for straying fry (about 80% of observation period in both cases). However, after the first post-hatching day, sticklebacks from North Uist breeding in laboratory tanks showed rather little interest in their fry, which rapidly left the nest and remained dispersed, hidden in the substrate. In the field, these fish nest in the open in fast-flowing streams and the fry get washed away from the nest as soon as they emerge; there is therefore little opportunity for retrieval (Wright and Wright, in press).

A difference was found in the quality of the father-fry interaction in sticklebacks from the two mainland sites (Huntingford and Wright, unpublished data). On the day of hatching, high-risk fry were more likely than low-risk fry to respond to their father's approach, and the responses they gave were stronger (remaining motionless rather than simply giving a startle response; Figure 6). By 2 days after hatching, more fry from both populations were reacting to their father, but the stronger responses of the high-risk fry were even more marked; these were associated with faster escape by the fry and (possibly as a consequence) with faster retrieval attempts by the fathers (Figure 7). Over the parental period, high-risk fry became more effective at avoiding the attentions of their father (making fast, well-oriented escape responses and remaining still afterwards), so that by the end of the period of paternal care they already possessed many of the behavioural features that protect wild-caught adults at this site from predatory fish. Throughout the paternal period, low-risk fry continued to make weak responses, jumping up towards the approaching father, whose success at retrieval did not decline; as fry that stray far from the nest run a risk of cannibalism in Inverleith Pond, this is probably an adaptive trait.

These results suggest that the population differences described in wild-caught fish from the three study sites develop as follows: sticklebacks from North Uist, where circumstances prevent a long period of paternal care, hatch with protective responses already well developed. A similar situation may exist in the related ten-spined stickleback (*Pungitius pungitius*). In this species breeding males build nests above the ground in clumps of vegetation (see Wootton, 1976). Soon after hatching, the fry disperse gradually from the nest and take cover in the weeds where they receive little by way of attention from their father (Benzie, 1965). So in this species fry do not experience retrieval and here too (as in three-spined sticklebacks from North Uist) anti-predator behaviour develops more-or-less fully in orphaned fish (Figure 5b).

Returning to three-spined sticklebacks, in contrast to the fish from North Uist, newly-hatched fry from the mainland site with abundant predators are only slightly more responsive to the approach of their father than are fry from the low-risk site.

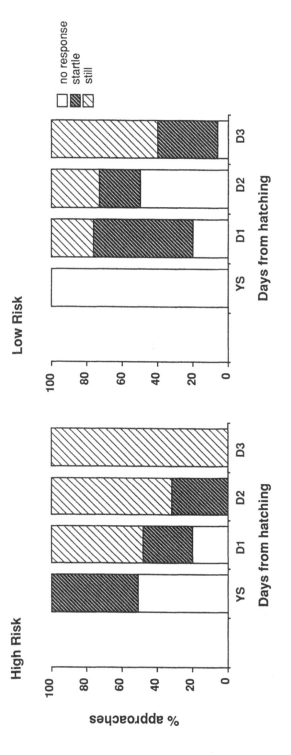

Figure 6 The proportion of fry from the two mainland sites showing different categories of response to approaches by their father on successive days after hatching. Analysing at the level of broods, the proportion of fry showing no response is significantly higher for low-risk fish on day YS and day 3 (Mann-Whitney U = 0, P < 0.05 in both cases). For both high-risk and low-risk fry, the proportion of no responses increases with age (U = 0, P < 0.05 in each case; Huntingford and Wright, unpublished data).

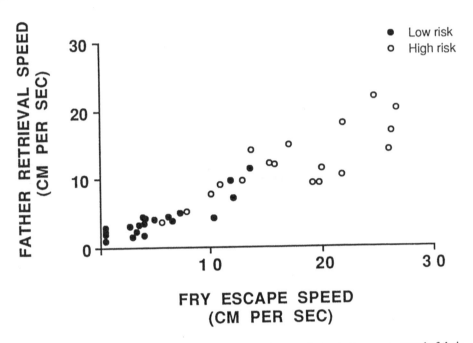

Figure 7 Speed of retrieval attempts by paternal stickleback in relation to the escape speed of their fry in fish from a high-risk site and a low-risk site. Escape and attack speeds are significantly correlated in both sets of fish (R = 0.80 and 0.88, N = 18 and 20, P < 0.01 for high- and low-risk fish respectively) and both are significantly lower for low-risk fish (Mann-Whitney U = 19 and 17, P << 0.001 for fry and father speed respectively; Huntingford and Wright, unpublished data).

This initial small difference is amplified and extended by the more active chasing that high-risk fry experience during the parental phase; by the time they are independent they already have a well developed repertoire of evasive responses to their father. Thus during interactions with their father, high-risk fry acquire exactly those traits that typify wild-caught fish from the same site, through a learning process that apparently generalises to later encounters with real predatory fish, but (incidentally) not to encounters with avian predators (Huntingford and Wright, unpublished data).

EFFECTS OF DIRECT EXPERIENCE OF PREDATORS

However, this is probably not the whole story; a number of lines of evidence suggest that differential experience during the period of parental care may interact with a predisposition in high-risk sticklebacks to learn from adverse experience. Looking at the effects of experience of predatory attack on the development of behaviour in two closely related species, Benzie (1965) subjected groups of 3 month old laboratory-reared three-spined and ten-spined sticklebacks to a 12 h encounter with a live hunting pike, during which a small proportion of fish were killed. The responses of these experienced fish to a subsequent standardised exposure to a pike were compared to those of inexperienced sticklebacks of the same age (Figure 8a). In three-spined sticklebacks, this experience produced a marked improvement in various aspects of anti-predator responses, with experienced fish being more likely to take fright when first meeting the pike, making better use of weed and being more effective at avoiding attack and capture. There is some evidence to suggest that such learning effects act in the wild; sticklebacks caught at a high risk site at an age of 3 months and subsequently acclimatised to laboratory conditions showed stronger responses to a model of a piscivorous bird than did fish from the same site and of the same age that had been reared in the laboratory (Figure 9). No such differences were observed when laboratory-reared and wild-caught fish from a low risk site were compared (both of which showed weak responses), suggesting that where encounters with predators are common, behavioural performance improves. Benzie (1965) found few significant effects of experience of predation on anti-predator responses of ten-spined sticklebacks (Figure 8b); there are a number of possible reasons for this, one being a species difference in the readiness with which behaviour is modified in the light of experience. At the population level, Magurran (1990) showed that the effect of experience of predation in minnows depends on predation risk in their site of origin; improved performance was most marked in fish from the population that naturally coexists with pike (Figure 2), suggesting that such fish might have an inherited predisposition to modify their behaviour in the light of adverse experience. To quantify population differences in learning ability in three-spined sticklebacks more formally, Huntingford and Wright (1989) compared rates of avoidance learning in wild-caught fish from the two mainland sites. The fish were housed in the centre of a tank divided into three sections, a home compartment with weed but no food separated by partitions with sliding doors (distinguished by coloured symbols) from two feeding compartments, each with a food patch visible from the door. Together the two patches contained just enough food to satiate the fish. Once per day the doors were opened, allowing the fish access to the food patches; they learned quickly to enter one feeding patch as soon as the door was opened and to swim across to the other patch once the first was depleted. In every case the fish had a preferred side that they usually entered first and in which they spent most of their foraging time. Once the criterion for learning to use the food patches was reached, avoidance training began; during a daily period of 5 minutes, each time the fish entered its preferred compartment and approached the feeding patch it received a simulated attack from a model predator. From the first day and throughout the training period, fish from both sites responded to the attack (by jumping and then either freezing for a variable period or escaping to the home compartment). The criterion for learning was 3 days in a row in which the fish never entered the previously-favoured-but-now-dangerous compartment at all. Although this would seem to be a difficult task (the dangerous patch had been highly favoured, the fish were given only 5 minutes training per day and, since they could see the undepleted food patch in the dangerous compartment, the criterion was quite stringent), in almost every case, the fish learned to avoid the dangerous compartment altogether. However, high-risk fish were significantly quicker to learn this avoidance task (and received fewer attacks in the process) than their low-risk counterparts (Huntingford and Wright, 1989). The fish used in this experiment were caught from the wild and so the high-risk fish might well have had experience of predatory attack, but the same learning differences were found

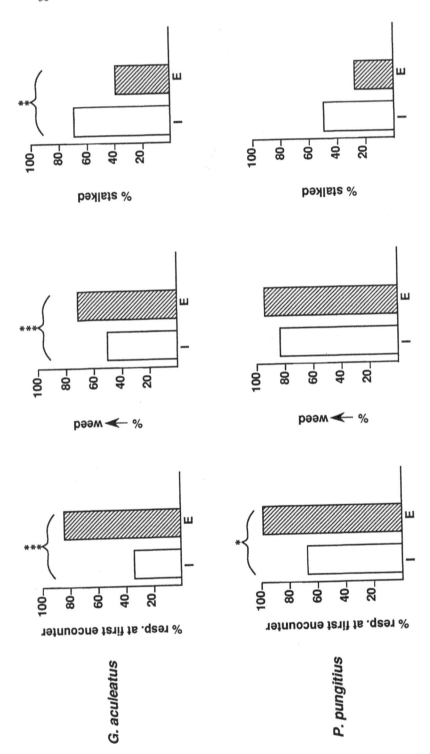

Figure 8 The proportion of inexperienced and experienced three-spined sticklebacks, *G. aculeatus*, and of ten-spined sticklebacks, *P. pungitius*, that took fright at the first encounter with a hunting pike, that took cover in the weed following an escape and that were stalked by the pike (redrawn from data in Benzie, 1965).

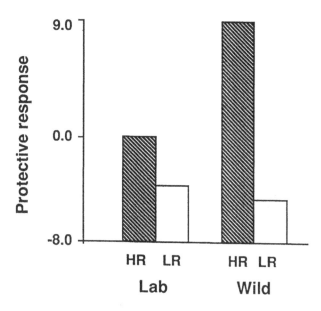

Figure 9 Multi-variate score summarising the strength of protective response shown by predator-naive and wild-caught 30 mm (standard length) stickleback from a high- and a low-risk site to exposure to a model avian predator. HR = high-risk fish; LR = low-risk fish (redrawn from Tulley and Huntingford, 1987a).

for predator-naive fish (Figure 10). While the cause of this difference in rate of learning is not clear, the indications are that, although the actual strength of attack was standardised, this was perceived as a greater risk and so represented a stronger negatively reinforcing stimulus for high risk sticklebacks (Huntingford and Wright, unpublished data).

CONCLUSIONS

There is now an extensive body of literature on small fish, and other prey species, describing the way in which population-specific patterns of anti-predator behaviour can be related to the nature and intensity of prevailing predation risk. To the extent that these patterns are inherited, they can be regarded as micro-evolutionary adaptations to local selective regimes, and so can be used to elucidate the way in which natural selection acts on behavioural phenotypes. In all the species of fish discussed in this article, it has been shown experimentally that experience of direct predatory attack is not necessary for the development of population-specific patterns of protective responses. One might, therefore, describe these patterns as inherited, but it is clear from the examples described above that their developmental origins are complex and variable, both within and between species. First of all, it is clear that inherited differences in response to a particular category of predator can be amplified by differential experience. Most obviously, predation itself is an important developmental influence. Thus in sticklebacks (Benzie, 1965),

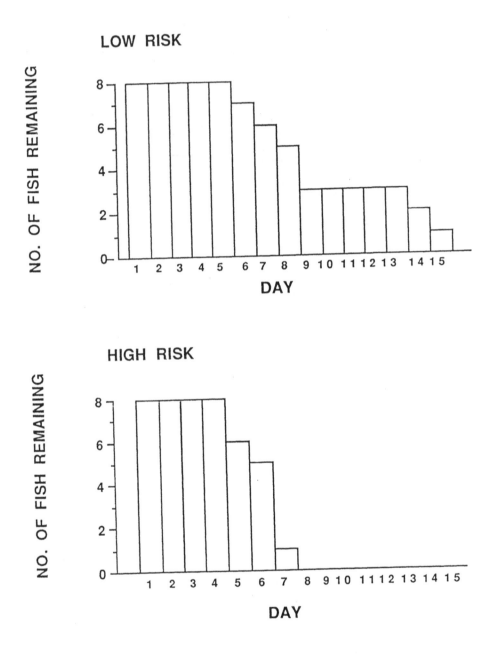

Figure 10 The number of predator-naive stickleback from the two study sites that had failed to reach the criterion for learning to avoid a previously-favoured but now-dangerous feeding patch on successive days of avoidance training (Mann-Whitney test on days to reach criterion: $H = 6.05$, n_1 and $n_2 = 8$, $P < 0.05$; redrawn from Huntingford and Wright, 1992).

salmon (Olla *et al.*, 1992), minnows (Magurran, 1990) and paradise fish (Czanyi *et al.*, 1989) experience during encounters with predators produces marked and long term alterations in behaviour. Experience that apparently has little to do with predation can also influence the development of anti-predator responses, which are promoted in sticklebacks (from some populations at least) by paternal retrieval attempts and in guppies by interaction with older conspecifics (Goodey and Liley, 1986). There is little information as to whether the amount of chasing received by young guppies is population-specific, although it seems that cannibalism on newly-born fry occurs more frequently in sites with low predation risk, (Magurran, pers. comm.) where schooling is weak and levels of aggression are high (Magurran and Seghers, 1991) Secondly, it is not simply that post-hatching experience amplifies inherited behaviour differences; the same experience can have differential effects depending on the genotype of the fish concerned. For example, in both minnows and sticklebacks, fish from high-risk sites show more marked behavioural changes following a standardised predatory encounter than did low-risk fish. For the marked differences in predation regime discussed in this article, the general level of predation prevalent at a given site is likely to be sufficiently predictable for inherited differences to produce fish with an appropriate level of overall response. However, local variation, both spatial and temporal, would favour an additional degree of flexibility. The ability to adapt behaviour in the light of experience during encounters with predators provides a mechanism for such fine-tuning. It is easy to see how a predisposition to learn from adverse experience would be advantageous in high-risk sites, but is more difficult to find adaptive explanations for poor learning in low-risk fish. There might possibly be a cost to maintaining the ability to learn at a high level, but we suggest the following alternative idea. Although sticklebacks from low-risk sites will escape from direct attack, this experience does not produce any lasting behavioural disturbance. At their site of origin, where predation is not a serious threat, this is probably adaptive as it leaves the fish free for other important activities. In the context of our learning tests, it means that our standardised attack is a relatively weak negatively reinforcing stimulus for low-risk fish, whose poor performance is therefore a by-product of a behavioural trait that is adaptive in other contexts.

ACKNOWLEDGMENTS

We would like to thank Jayne Tierney and Roddy MacDonald for help in conducting a number of these experiments, Mr R. Campbell for drawing our attention to the spine-deficient sticklebacks from North Uist (and to the North Uist Estate for permission to collect them) and many colleagues for helpful comments at various stages in the study. This work was funded by SERC grant no. GR/F/10743.

References

Bell, M. (1990). Stickleback fishes: bridging the gap between population biology and paleobiology. *Trends Ecol. Evol.*, **3**, 320–325.
Benzie, V. L. (1965). Some aspects of the anti-predator responses of two species of sticklebacks. Ph.D. thesis. University of Oxford.

Breden, F., Scott, M.A. and Michel, E. (1987). Genetic differentation for anti-predator behaviour in the Trinidad guppy, *Poecilia reticulata*. *Anim. Behav.*, **35**, 618–620.

Campbell, R. N. and Williamson, R. B. (1979). The fishes of inland waters in the Outer Hebrides. *Proc. Roy. Soc. Edinburgh B*, **77**, 377–394.

Czanyi, V., Csizmadia, G. and Miklosi, A. (1989). Long-term memory and recognition of another species in the paradise fish. *Anim. Behav.*, **37**, 908–911.

Dowdey, T. G. and Brodie, E. D. (1989). Antipredator strategies of salamanders: individual and geographical variation in responses of *Eurycea bislineata* to snakes. *Anim. Behav.*, **38**, 707–711.

Foster, S. A. (1988). Diversionary displays of paternal stickleback. Defences against cannibalistic groups. *Behav. Ecol. Sociobiol.*, **22**, 335–340.

Giles, N. (1983). The possible role of environmental calcium levels during the evolution of phenotypic diversity in Outer Hebrides populations of three-spined stickleback, *Gasterosteus aculeatus*. *J. Zool., London*, **199**, 535–544.

Giles, N. and Huntingford, F. A. (1984). Predation risk and inter-population variation in anti-predator behaviour in the three-spined stickleback, *Gasterosteus aculeatus*. *Anim. Behav.*, **32**, 264–275.

Goodey, W. and Liley, N. R. (1986). The influence of early experience on escape behaviour in the guppy. *Can. J. Zool.*, **64**, 885–888.

Herzog, H. A. and Schwartz, J. M. (1990). Geographical variation in anti-predator behaviour of neonate garter snakes, *Thamnophis sirtalis*. *Anim. Behav.*, **40**, 579–601.

Hoogland, R., Morris, D. and Tinbergen, N. (1957). The spines of sticklebacks (*Pygosteus* and *Gasterosteus*) as a means of defence against predators. *Behaviour*, **10**, 205–236.

Huntingford, F. A. (1976). The relationship between anti-predator behaviour and aggression among conspecifics in the three-spined stickleback, *Gasterosteus aculeatus*. *Anim. Behav.*, **24**, 245–260.

Huntingford, F. A. (1982). Do inter- and intra-specific aggression vary in relation to predation pressure in sticklebacks? *Anim. Behav.*, **30**, 909–916.

Huntingford, F. A. and Wright, P. J. (1989). How sticklebacks learn to avoid dangerous feeding patches. *Behav. Processes*, **19**, 181–189.

Ibrahim, A. I. and Huntingford, F. A. (1989). Laboratory and field studies of the effect of predation risk on foraging in three-spined sticklebacks (*Gasterosteus aculeatus*). *Behaviour*, **109**, 46–57.

Krebs, J. R. and Davies, N. B. (1987). *An Introduction to Behavioural Ecology*, 2nd edn. Oxford: Blackwell Scientific.

Lima, S. L. and Dill, L. M. (1990). Behavioural decisions made under the risk of predation: a review and a prospectus. *Can. J. Zool.*, **68**, 619–640.

Loughry, W. J. (1989). Discrimination of snakes by two populations of black tailed prairie dogs. *Behav. Ecol. Sociobiol.*, **19**, 828–834.

Magnhagen, C. (1990). Reproduction under predation risk in the sand goby, *Pomatoschistus minutus*, and the black goby, *Gobius niger*: the effects of age and longevity. *Behav. Ecol. Sociobiol.*, **26**, 331–335.

Magurran, A. E. (1986). Predator inspection behaviour in minnow shoals: differences between populations and individuals. *Behav. Ecol. Sociobiol.*, **19**, 267–273.

Magurran, A. E. (1990). The inheritance and development of minnow anti-predator behaviour. *Anim. Behav.*, **39**, 828–834.

Magurran, A. E. and Pitcher, T. J. (1987). Provenance, shoal size and the sociobiology of predator evasion behaviour in minnow shoals. *Proc. Roy. Soc., Lond. B*, **229**, 439–465.

Magurran, A. E. and Seghers, B. (1990). Population differences in predator recognition and attack cone avoidance in the guppy *Poecilia reticulata*. *Anim. Behav.*, **40**, 443–452.

Magurran, A. E. and Seghers, B. (1991). Variation in schooling and aggression amongst guppy (*Poecilia reticulata*) populations in Trinidad. *Behaviour*, **118**, 214–234.

Milinski, M. and Heller, R. (1978). Influence of a predator on the optimal foraging behaviour of sticklebacks *Gasterosteus aculeatus*. *Nature, Lond.*, **275**, 642–644.

Olla, B. I., Davis, M. W. and Ryer, C. H. (1992). Foraging and predator avoidance in hatchery-reared Pacific salmon; achievement of behavioural potential. In: *Behavioural Concepts in Aquaculture* (Ed. by J. E. Thorpe, and F. A. Huntingford). World Aquaculture Series.

Pitcher, T. J., Green, D. and Magurran, A. E. (1986). Dicing with death: predator inspection behaviour in minnow shoals. *J. Fish Biol.*, **28**, 439–448.

Reznick, D. N. and Endler, J. A. (1982). The impact of predation on life history evolution in Trinidadian guppies (*Poecilia reticulata*). *Evolution*, **36**, 160–177.

Riechert, S. E. and Hedrick, A. V. (1990). Levels of predation and genetically-based anti-predator behaviour in the spider *Agelenosis aperta*. *Anim. Behav.*, **40**, 679–687.

Seghers, B. H. (1974). Schooling behaviour in the guppy (*Poecilia reticulata*): an evolutionary response to predation. *Evolution*, **28**, 286–289.

Towers, S. R. and Coss, R. G. (1990). Confronting snakes in the burrow: snake-species discrimination and anti-snake tactics of two California ground squirrel populations. *Ethology*, **87**, 177–192.

Tulley, J. J. and Huntingford, F. A. (1987a). Age, experience and the development of adaptive variation in anti-predator responses in three-spined stickleback, *Gasterosteus aculeatus*. *Ethology*, **75**, 285–290.

Tulley, J. J. and Huntingford, F. A. (1987b). Parental care and the development of adaptive variation in anti-predator responses in sticklebacks. *Anim. Behav.*, **35**, 1570–1572.

Ukegbu, A. A. and Huntingford, F. A. (1986). Brood value and life expectancy as determinants of parental investment in the three-spined stickleback (*Gasterosteus aculeatus*). *Ethology*, **75**, 285–290.

Wootton, R. J. (1976). *The Biology of Sticklebacks*. London: Academic Press.

Wright, P. J. and Huntingford, F. A. (1992). Inherited population differences in avoidance conditioning in three-spined sticklebacks (*Gasterosteus aculeatus*). *Behaviour*, **122**, 264–273.

Wright, P. J. and Wright, K. (In press). A field study of the reproductive behaviour of spine-deficient sticklebacks from North Uist, Outer Hebrides. *Scottish Naturalist*.

LEARNING INTERACTIONS BETWEEN PREY AND PREDATOR FISH

VILMOS CSÁNYI and ANTAL DÓKA

Department of Ethology, ELTE University, Budapest, Hungary

INTRODUCTION

The predator–prey relationship is one of the strongest selection factors which contributes to the evolution of behavior. Among the variety of predator–prey relations we intend to deal with defensive behavior in detail. Genetic factors, play a major role in many forms of this behavior. Despite this fact one can also find many learning-based defensive behavior in fish. Literature dealing with learning in fishes is rather dispersed; we do not aim to present a complete picture but to show at least one example for each of the learning mechanisms relevant from the point of view of adaptation.

In our opinion four different classes of the predator–prey interactions could be distinguished and learning might contribute to each of them (Figure 1). These four classes are as follows:

1. Interactions at the individual level.

This category involves recognition and in most cases inspection of the predator, and also discrimination between a hungry and satiated one. Recognition is usually followed by some defense response such as flight, tonic immobility or other kind of avoidance. Individuals also frequently need to make decision about approaching desirable territories or food under the threat of predators.

2. Interactions at the group level.

There are many possible advantages for living in a group, e.g. diminishing the effect of the predator threat. Shoaling by fish can lead to predator confusion, to dilution of the attack and vigilance results in early predator detection. Alarm displays, and releasing alarm substances are also effective means of the social defence. Mobbing also might decrease the effectiveness of stalking and deters the attack of the predator.

3. Interactions among the prey.

Effects of the various warning mechanisms such as alarm displays and the release of alarm pheromones are enhanced by social facilitation mechanisms and result in earlier, faster, and more effective flights. In some cases the contribution of learning to predator avoidance can be so highly developed in fish that even their ability for pre-cultural behavior was suggested.

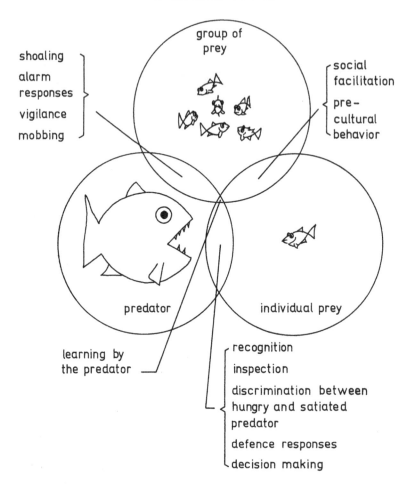

Figure 1   Different classes of predator–prey interactions.

4. Learning by the predator.

The other main actor of the predator–prey interactions, the predator itself can also learn from experience and increases its own chance in the game. Learning by the predator will be discussed briefly, only to complete the above classification.

INTERACTIONS AT THE INDIVIDUAL LEVEL

Recognition of the Predator

The most important components of the environment for an animal are other animals, such as conspecifics, predators or prey. Many mammals (Owings and Coss, 1977; Owings and Owings, 1979; Hirsch and Bolles, 1980) and birds (Curio, 1975) were found to be able to recognize their predators without prior experience, and respond appropriately upon their first exposure to them. There are a few studies showing that learning may improve predator avoidance behavior (Hinde, 1954; Kruuk, 1976; Curio et al., 1978).

There are numerous of observations which indicate that predator recognition in fish has strong genetic components. Minnows (*Phoxinus phoxinus*) from a population that lived without pike (*Esox lucius*) for many thousand of years are capable of performing synchronized anti-predator behavior in their first encounter with a pike (Pitcher, 1986). Local variation in predation pressure appears to affect significantly many behaviors in the Trinidadian guppy (*Poecilia reticulata*). Seghers (1970, 1973) reports differences in micro-distribution, shoaling, feeding and fright responses between guppies sympatric with some large cichlids and guppies allopatric with the large predators. Differences in overhead fright response measured experimentally persisted in laboratory-reared young from various sites indicating that the basis of this behavioral variation may be inherited (Seghers, 1974).

Observations on the anti-predator behavior in the three-spined stickleback *Gasterosteus aculeatus* showed that there are significant differences between populations in response to both fish and bird predators, and these differences relate to known and estimated predation risk at the study sites (Giles and Huntingford, 1984).

Predator Inspection

When first encountering a stalking pike, individual minnows leave the school and approach to within 4–6 body lengths of the predator (Magurran and Pitcher, 1987). There they wait for a few seconds, slowly turn, and return to the shoal. Pitcher, Green and Magurran (1986) term this behavior a "predator inspection visit". The same behavior is shown to models, and initially to other intruders such as large non-predatory fish. It also has been described for sticklebacks (Reist, 1983; Giles and Huntingford, 1984). In paradise fish (*Macropodus opercularis*) we observed this behavior in individuals meeting species new to them. We called this behavior "approach" (Csányi, 1985; Csányi et al., 1989).

Certain features of a predator appear to be recognized by predator-naive fish (Csányi, 1985; Tulley and Huntingford, 1987). However, some of the information that allows dangerous fish to be identified is obtained during predator inspection visits, when some kind of learning occurs (Magurran and Pitcher, 1986).

Magurran and Girling (1986) showed that predator inspection functions, at least in part, in predator recognition. Predatory and nonpredatory fish may be similar in appearance and the minnow can make the distinction only after a close range examination of potential predators. This can be a risky affair, and several lines of evidence suggest that predator inspection in minnows is strongly influenced by their assessment of risk. Pitcher et al., (1986) found that feeding minnows (a) increase predator inspections as a pike approaches, (b) inspect a stationary predator more than a more dangerous, moving one, and (c) approach a predator more closely when inspecting in a group. Magurran and Pitcher (1987) also found inspections to cease after a successful attack on another fish which clearly indicates some learning during the process. Habituation of predator inspection was found in the three-spined stickleback (Huntingford and Coulter, 1989). Habituation to predator models was found by Welch and Colgan (1990). In a recent series of experiments Milinski and co-workers found that in sticklebacks individual fish had partners with whom they repeatedly performed pairwise predator inspection visits, this finding raised the possibility of cooperation (Milinski *et al.*, 1990).

Beyond predator recognition, "inspectors" may gain information concerning the state (or motivation) of the predator (Magurran and Girling, 1986). Predator inspection may also function in pursuit deterrence (Magurran and Pitcher, 1987).

Discrimination Between Hungry and Satiated Predators

The ability to discriminate between hungry and satiated predators and a flexible response to the differential threat can help to lower the cost of anti-predator behavior. Guppies (*Poecilia reticulata*) were found by Licht (1989) to distinguish between hungry and satiated predators and adjust their antipredatory behavior accordingly. We found little information on the role of learning in the discrimination by the prey between hungry and satiated predators. Further work would be worthwhile.

Defense Responses

Defense by closed genetic programs

The most common form of defensive behavior is flight. The rock-dwelling African cichlid fish (*Melanochromis chipokae*) flees back to safe rocks when a predator appears (Dill, 1990). An other widespread passive form of defense is tonic immobility or freezing (Ratner, 1967; Gallup, 1977; Kabai and Csányi, 1978; Csányi *et al.*, 1985). Behavior genetics studies on the paradise fish have shown that many elements of defensive behavior, or whole sequences can be determined by genetic factors (Csányi and Gervai, 1985, 1986; Csányi and Gerlai, 1988; Gerlai and Csányi, 1989, 1990; Gervai and Csányi, 1989; Gerlai *et al.*, 1990). Genetically based inter-population differences in stickleback anti-predator behavior in response to predation risk are now well established (Huntingford, 1982; Giles and Huntingford, 1984).

Defense modified by learning

In many cases the genetically programmed defense behavior patterns can be modified by learning. In conditioning experiments with the paradise fish, we found that the duration and frequency of the passive freezing responses was influenced by learning (Altbäcker and Csányi, in press). Evidence for the modification of more complex escape patterns was also found. The results with mosquitofish, *Gambusia affinis* indicate that with proper conditioning and with the benefit of other cues available in the natural environment, mosquitofish are able to return quickly and accurately to the home shore. Such behavior would have great survival value in mosquitofish populations subject to intense predation by larger species of fish (Goodyear and Ferguson, 1969; Goodyear, 1973).

Decision Making

Habitats and patches may vary not only in terms of their foraging profitability, but also in terms of predation risk (Fraser and Huntingford, 1986; Hart, 1986; Huntingford *et al.*, 1988). The extent of the threat of predation influences decisions. For example Helfman (1989) has shown that the strength of predator avoidance and escape in the threespot damselfish, *Stegastes planifrons* varies directly with the size and threatening posture of predatory Atlantic trumpetfish, *Aulostomus maculatus*. Decision making involves many complex behavioral patterns and various learning mechanisms certainly play an important role (Lima and Dill, 1990).

A Case Study: Experiments with the Paradise Fish

The natural habitats of the paradise fish are the ponds and rivulets of South-East Asia. We have maintained a genetically controlled, outbred colony of the paradise fish since 1978 (Csányi *et al.*, 1984). Members of this colony are very similar in their behavior to the paradise fish caught in the wild (Csányi *et al.*, 1985). The predators (pike, catfish (*Silurus glanis*)) were also raised in our laboratory in order to adapt them to aquaria and manipulations. They were fed continuously with paradise fish of an appropriate size. Goldfish (*Carasius auratus*) were acquired from petshops.

Meeting and recognition predators

First we examined the reaction of naive paradise fish to their first exposure to an individual of another fish species (Csányi, 1985). During the encounters the paradise fish always approached the other fish and examined it thoroughly, swimming around it several times and carefully avoiding its head. If the encounters were repeated a few hours or days later it was found that the exploratory reaction observed on the first occasion had diminished. During repeated further meetings the percentage time showing approach decreased further.

Using the decrease of the "approach" reaction between two subsequent meetings we studied how long a paradise fish can remember fish of other species. It was found that the decrease of the frequency of approach behavior between the first and second encounters with a harmless goldfish is significant even if the second encounter took place three months, later (Csányi *et al.*, 1989).

Smell plays an important part in the recognition of other species (Davis and Pilotte, 1975). If the olfactory nerves of the paradise fish are cut by appropriate surgery, the approach reaction does not decrease as fast during repeated encounters as in the controls. For a significant decrease ten to fifteen repeats are necessary. It is therefore reasonable to conclude that the memory formed during the encounters contains not only visual but also olfactory information (Miklósi and Csányi, 1989).

Paradise fish which were raised individually in isolation showed the same "approach" reaction as normally raised fish in they first encounters with a goldfish. Our conclusion was that the curiosity of this animal, its approach reaction, does not require experience to develop (Csányi, 1985).

We examined the reaction of the paradise fish to repeated attacks by a predator and found that the "approach" reaction of the chased fish was considerably diminished compared to the inexperienced control fish. Instead of "approach" the fish tried to *"escape"* through the glass wall of the aquarium by a constant swimming perpendicular to the glass wall. We suppose that the fear caused by the attacking pike induced some learning process which was expressed as an "escape" attempt in the next encounter (Csányi, 1985).

This conclusion was given further support when we successfully transformed a harmless goldfish into an object to be avoided by using electric shocks. As the main motivating factors for avoidance are pain and fear, mild electric shocks applied in the presence of a goldfish make it a negative stimulus and it is avoided on the next encounter.

An interesting finding was that only members of other species can be transformed in this way (Csányi, 1985).

Avoidance strategies

From these ethological observations we concluded that learning plays an important role in predator avoidance in the paradise fish. It therefore seemed worthwhile to study avoidance learning in the paradise fish by the methods of comparative psychology.

Whereas in the traditional shuttle aquarium of the psychologists mild shocks usually serve as punishment during an avoidance training session, we used a live hungry pike as punishment. (Csányi and Altbäcker, in press).

Three groups of paradise fish were observed in a large shuttle aquarium. Members of the control group did not encounter the pike. Members of the other two groups met a hungry or a satiated pike respectively. The hungry pike was very active, it chased the paradise fish as soon as they entered the large compartment, while the satiated one was very quiet. Members of the control and the group which met the satiated pike actively shuttled between the two compartments, and spent roughly the same time in each compartment. The behavior of the group which encountered the hungry pike was entirely different. Entry latency rapidly increased from test to test, the number of entries and the time spent in the larger compartment steadily decreased. Clearly the paradise fish which were attacked by the pike quickly learned to avoid the larger compartment, while there was no effect in the presence of the satiated pike. These findings are similar to those found by Brookshire and Hognander (1968) in avoidance learning experiments in which paradise fish were punished by electric shock if they entered one of the compartments of a shuttle tank. Similar experiments were done with other predators such as perch and catfish and the results were the same: the presence of a predator as a punishment conditioned the avoidance of a particular compartment by the paradise fish.

It is also worth mentioning an experiment which used harmless carp of different sizes (Altbäcker and Csányi, in press) in a shuttle aquarium, since here the early reactions of the paradise fish were dependent on the size of the carp. Small carp which were similar in size to the paradise fish (6–7 cm) were examined by the paradise fish for a very long time, and the number and duration of the visits to the large compartment increased considerably. Large carp (30 cm) set off the same avoidance reaction as a hungry pike, although the carp never attacked the paradise fish. After three or four trials, avoidance decreased and at the fifth or sixth trial it ceased completely, the paradise fish ignoring the presence of the big fish. This observation is very important because it shows clearly the role of experience and learning in the relationship of the paradise fish with other species. On the first few encounters the paradise fish were obviously afraid of their large, strange companion but after repeated meetings and during careful exploration the paradise fish habituated to the presence of the carp because it made no threatening or unpleasant actions.

In an analysis of the behavior units of both the paradise fish and the attacking pike in a shuttle tank it was found that at the start of the attack the paradise fish tended to react immediately with "display", which seems to *inhibit* the attack of the pike. If the attack is unexpected then leaping is the only effective defense. Attacks usually started when the paradise fish emitted "escape" or "orientation" units or simply swam. In repeated encounters the frequency of these decreased and at the same time latency to enter the pike's compartment increased.

In further experiments (Csányi and Altbecker, in press) we examined the individual learning curves of the paradise fish rather than the group *average* (as was advised and carried out very successfully by Krechevsky (1932)), it was clearly seen that the process of conditioning is not a gradual one, but an "all or nothing" type of phenomenon. During the conditioning process the paradise fish explored very intensively, they also interacted with the predator very frequently, were chased and perhaps bitten, but despite all this exploration remained high. Then, in a subsequent trial, some fish remained in the starting compartment. It was concluded in this case the fish had learned the task, and fulfilled the criteria of avoidance. Nevertheless, avoidance of the predator is not final and the animals, even those which had been attacked on several occasions, sometimes visited the predator again. The frequency of these visits is not high, and thus the *average* latency time remains high and the average curve of this avoidance learning is near to 90 percent of the possible maximum latency.

These findings are not in close agreement with the traditional account of gradual learning. It seems that unusual stimuli induce *exploration* in the paradise fish. Motivation for this is extremely high and decreases, at least temporarily, only because of gradually developing fear. During exploration a memory is formed, but this will only take over control of the paradise fish's behavior (i.e. make it avoid the predator) if the motivation for exploration has already considerably decreased. The memory process itself is probably very quick, just 2 minutes were enough for the formation of a permanent memory trace connected with members of other species (Csányi et al., 1989); thus particular instances of learning are most probably "all or nothing" events. After the memory traces are established, behavior of the paradise fish in the performance phase of learning is due to a *decision* process which is probably controlled by the exploration drive and the fear of the animal.

Key stimuli in learning

In the experiments discussed in this section we investigated whether there were key stimuli connected to predators important in predator recognition and in the avoidance learning process in the paradise fish.

We experimented with a live goldfish and a series of dummies which looked more and more like a fish. We carried out the usual conditioning experiment in a small shuttle tank, except that after the habituation period we also put a dummy or the goldfish into the dark compartment. In some of the experimental groups we applied electric shocks (50 mA) as punishment in addition to the dummy (Csányi, 1986a).

Neither the dummies, the presence of a goldfish, nor the mild electric shocks alone caused significant changes in latency. Nor did latency increase if the dummy-head with a "single eye" or without eyes was applied together with electric shocks. Latency significantly increased, however, if punishment was applied in the presence of other dummies. We got the largest increase in the group conditioned in the presence of a live goldfish, and latency also increased remarkably, although to a somewhat lesser extent, in the groups where the paradise fish met a dummy with two lateral eye-like spots. The increase in latency was much smaller in the presence of a dummy with two vertically arranged spots.

In some experiments the effects on the exploratory reaction of the paradise fish, of an eyeless fish-shaped body, of two or four eyes, of eyes without a body and of combinations of these were studied (Altbäcker and Csányi, 1989a). Eyes without a body did not cause any reaction. Very mild exploration was activated by fish shaped bodies but display reactions were set off only by the body with eyes and the most effective stimulus was the body with two horizontally positioned eyes.

In a further experiment we compared the behavior of different paradise fish groups in encounters with a pike or a catfish or a "modified" catfish. The catfish was "modified" by attaching two glass eyes to it with a thin wire. It was found that the defense reaction characteristic in the presence of a pike was activated by the large eyes of the pike, since reactions to the "modified" catfish were similar to those caused by the pike (Altbäcker and Csányi, 1990).

Judging from the results of this experiment we can maintain that eyes or lateral eye-like spots on a body have a *key stimulus* character and are used in learning by the paradise fish. A remarkable amount of learning can be observed in the presence of the dummies with the key stimulus and if the dummy is "moving", like the goldfish for example, a single trial is virtually sufficient to make the paradise fish avoid the dark compartment. Motion is very likely to be an additional factor in the key stimulus, or may be a key stimulus itself.

It is therefore quite certain that avoidance learning occurs if fear or pain is inflicted upon the animal in the presence of certain key stimuli, but what does the paradise fish learn? Does it learn to avoid the dangerous place or the source of the key stimulus? We tried to find an answer to this question with the next experiment.

Paradise fish were put in one of two differently arranged aquaria by random assignment. Both tanks had a dark and light compartment but one of the tanks was safe, and punishment was never applied in it. In the other aquarium some fish got strong (100 mA) electric shocks as punishment, and some fish got punishment with a fish-like dummy. Each fish was tested in both tanks (Csányi and Lovász, 1987). It was found that strong shocks alone or combined with the dummy were very effective in making the paradise fish avoid the dark compartment. There was a characteristic difference between the groups punished by shock alone or in the presence of the dummy. Those groups which were punished by shocks alone extended their avoidance behavior to the dark compartment of the safe tank also, while those which were punished in the presence of the dummy did not avoid the dark compartment in the safe tank. It is therefore not the place, but the carrier of the key stimulus that elicits avoidance.

Interactive learning

On the bases of our observations and experiments on the paradise fish we assume that the main part of the mechanism of predator avoidance is the organized system of the memory traces in the animal brain acquired during daily experiences. The system of the memory traces is intimately connected to certain behavior mechanisms which are mainly genetically determined. One of such is *curiosity* which is expressed as "approach" behavior. A new environment or a new, not too unpleasant stimulus (e.g. a mild shock) in the familiar environment, activates *exploratory behavior*. The paradise fish is especially curious about living things in its environment. The "exploratory behavior" enables the animal to gain experience and knowledge about the new objects in its environment and to store these experiences during the learning process. Accumulation of these daily experiences forms very important individual differences among the members of a local population.

Besides conspecifics and prey organisms, predators are undoubtedly the most important living things in the environment of the paradise fish. Key stimuli promote the recognition, recording and learning of the features of the predator. Such stimuli are the eyes on a body, motion, and presumably smell. A carrier of such key stimuli immediately arouses the interest of the paradise fish. Different learning mechanisms are likely to be activated during exploration. If no unpleasant stimulus or pain occurs during exploration, the paradise fish habituates, that is, it gets used to the presence of the carrier of the key stimulus and does not attend to it any more. But if it is attacked by the carrier of the key stimulus, the fish will try to avoid it later on. During our observations and experiments, attack by the predator and shocks in the presence of a harmless goldfish or a fish dummy activated the avoidance reaction in the paradise fish. On the basis of all this an "interactive model of avoidance learning" can be formulated (Csányi, 1986a, 1986b), in which, besides the species-specific behavior units and the un-conditioned or conditioned stimuli (Bolles, 1970, and Bolles and Fanselow, 1980), the key stimuli and the exploratory behavior of the animal also play an important role (Figure 2).

The hypothesis of the paradise fish showing interactive learning enables us to draw several important general conclusions. It follows from the interactive learning theory that the animal always keeps an eye on its environment and if it meets "new" strange objects or living things it explores them and its experiences determine the way it acts at the next encounter. Memory traces *representing* the predator form a close link with effective escape reactions, while those representing harmless fishes link to neutral behavior units. Interactive learning continuously refreshes, and if

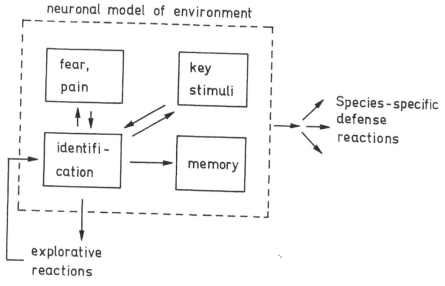

Figure 2   Sketch of the interactive learning hypothesis.

necessary rearranges, this system of memory traces and responses. It is reasonable to suppose that the environmental model includes not only information on the predators but also individual relationships with conspecifics. The adaptive value of the interactive learning mechanism is that a description or dynamic model of the environment is built up in the brain of the paradise fish during its lifetime and this is what enables it to predict the reactions and events of the environment (Csányi, 1987, 1988, 1989). This dynamic model is constructed by individual experiences and represents the major factor of the paradise fish's adaptation to its environment.

Paradise fish fry fall prey to smaller predators and occasional attacks activate the learning mechanism in tens or hundreds of individuals growing up in the group. The number of bad learners slowly decreases because they are eaten, and the individuals that have built up the most appropriate environmental model on the basis of their experiences survive and are the most successful in the struggle for life.

INTERACTIONS AT THE GROUP LEVEL

Shoaling

Shoaling behavior is common among teleosts (Keenleyside, 1955; Godin, 1986). Clumping of the chased fish, predator confusion, early predator detection and social communication of alarm are well studied anti-predator mechanisms of shoaling fish (Pitcher, 1986).

Individuals may be especially vulnerable to predation if they appear different from other group members (the "oddity effect") For an example Hobson (1968) showed that predators attacking large shoal of prey fish captured only injured individuals. It is therefore of interest that both marine (Wolf, 1985) and freshwater fish (Allan and Pitcher, 1986) tend to leave shoals comprised largely of another species when these are threatened by predators. No direct evidence for involvement of learning in shoaling behavior was found but it would be rather odd if this behavior could properly function without some learning mechanisms.

Alarm Displays

The recognition of predators in groups is not always triggered by their direct visual, olfactory or other traits. In many species the contact with a predator elicits characteristic alarm reactions, which are able to activate defensive mechanisms in conspecifics.

Displays of alarm behavior have been reported for a schooling fish, *Pristella riddlei* (Keenleyside, 1955) and from parents of fry to cichlid species (*Hemichromis bimaculatus*, Noble and Curtis, 1939). The pattern on the dorsal fin of *Pristella riddlei* is certainly a social releaser which attracts the members of the species. In case of disturbance the frequency of flicking movements of the dorsal fin increases and this acts as an alarm signal. Minnows that detect an approaching predator (real or model) perform a low-intensity skitter, often returning to near their original position in a curved path. The skitter may act as a warning to shoal members (Magurran *et al.,* 1985).

Essentially similar mechanisms are activated when the attacked individuals secrete species-specific alarm substances and the detection of these is what induces defensive mechanisms in conspecifics (Heczko and Seghers, 1981). In zebra danios

(*Brachydanio rerio*), the alarm substance acts to initiate learning of a co-occurring novel and behaviorally neutral chemosensory stimulus. The new learned chemosensory stimulus can be a base of a cultural tradition (Suboski and Templeton, 1989)

Vigilance

The prey's vigilance under the risk of predation as a social phenomenon is well studied (Lazarus, 1979). The common observation is that individuals in a foraging group spend less time being vigilant with an increase in group size. The results with fish are also very suggestive (Magurran *et al.*, 1985; Godin *et al.*, 1988). Pitcher *et al.* (1986) and Magurran and Higham (1988) demonstrated that information gained by the inspecting minnows may be transferred to other group members. Magurran and Higham (1988) suggested that inspectors may actually "manipulate" other group members, while Milinski (1987) suggest that sticklebacks inspecting a predator engage in evolutionary stable reciprocation. Godin *et al.* (1988) also suggested that social transmission of the flight response occurred within glowlight tetra (*Hemigrammus erythrozonus*) shoals.

Learning mechanisms could be involved in vigilance most probably through the recognition of the predator.

Mobbing

Mobbing of predators by fish has been reported (e.g. in bluegill, *Lepomis macrochirus*), Dominey, 1983, in zebra danio (*Brachydanio rerio*) Suboski, 1988; Suboski *et al.*, 1990). Studies for learning involvement are still missing.

INTERACTIONS AMONG THE PREY

Social Facilitation

Several cases of socially transmitted alarm reactions are known in fish. In *Rasbora heteromorpha* the mere sight of alarmed conspecifics is enough for the same type of alarm reaction (Verheijen, 1965). Unfortunately no effort was made to examine whether stable conditioning to some visual stimuli could be established by this technique.

A similar observations were made on shoal of small roach (*Rutilus rutilus*) which acquired a conditioned avoidance reaction against a predator (*Perccottus glehni*) (without direct contact with the predator), through observation of the consumption of other members of the school. The lack of control is a problem with this experiment (Leshcheva, 1968).

Patten (1977) obtained similar results for coho salmon (*Oncorhynchus kisutch*) fry. Naive fry exposed to predators fell prey at a significantly higher rate than did naive fry exposed to the predators but in the company of conspecific survivors of earlier exposure to sculpins (*Scorpaena guttata*).

The experiment with the rasboras clearly shows social facilitation and in the one on roach a conditioned avoidance is established by some (visual and perhaps other) stimuli. If stable avoidance behavior can be established by such conditioning then we are dealing with a kind of precultural behavior which will be discussed below.

Precultural Behavior

Cultural behavior was defined very rigorously by Mundinger (1980), but if we apply a somewhat less stringent definition by which (pre)-cultural behavior involves the non-genetic transmission of social information across age classes or generations, then some observations indicate its presence among various species of fish.

Small juveniles of French grunts (*Haemulon flavolineatum*), a coral-reef fish, exhibit social traditions of daytime schooling sites and twilight migration routes. Individuals transplanted to new shoaling sites, and allowed to follow residents at the new sites, used new migration routes and returned to new sites in the absence of resident fish. Control fish with no opportunity to learn showed no such directionality or return. This is a clear demonstration of apparent pre-cultural behavior in free-living fish (Helfman and Schultz, 1984).

LEARNING BY THE PREDATOR

Triplett (1900–1) using a tank divided into two compartments by a glass partition presented minnows to two perches (*Perca americana*); the perches tried to attack the minnows across the partition, certainly without success. After several days of reiteration the intensity of attacks decreased; later, when the glass screen was removed the perches did not attack the minnows but swam amongst them for a long period.

Foraging efficiency of sticklebacks on *Gammarus* and *Artemia* as prey was studied by Croy and Hughes (1991a, 1991b). Increased hunger and associated feeding motivation increased the foraging efficiency of fish already experienced with the prey. Foraging efficiency also increased as naive fish gained experience with prey.

CONCLUSIONS

The study of learning mechanisms in fish has just been started, however, it seems to be a very promising field. In fish, as with other animals, more or less closed genetic programs are functioning. Those which are related to predator interactions are usually supported by some learning mechanisms. The simplest interpretation of the genetic program–learning relationship could be formulated as follows:

1. It might be supposed that the genetic programs include several behavioral options for a given situation and learning helps the animal to select the most appropriate one in the given environment. Recognition of the dangerous situation by fish is based partly on genetically constrained perception of key stimuli and partly on memory. The recognized predator elicits defensive mechanisms in the prey fish; carrying out of them is preceded by a decision. The decision might involve the evaluation of the threat posed by the predator, the need for feeding, and the choice of the safest escape route.

2. Defense against predators is realized on both organizational levels: the individual as well as the group level. The most important individual actions are flight, freezing or more complex, frequently learned escape reactions. In shoaling fish defensive behavior occurs not only when the predator is directly noticed but they can also be triggered by the various alarm reactions of conspecifics which are the components of the group-level defense. The most important mechanism of group-level defense is social facilitation which manifests its effects through the spread of escape reactions and is the base of vigilance and mobbing. There are observations which support the assumption that cooperation and perhaps precultural forms of behavior are also involved in defense at the group level of certain fish.

ACKNOWLEDGMENT

This work was supported by a grant (No. 368-0813\1991) from the Hungarian OTKA Scientific Foundation.

References

Allan, J. R. and Pitcher, T. J. (1986). Species segregation during predator evasion in cyprinid fish shoals. *Freshwater Biol.*, **16**, 653–659.
Altbäcker, V. and Csányi, V. (1990). The role of eye-spots in predator recognition and antipredatory behaviour of the paradise fish (*Macropodus opercularis*). *Ethology*, **85**, 51–57.
Altbäcker, V. and Csányi, V. (In press). Experience and avoidance of other livings by the paradise fish (*Macropodus opercularis*). *Acta. Biol. Hung.*
Bolles, R. C. (1970). Species-specific defense reaction and avoidance learning. *Psychol. Review*, **77**, 32–48.
Bolles, R. C. and Fanselow, M. S. (1980). A perceptual-defensive-recuperative model of fear and pain. *Behav. Brain Sci.*, **3**, 291–323.
Brookshire, K. H. and Hognander, O. C. (1968). Conditioned fear in the fish. *Psychol. Repts.*, **22**, 78–81.
Croy, M. I. and Hughes, R. N. (1991a). The role of learning and memory in the feeding behaviour of the fifteen-spined stickleback, *Spinachia spinachia* L. *Anim. Behav.*, **41**, 149–159.
Croy, M. I. and Hughes, R. N. (1991b). The influence of hunger on feeding behaviour and on the acquisition of learned foraging skills by the fifteen-spined stickleback, *Spinachia spinachia* L. *Anim. Behav.*, **41**, 161–170.
Csányi, V. (1985). Ethological analysis of predator avoidance by the Paradise fish (*Macropodus opercularis*). I. Recognition and learning of predators. *Behaviour*, **92**, 227–240.
Csányi, V. (1986a). Ethological analysis of predator avoidance by the Paradise fish (*Macropodus opercularis*). II. Key stimuli in avoidance learning. *Anim. Learn. Behav.*, **14**, 101–109.
Csányi, V. (1986b). How is the Brain Modelling the Environment? A Case Study by the Paradise Fish. In: *Variability and Behavioural Evolution* (Ed. by G. Montalenti and G. Tecce). Proceedings, Accademia Nazionale dei Lincei, Roma, 1983, *Quaderno*, **259**, 142–157.
Csányi, V. (1987). The Replicative Evolutionary Model of Animal and Human Minds. *World Future: J. Gen. Evol.*, **24**, 174–214.

Csányi, V. (1988). Contribution of the Genetical and Neural Memory to Animal Intelligence. In: *Intelligence and Evolutionary Biology* (Ed. by H. Jerison and Irene Jerison), pp. 299–318. Berlin: Springer-Verlag.

Csányi, V. (1989). *Evolutionary Systems and Society: A General Theory.* p. 304. Durham: Duke University Press.

Csányi, V. and Altbäcker, V. (1991). Variable learning performance: the level of behaviour organization. *Acta. Biol. Hung.*, **41**, 321–332.

Csányi, V. and Altbäcker, V. (In press). Predator in the shuttlebox: An ethological analysis of passive avoidance conditioning of the paradise fish (*Macropodus opercularis*). *Ethology*.

Csányi, V. and Gerlai, R. (1988). Open-field behaviour and the behaviour-genetic analysis of the Paradise fish (*Macropodus opercularis*). *J. Comp. Psych.*, **102**, 226–236.

Csányi, V. and Gervai, J. (1985). Genotype-environment interaction in passive avoidance learning of the Paradise fish (*Macropodus opercularis*). *Acta. Biol. Hung.*, **36**, 259–267.

Csányi, V. and Gervai, J. (1986). Behaviour-genetic analysis of the Paradise fish (*Macropodus opercularis*) II: Passive avoidance conditioning of inbred strains. *Behav. Genet.*, **16**, 553–557.

Csányi, V. and Lovász, F. (1987). Key stimuli and the recognition of the physical environment by the paradise fish (*Macropodus opercularis*). *Anim. Learn. Behav.*, **15**, 379–381.

Csányi, V., Tóth, P., Altbäcker, V., Dóka, A. and Gervai, J. (1984). Behaviour elements of the Paradise fish (*Macropodus opercularis*) I: Regularities of defensive behaviour. *Acta. Biol. Hung.*, **36**, 93–114.

Csányi, V., Tóth, P., Altbäcker, V., Dóka, A. and Gervai, J. (1985). Behaviour elements of the Paradise fish (*Macropodus opercularis*) II: A functional analysis. *Acta. Biol. Hung. Acad. Sci.*, **36**, 115–130.

Csányi, V., Csizmadia, G. and Miklósi, Å. (1989). Long-term memory and recognition of another species in the paradise fish. *Anim. Behav.*, **37**, 908–911.

Curio, E. (1975). The functional organization of anti-predator behaviour in the pied flycatcher: A study of avian visual perception. *Anim. Behav.*, **23**, 1–45.

Curio, E., Ernst, K. and Vieth, W. (1978). The adaptive significance of avian mobbing II: Cultural transmission of enemy constraints. *Zeitschr. Tierpsychol.*, **48**, 184–202.

Davies, R. E. and Pilotte, N. J. (1975). Attraction to conspecific and non-conspecific chemical stimuli in male and female *Macropodus opercularis*. *Behav. Biol.*, **13**, 191–196.

Dominey, W. J. (1983). Mobbing in colonially nesting fishes, especially the bluegill, *Lepomis macrochirus*. *Copeia*, **4**, 1086–1088.

Fraser, D. F. and Huntingford, F. A. (1986). Feeding and avoiding predation hazard: the behavioral response of the prey. *Ethology*, **73**, 56–68.

Gallup, G. G. Jr. (1977). Tonic Immobility: the role of fear and predation. *Psychol. Rec.*, **27**, 41–61.

Gerlai, R. and Csányi, V. (1989). Diallel Genetic Analysis of the Elements of Paradise Fish's (*Macropodus opercularis*) Ethogram. *Acta. Biol. Hung.*, **40**, 57–66.

Gerlai, R. and Csányi, V. (1990). Genotype-environment interaction and the correlation structure of behaviour elements in the paradise fish (*Macropodus opercularis*). *Phys. Behav.*, **47**, 343–356.

Gerlai, R. Crusio, W. E. and Csányi, V. (1990). What behaviour could be advantageous in novel and familiar environments? A diallel genetic analysis of the paradise fish (*Macropodus opercularis*). *Behav. Genet.*, **20**, 487–498.

Gervai, J. and Csányi, V. (1989). Behaviour genetic analysis of the Paradise fish (*Macropodus opercularis*) IV: Behavioural unit analysis of the response to novelty using recombinant inbred strains. *Acta. Biol. Hung.*, **40**, 57–66.

Godin, J.-G. J. (1986). Antipredator function of shoaling in teleost fishes: A selective review. *Nat. Can. (Rev. Écol.Syst.)*, **113**, 241–250.

Godin, J.-G. J. Classon, L. J. and Abrahams, M. V. (1988). Group vigilance and shoal size in a small characin fish. *Behaviour*, **104**, 29–40.

Goodyear, C. P. (1973). Learned orientation in the predator avoidance behavior of Mosquitofish, *Gambusia affinis. Behaviour*, **45**, 191–224.

Goodyear, C. P. and Ferguson, D. E. (1969). Sun-compass orientation in the mosquitofish, *Gambusia affinis. Anim. Behav.*, **17**, 636–640.

Giles, N. and Huntingford, F. A. (1984). Predation risk and interpopulation variation in anti-predator behaviour in the three-spinned stickleback. *Anim. Behav.*, **32**, 264–275.

Hart, P. J. B. (1986). Foraging in Teleost Fishes. In: *The Behaviour of Teleost Fishes* (Ed. by T. J. Pitcher), pp. 211–235. London: Croom Helm.

Heczko, E. J. and Seghers, B. H. (1981). Effects of alarm substance on schooling in the common shiner (*Notropus cornutus*). *Environm. Biol. of Fishes*, **6**, 25–29.

Helfman, G. S. (1989). Threat-sensitive predator avoidance in damselfish-trumpetfish interactions. *Behav. Ecol. Sociobiol.*, **24**, 47–58.

Helfman, G. S. and Schultz, E. T. (1984). Social transmission of behavioural traditions in a coral reef fish. *Anim. Behav.*, **32**, 379–384.

Hinde, R. A. (1954). Factors governing the changes in strength of a partially inborn response, as shown by the mobbing behaviour of the chaffinch (*Fringilla coelebs*): I. Nature of the response, and an examination of its course. *Proc. Roy. Soc. B.*, **142**, 306–331.

Hinde, R. A. (1970). *Animal Behaviour: a Synthesis of Ethology and Comparative Psychology*, 2nd edn. New York: McGraw-Hill.

Hirsch, S. M. and Bolles, R. C. (1980). On the ability of prey to recognize predators. *Zeitschrift für Tierpsychologie*, **54**, 71–84.

Hobson, E. S. (1968). Predatory behavior of some shore fishes in the Gulf of California. *Res. Rep. U.S. Fish. Wildl. Serv.*, **73**, 1–92

Hoogland, R., Morris, D. and Tinbergen, N. (1957). The spines of sticklebacks (*Gasterosteus* and *Pygosteus*) as a means of defense against predators (Perch and Esox). *Behaviour*, **10**, 207–236.

Huntingford, F. (1982). Do inter- and intra-specific aggression vary in relation to predation pressure in sticklebacks? *Anim. Behav.*, **30**, 909–916.

Huntingford, F. A. and Coulter, R. M. (1989). Habituation of a predator inspection in the three-spined stickleback, *Gasterosteus aculeatus* L. *J. Fish Biol.*, **35**, 153–154.

Huntingford, F. A., Metcalfe, N. B. and Thorpe, J. E. (1988). Feeding motivation and response to predation risk in Atlantic salmon parr adopting different life history strategies. *J. Fish Biol.*, **32**, 777–782.

Keenleyside, M. H. A. (1955). Some aspects of the schooling behaviour of fish. *Behaviour*, **8**, 183–248.

Krechevsky, I. (1932). Hypothesis in Rats. *Psychol. Review*, **39**, 516–532.

Kruuk, H. (1976). The biological function of gull's attraction towards predators. *Animal Behaviour*, **24**, 146–153.

Lazarus, J. (1979). The early warning function of flocking in birds: an experimental study with captive quelea. *Anim. Behav.*, **27**, 855–865.

Lescheva, T. S. (1968). Formation of defensive reflexes in roach (*Rutilus rutilus* L.) larvae through imitation. *J. Ichthyology*, **8**, 838–841.

Licht, T. (1989). Discriminating between hungry and satiated predators: the response of guppies (*Poecilia reticulata*) from high and low predation sites. *Ethology*, **82**, 238–243.

Lima, S. L. and Dill, L. M. (1990). Behavioral decisions made under risk of predation: a review and prospectus. *Can. J. Zool.*, **68**, 619–640.

Magurran, A. E. and Higham, A. (1988). Information transfer across fish shoals under predation threat. *Ethology*, **78**, 153–158.

Magurran, A. E. and Girling, S. L. (1986). Predator model recognition and response habituation in shoaling minnows. *Anim.Behav.*, **34**, 510–518.

Magurran, A. E. and Pitcher, T. J. (1987). Provenance, shoal size and the sociobiology of predator-evasion behaviour in minnow shoals. *Proc. Roy. Soc. London, Ser. B.* **229**, 439–465.

Magurran, A. E., Oulton, W. and Pitcher, T. J. (1985). Vigilant behaviour and shoal size in minnows. *Zeitschrift für Tierpsychologie*, **67**, 167–178.

Miklósi, Á. and Csányi, V. (1989). The influence of olfaction on exploratory behaviour in the paradise fish (*Macropodus opercularis* L.). *Acta. Biol. Hung.*, **40**, 195–202.

Milinski, M. (1979). Can an experienced predator overcome the confusion of swarming prey more easily? *Anim. Behav.*, **27**, 1122–1126.

Milinski, M. (1987). Tit for tat in sticklebacks and the evolution of cooperation. *Nature*, **325**, 433–435.

Milinski, M., Pfluger, D., Killing, D. and Kettler, R. (1990). Do sticklebacks cooperate repeatedly in reciprocal pairs? *Behav. Ecol. Sociobiol.*, **27**, 17–21.

Mundinger, P. C. (1980). Animal cultures and a general theory of cultural evolution. *Ethol. Sociobiol.*, **1**, 183–223.

Noble, G. K. and Curtis, B. (1939). The social behavior of the jewel fish, *Hemicromis bimaculatus*. *Bulletin of the American Museum of Natural History*, **76**, 1–46.

Owings, D. H. and Coss, R. G. (1977). Snake mobbing by California ground squirrels: Adaptive variation and ontogeny. *Behaviour*, **62**, 50–69.

Owings, D. H. and Owings, S. C. (1979). Snake directed behaviour by blacktailed prairie dogs (*Cynomys ludovicianus*). *Zeitschrift für Tierpsychologie*, **49**, 35–54.

Patten, G. G. (1977). Body size and learned avoidance as factors affecting predation on coho salmon, *Oncorhynchus kisutch*, fry by torrent sculpin, *Cottus rhotheus*. *Fisheries Bulletin*, **75**, 457–459.

Pitcher, T. J. (1986). Functions of Shoaling Behaviour in Teleosts. In: *The Behaviour of Teleost Fishes* (Ed. by T. J. Pitcher), pp. 294–338. London: Croom Helm.

Pitcher, T. J., Green, D. A. and Magurran, A. E. (1986). Dicing with death: predator inspection behaviour in minnow shoals. *J. Fish. Biol.*, **28**, 438–448.

Ratner, S. C. (1967). Comparative Aspects of Hypnosis. In: *Handbook of Clinical and Experimental Hypnosis* (Ed. by J. E. Gordon) New York: Macmillan.

Reist, J. D. (1983). Behavioural variation in pelvic phenotypes of brook stickleback *Culaea inconstans* in response to predation by the northern pike, *Esox lucius*. *Biology of Fishes*, **8**, 255–268.

Seghers, B. H. (1970). Behavioural adaptations of natural populations of the guppy, *Poecilia reticulata* to predation. *Amer. Zool.*, **10**, 489–490.

Seghers, B. H. (1973). An analysis of geographic variation in the anti-predator adaptations of the guppy, *Poecilia reticulata*. Ph.D. thesis. University of British Columbia.

Seghers, B. H. (1974). Geographic variation in the responses of guppies (*Poecilia reticulata*) to aerial predators. *Oceologia (Ber.)*, **14**, 93–98.

Suboski, M. D. (1988). Acquisition and social communication of stimulus recognition by fish. *Behavioural Processes*, **16**, 213–244.

Suboski, M. D. and Templeton, J. J. (1989). Life skills training for hatchery fish: Social learning and survival. *Fisheries Res.*, **7**, 343–352.

Suboski, M. D., Bain, S., Carty, A. E., McQuoid, L. M., Seelen, M. I. and Seifert, M. (1990). Alarm reaction in acquisition and social transmission of simulated-predator recognition by zebra danio fish (*Brachydanio rerio*). *J. Comp. Psychol.*, **104**, 101–112.

Triplett, N. (1900–1901). The educability of the perch. *Amer. J. Psychol.*, **2**, 354–360.

Tulley, J. J. and Huntingford, F. A. (1987a). Paternal care and the development of adaptive variation in anti-predator responses in sticklebacks. *Anim. Behav.*, **35**, 1570–1572.

Tulley, J. J. and Huntingford, F. A. (1987b). Age, experience and the development of adaptive variation in antipredator responses in three-spined stickleback *Gasterosteus aculeatus* L. *Ethology*, **75**, 285–290.

Verheijen, F. J. (1956). Transmission of a flight reaction amongst a school of fish and the underlying sensory mechanisms. *Experientia*, XII/5, 202–204.

Welch, C. and Colgan, P. (1990). The effect of contrast and position on habituation to models of predators in eastern banded killifish (*Fundulus diaphanus*). *Behav. Proc.*, **22**, 61–71.

Wolf, N. G. (1985). Odd fish abandon mixed-species groups when threatened. *Behav. Ecol. Sociobiol.*, **17**, 47–52.

CONFLICTING DEMANDS IN GOBIES: WHEN TO EAT, REPRODUCE, AND AVOID PREDATORS

CARIN MAGNHAGEN

Dept. of Zoology, Uppsala University, Box 561, S-751 22 Uppsala, Sweden

Trade-offs between different types of behaviour can be found when an increase in one type of behaviour, performed, for example, to enhance survival or reproductive output, leads to a decrease in some other fitness-related trait. In these cases there are conflicts between different demands, and therefore animals will have several factors to consider in order to optimize their behaviour. For example, foraging behaviour can increase the animal's susceptibility to predators, and foraging rate and foraging habitat are commonly found to be influenced by predation risk (reviewed by Lima and Dill, 1990). Also, reproduction may make an animal more vulnerable to predation than it is otherwise (Magnhagen, 1991), either because of a higher visibility caused by egg-carrying (Berglund and Rosenqvist, 1986; Svensson, 1988) or breeding coloration (Endler, 1982; Wootton, 1984), or because of conspicuous displays that may attract predators as well as mates (Giles, 1984; Helfman, 1986). However, very few studies have investigated how reproductive behaviour may be influenced by predation risk (but see Sih, 1989; Sih *et al.*, 1990; Magnhagen, 1990). Furthermore, trade-offs between reproduction and foraging are likely to be influenced by risk of starvation, probability of future reproduction and reproductive value of the current brood.

Lifetime reproductive success in an animal is often connected with its success of survival. Being killed by a predator inevitably eliminates future fitness, and predation may therefore be a strong selective force for the evolution of adaptations, such as cryptic coloration and behaviour (Sih, 1987). However, when there is a trade-off between different behaviours, the optimal choice of action differs under different circumstances. For a foraging animal, the amount of risk it should take in order to get food should be negatively correlated with its metabolic state (measured as energy reserves or hunger levels), since at low energy levels there is also a probability of dying from starvation (Mangel and Clark, 1986; McNamara and Houston, 1986). Similarly, when a trade-off has to be made between reproduction and predator avoidance, the degree of risk-taking should be correlated with the probability of future reproduction, and during the parental phase also with the value of the current brood (Montgomerie and Weatherhead, 1988). The optimal behaviour may be influenced by many factors, but should, with these predictions, differ between species with different longevities and within species between individuals of different age (if mortality is age-dependent). Thus, short-lived species and older individuals should take relatively high risks in order to reproduce (Clark and Ydenberg, 1990; Magnhagen, 1990).

This article considers how fishes of the family Gobiidae trade different types of behaviours and which factors influence behavioural "decisions" in these fishes when there are conflicts between different demands, such as the demands for foraging, reproduction and predator avoidance.

NATURAL HISTORY

The typical goby is a small, benthic, marine fish, although freshwater species do occur. The three species I have studied are very abundant in shallow softbottom areas along the European coast. Breeding occurs in spring and summer. All of these species spawn repeatedly and have paternal care. Males build nests, using a mussel shell or a stone, depending on species, and attract females to their nest with a display. Females lay their eggs in the nest and males subsequently guard the eggs and fan them until hatching (Miller, 1984). These gobies usually feed on small invertebrates, often infauna such as polychaetes and amphipods. Sand goby, *Pomatoschistus minutus*, and common goby, *P. microps,* are closely related and very similar both in morphology and ecology. They build nests out of mussel shells covering them completely with sand and leaving just a small entrance opening (Vestergaard, 1976; Hesthagen, 1977). These two species are short-lived, with a life span of approximately 1 year, reproducing during only one season. The black goby, *Gobius niger,* lives for up to five years and thus has several breeding seasons. This species uses either shells or rocks as nest material, digging a hole underneath or placing the eggs completely in the open (Vaas *et al.,* 1975). The male usually cleans an area around the nest from vegetation and other objects and defends this area. All three species are subjected to predation by both fish (Pihl, 1982) and birds (Dornboos, 1987). Nest building, courtship (Figure 1), (Potts, 1984), and searching for food should make gobies conspicuous and, hence, vulnerable to predation, and one would therefore assume that they have evolved a capacity to adjust their behaviour according to current predation risk.

FORAGING OR PREDATOR AVOIDANCE?

The risk of being eaten by a predator may influence the foraging behaviour of an animal in several ways. To avoid predators it could, for example, use a refuge

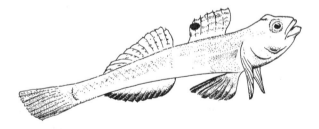

Figure 1   A male sand goby trying to attract a female raises his fins to make himself look larger and to display his dark breeding coloration, which probably makes him more conspicuous to predators as well as to potential mates.

(Werner *et al.*, 1983) or reduce its activity (Giles, 1987; Prejs, 1987). Both of these strategies can be costly in terms of lost feeding opportunities. A hungry animal therefore may have to take higher risks in order to get the food it requires than a less hungry one (Mangel and Clark, 1986; McNamara and Houston, 1986).

Table 1 Important features of the populations of gobies mentioned in this article

Species	Pomatoschistus minutus (Pallas)	Pomatoschistus microps Krøyer	Gobius niger L.
Common name	Sand goby	Common goby	Black goby
Lifespan	1 year	1 year	5 years
Max. body length	7 cm	5 cm	15 cm
Habitat	Open sand bottoms	Open sand-mud bottoms	Vegetation or open areas
Nest	Covered mussel shell	Covered mussel shell	Shell or rock

In an aquarium experiment, the food intake of sand gobies and black gobies was found to be clearly suppressed in the presence of a predator (cod, *Gadus morhua*), (Magnhagen, 1988). However, when the gobies were starved before the experiment, their food intake was higher than when they had continuous access to food, even in the presence of a predator (Figure 2). Photocells were used to measure the swimming

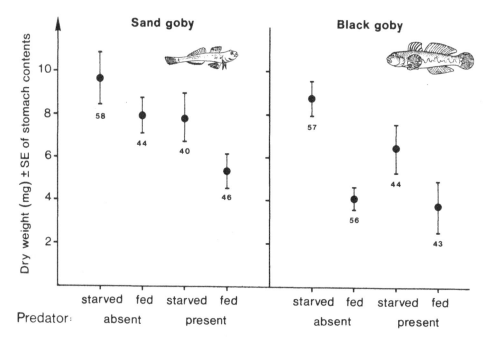

Figure 2 Mean weight (mg) ± SE of stomach contents from sand gobies and black gobies feeding for 24 h in the presence or absence of a cod (Magnhagen, 1988).

activity of sand gobies when feeding on the amphipod *Corophium volutator*, buried in the sediment. Individual activity levels were lower in the presence of the predator, and, in addition, starved gobies were more active than those that were previously fed (Figure 3). Thus it is likely that the decrease in food intake found in gobies in the presence of a predator was a consequence of a reduced activity.

A decrease in food consumption due to a choice of safe habitats that are less profitable than such with a high predation risk has been observed in fish such as bluegill sunfish, *Lepomis macrochirus* (Werner *et al.*, 1983), and armoured catfish, Loricariidae (Power, 1984). Sand gobies inhabit open sand bottoms and usually do not have the option to seek a refuge. However, these gobies are able to burrow, and are also very cryptic against the sediment. Thus they are probably easier to detect for a predator when moving. Decreasing their activity would then lower their conspicuousness and at the same time decrease their prey encounter rate. However, when starved, they have to take higher risks in order to find food, in accordance with the predictions that animals with a low metabolic state should take higher risks to avoid starvation (Mangel and Clark, 1986; McNamara and Houston, 1986), and thus starved gobies showed higher activity levels than fed individuals.

Atlantic salmon, *Salmo salar*, reduced their movements after exposure to a predator model (Metcalfe *et al.*, 1987) and as a result suffered a reduced feeding rate. This was also found in coho salmon, *Oncorhynchus kisutch*, where hunger was also influencing movements and thus risk-taking (Dill and Fraser, 1984). In addition, diet selection was influenced by predation risk. A decrease in swimming activity in the common goby as prey density increased (Magnhagen, 1986) was suggested to be an adaptation for predator avoidance, lowering the conspicuousness of the foraging fish,

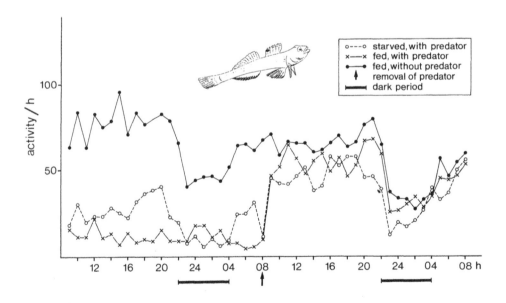

Figure 3   Activity of sand gobies measured as number of passages per hour through an infrared beam. Data points show the means from 5 replicate experiments with 3 fish in each (Magnhagen, 1988).

but also decreasing the possibility for prey size selection. Consequently, a preference for large *Corophium* over smaller was found at low prey densities, but prey size selection became random when small prey increased in abundance (Magnhagen, 1985), probably compensating for a reduced foraging activity.

REPRODUCTION UNDER PREDATION RISK

Predation risk can in some cases be considered as a cost of reproduction (Bell and Koufopanou, 1986; Magnhagen, 1991). Risks taken now can decrease the probability of future reproduction, and for this, among other reasons, there is a trade-off between current and future mating efforts. Sand gobies and black gobies were allowed to build nests and to spawn in the presence or absence of a piscivorous cod behind a glass wall (Magnhagen, 1990) and the number of nests built and number of nests with eggs was then counted. The sand goby, which is a short-lived species with only one breeding season, showed no change in spawning frequency in response to the presence of the predator. Neither the number of nests built nor the number of spawnings differed between treatments (Figure 4). In contrast, the black goby, which has a maximal lifespan of five years, totally refrained from spawning when the cod was visible (Figure 4). The black gobies used in this experiment were 2–3 years old. However, there was a big difference between fish of different ages. Of seven young pairs (2–3 yrs), only one reproduced in the presence of a predator. Of older pairs (4–5 yrs), six of seven spawned (Magnhagen, 1990). Both of these results, i.e. from the between species

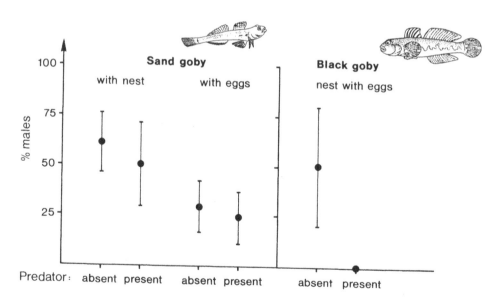

Figure 4 a) Percentage of sand goby males building nests and percentage of nests with eggs ($\bar{x} \pm SD$), b) percentage of black gobies spawning ($\bar{x} \pm SD$) in the presence and absence of a predator (Magnhagen, 1990).

comparison and the between age comparison, probably reflect the differences in expected future reproduction. Both the sand goby, with only one breeding season, and the old black gobies (assuming an age-dependent mortality) had a lower probability of future reproduction compared to the young black gobies, and should therefore take higher risks in order to maximize lifetime reproductive success.

Adaptations to decrease predation risk during reproduction can take several forms. Yet studies of such adaptations in fish are rare. However, in the guppy, *Poecilia reticulata*, a whole suite of evolutionary responses to predation risk have been discovered.

These responses include breeding coloration (Endler, 1982), female preference for brightly coloured males (Breden and Stoner, 1987), timing of mating (Endler, 1987) and mating behaviour (Luyten and Liley, 1985). In the sand goby, even though spawning rate did not decrease under predation risk, other adaptive behaviours could be expected, for instance a change in mate choice. Usually, larger males are preferred by females over smaller males (Forsgren, 1992). Under predation risk, however, this preference was found to disappear, and mate choice became random. This may be a way for females to decrease their mate searching time and thus the probability of being detected by the predator. Another explanation for a decreased scope for selectivity for mates in female gobies could be a decreased courtship rate in males under predation risk, which would make comparisons between potential mates more difficult. This has previously been found in guppies, *P. reticulata* (Farr, 1975).

NEST GUARDING UNDER PREDATION RISK

During the parental phase, the amount of investment (including risk-taking) put into the current brood should depend, among others, on two things: the reproductive value of the current brood and the probability of future reproduction for the parent (Pressley, 1981; Sargent and Gross, 1985). In a field experiment we investigated risk-taking during nest-guarding in male common gobies. The prediction was that the males would accept an increased risk as their clutch developed and thus increased in reproductive value and as the males themselves got a decreased probability of future reproductive opportunities (Magnhagen and Vestergaard, 1991). When a predator (eelpout, *Zoarces viviparus*) was placed in a glass in front of the nest, the male common gobies, after being chased away, spent a decreasing amount of time away from their nest as the eggs developed and thus increased in reproductive value (Figure 5). When testing for habituation, no such effect was found. Instead, the faster return rates of males with the progress of the brood cycle, likely reflects an acceptance of an increased risk of being eaten. In this case there may be a trade-off between the risk of being eaten and the risk of the nest being occupied by another male, which means a complete loss of the brood. The increased risk-taking may reflect the increase in brood value with time.

Similarly, threespine sticklebacks, *Gasterosteus aculeatus*, decreased time away from their nest, after fleeing from a predator, as the age of their eggs increased (Pressley, 1981). These males were also found to increase their aggressive behaviour with the development of the brood, as was also the case in bluegill sunfish (Coleman et al., 1985).

Reproductive effort and risk-taking should increase with a decrease in expected future reproductive success. In the common gobies most individuals will die after the

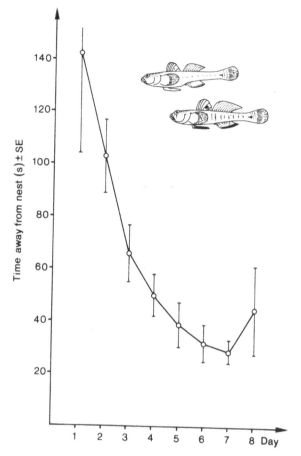

Figure 5 Time (s) before returning to the nest in male common gobies after being chased away in the presence of a predator. The data points are the average (± SE) of between 11 and 23 individuals, tested in May, once a day from receiving eggs in their nest until the hatching of the eggs (Magnhagen and Vestergaard, 1991).

breeding season, thus they should be expected to invest heavily in their last brood. There is an indication of this effect in our study (Magnhagen and Vestergaard, 1991). In August the males always came back to their nests just as fast in the presence of the predator as in its absence, which was not the case in May nor in the beginning of the breeding cycle in June (Magnhagen and Vestergaard, 1991). Thus, males seemed to accept higher risks when their probability of future reproduction was decreasing. In contrast to other studies, Ukegbu and Huntingford (1988) found that male threespine sticklebacks took fewer risks to defend their nest late in the season. This was explained by a lower value of late-hatching broods, since these young have a low probability to grow enough to reproduce the next season. This should also be the case in the common goby, since the offspring hatching late in the season should be smaller when the reproductive season starts the following spring and may therefore experience a lower reproductive success compared to conspecifics hatched at an earlier date. Actually, more nests were abandoned late in the season than otherwise, which can imply a lower value of these broods. However, the male common gobies staying with their brood all seem to modify their nest guarding behaviour both in accordance to the developmental stage of the brood and to their own prospects of future reproduction.

FORAGING AND NEST GUARDING

In female gobies reproductive success depends on the number of eggs produced, and also on the survival of these eggs, i.e. it is important to choose a male who is good at taking care of the brood. The fitness of a female should benefit from a high energy intake both because it leads to a high growth rate and a large female can produce more eggs than a small, and also because much energy goes straight into egg production. In female common gobies 28–44% of the dietary energy content is devoted to reproductive investment (Rogers, 1988). In contrast, the fitness of male gobies depends on how many females he can attract to the nest in order to fill it with eggs but also on the survival of these eggs. To maximize egg survival the males should spend as much time as possible guarding the eggs. Hence, there is a trade-off between egg guarding and foraging. During the reproductive season, male common gobies had a lower food consumption than females, according to stomach analyses of fish caught in the field, and consequently a lower growth rate over the season was found (Figure 6), (Magnhagen, 1986). The same was found in the pipefish *Syngnathus typhle*, where males eat less than females during reproduction (but not during the non-reproductive season) and also grow more slowly during that period (Svensson, 1988). In the painted greenling, *Oxylebius pictus*, frequent

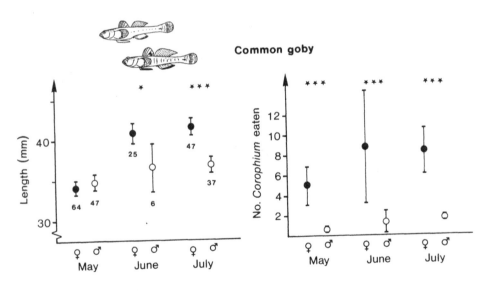

Figure 6 Total body length (mm) and number of *Corophium volutator* eaten ($\bar{x} \pm$ 95% CI) of female and male common gobies caught in the field.

guarders could only maintain their body weight, whereas infrequent guarders were able to increase their body weight (DeMartini, 1987). This was explained as a result of the lowered food intake from a restricted foraging area in the frequent guarders, as is probably also the case in male common gobies.

In the black goby there was no difference between the sexes in food intake, and males were slightly larger than females at the end of the breeding season (Figure 7). These differences between the species again may reflect the differences in the prospects of future reproduction due to differential longevity. It may be more important for the long-lived black goby to invest in body growth and survival in order to increase future reproductive success, and consequently invest less in the current brood, as compared to male common gobies with no expectations of survival to the next breeding season. Another explanation would be that a large size is more important for the ability to guard a nest in male black gobies than in common gobies, due to differences in nest building. A black goby is defending eggs that is placed more openly than is a common goby that has a nest that is hidden in the sand and very protected. Such a protected nest would be easier to defend than a more open one, and thus selection for a large size would not be as strong in male common gobies as in black gobies.

CONCLUSION

Several conflicting demands may simultaneously influence the behaviour of animals (Figure 8). In gobiid fishes I have shown trade-offs between foraging,

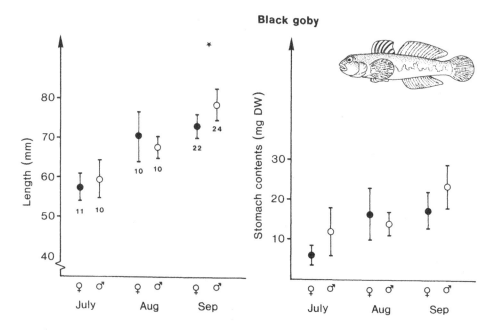

Figure 7 Total body length (mm) and stomach contents (mg dry weight) ($\bar{x} \pm 95\%$ CI) of female and male black gobies caught in the field.

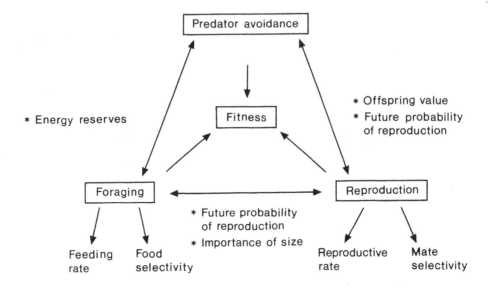

Figure 8 An animal's fitness depends on several conflicting demands. For example, predation risk can influence fitness both directly and through influencing foraging and reproductive behaviour in different ways. Trade-offs between conflicting demands (⟷) are influenced by several factors (examples marked by *).

spawning, nest guarding and predator avoidance, and also between foraging and nest guarding. What action an animal takes should depend both on intrinsic factors such as age, longevity and body condition, and on environmental factors such as food abundance, predation pressure and time of season. The interactions of these factors will lead to alternative behavioural "decisions" under different circumstances, in order to maximize fitness.

ACKNOWLEDGMENTS

My studies were supported financially by Swedish Natural Science Research Council and Magnus Bergvalls Foundation. The manuscript was scrutinized and improved by Ingrid Ahnesjö, Manfred Milinski, Ben Seghers, Staffan Ulfstrand and David Wiggins.

References

Bell, G. and Koufopanou, V. (1986). The cost of reproduction. In: *Oxford Surveys in Evolutionary Biology* (Ed. by R. Dawkins and M. Ridley), pp. 83–131. Oxford: Oxford University Press.
Berglund, A. and Rosenqvist, G. (1986). Reproductive costs in the prawn *Palaemon adspersus*: effect on growth and predator vulnerability. *Oikos*, **46**, 349–354.
Breden, F. and Stoner, G. (1987). Male predation risk determines female preference in the Trinidad guppy. *Nature*, **329**, 831–833.

Clark, C. W. and Ydenberg, R. C. (1990). The risks of parenthood. I. General theory and applications. *Evol. Ecol.*, **4**, 21–34.

Coleman, R. M., Gross, M. R. and Sargent, R. C. (1985). Parental investment decision rules: a test in bluegill sunfish. *Behav. Ecol. Sociobiol.*, **18**, 59–66.

DeMartini, E. E. (1987). Paternal defence, cannibalism and polygamy: factors influencing the reproductive success of painted greenling (Pisces, Hexagrammidae). *Anim. Behav.*, **35**, 1145–1158.

Dill, L. M. and Fraser, A. H. G. (1984). Risk of predation and the feeding behaviour of juvenile coho salmon (*Oncorhynchus kisutch*). *Behav. Ecol. Sociobiol.*, **16**, 65–71.

Doornbos, G. (1987). Piscivorous birds on the saline lake Grevelingen, the Netherlands: abundance, prey selection and annual food consumption. *Neth. J. Sea Res.*, **18**, 457–479.

Endler, J. A. (1982). Convergent and divergent effects of natural selection on color patterns in two fish faunas. *Evolution*, **36**, 178–188.

Endler, J. A. (1987). Predation, light intensity and courtship behaviour in *Poecilia reticulata* (Pisces: Poeciliidae). *Anim. Behav.*, **35**, 1376–1385.

Farr, J. A. (1975). The role of predation in the evolution of social behavior of natural populations of the guppy, *Poecilia reticulata* (Pisces, Poeciliidae). *Evolution*, **29**, 151–158.

Forsgren, E. (1992). Predation risk affects mate choice in a gobiid fish. *Am. Nat*, **140**, 1041–1049.

Giles, N. (1984). Implications of parental care of offspring for the anti-predator behaviour of adult male and female three-spined sticklebacks, *Gasterosteus aculeatus* L. In: *Fish Reproduction: Strategies and Tactics* (Ed. by G. W. Potts and R. J. Wootton), pp. 119–153. London: Academic Press.

Giles, N. (1987). Predation risk and reduced foraging activity in fish: experiments with parasitized and non-parasitized three-spined sticklebacks, *Gasterosteus aculeatus* L. *J. Fish Biol.*, **31**, 37–44.

Helfman, G. S. (1986). Behavioral responses of prey fishes during predator–prey interactions. In: *Predator–Prey Relationships: Perspectives and Approaches from the Study of Lower Vertebrates* (Ed. by M. E. Feder and G. V. Lauder), pp. 135–156. Chicago, London: University of Chicago Press.

Hesthagen, I. H. (1977). Migrations, breeding, and growth in *Pomatoschistus minutus* (Pallas) (Pisces, Gobiidae) in Oslofjorden, Norway. *Sarsia*, **63**, 17–26.

Lima, S. L. and Dill, L. M. (1990). Behavioral decisions made under the risk of predation: a review and prospectus. *Can. J. Zool.*, **68**, 619–640.

Luyten, P. H. and Liley, N. R. (1985). Geographic variation in the sexual behavior of the guppy, *Poecilia reticulata* (Peters). *Behaviour*, **95**, 164–179.

Magnhagen, C. (1985). Random prey capture or active choice? An experimental study on prey size selection in three marine fish species. *Oikos*, **45**, 205–216.

Magnhagen, C. (1986). Activity differences influencing the food selection in the marine fish *Pomatoschistus microps*. *Can. J. Fish. Aquat. Sci.*, **43**, 223–227.

Magnhagen, C. (1988). Changes in foraging as a response to predation risk in two gobiid fish species, *Pomatoschistus minutus* and *Gobius niger*. *Mar. Ecol. Prog. Ser.*, **49**, 21–26.

Magnhagen, C. (1990). Reproduction under predation risk in the sand goby, *Pomatoschistus minutus*, and the black goby, *Gobius niger*: the effect of age and longevity. *Behav. Ecol. Sociobiol.*, **26**, 331–335.

Magnhagen, C. (1991). Predation risk as a cost of reproduction. *TREE*, **6**, 183–186.

Magnhagen, C. and Vestergaard, K. (1991). Risk taking in relation to reproductive investments and future reproductive opportunities; field experiments on nest guarding common gobies, *Pomatoschistus microps*. *Behavioral Ecology*, **2**, 351–359.

Mangel, M. and Clark, C. W. (1986). Towards a unified foraging theory. *Ecology*, **67**, 1127–1138.

McNamara, J. M. and Houston, A. I. (1986). The common currency for behavioral decisions. *Am. Nat.*, **127**, 358–378.

Metcalfe, N. B., Huntingford, F. A. and Thorpe, J. E. (1987). The influence of predation risk on the feeding motivation and foraging strategy of juvenile Atlantic salmon. *Anim. Behav.*, **35**, 901–911.

Miller, P. J. (1984). The tokology of gobioid fishes. In: *Fish Reproduction: Strategies and Tactics* (Ed. by G. W. Potts and R. J. Wootton), pp. 119–153. London: Academic Press.

Montgomerie, R. D. and Weatherhead, P. J. (1988). Risks and rewards of nest defence by parent birds. *Quart. Rev. Biol.*, **63**, 167–187.

Pihl, L. (1982). Food intake of young cod and flounder in a shallow bay on the Swedish west coast. *Neth. J. Sea Res.*, **15**, 419–432.

Potts, G. W. (1984). Parental behaviour in temperate marine teleosts with special reference to the development of nest structures. In: *Fish Reproduction: Strategies and Tactics* (Ed. by G. W. Potts and R. J. Wootton), pp. 223–244. London: Academic Press.

Power, M. E. (1984). Depth distribution of armored catfish: predator-induced resource avoidance? *Ecology*, **65**, 523–528.

Prejs, A. (1987). Risk of predation and feeding rate in tropical freshwater fishes: field evidence. *Oecologia (Berlin)*, **72**, 259–262.

Pressley, P. H. (1981). Parental effort and the evolution of nest-guarding tactics in the threespine stickleback, *Gasterosteus aculeatus* L. *Evolution*, **35**, 282–295.

Rogers, S. I. (1988). Reproductive effort and efficiency in the female common goby, *Pomatoschistus microps*, Krøyer. *J. Fish Biol.*, **33**, 109–119.

Sargent, R. C. and Gross, M. R. (1985). Parental investment decision rules and the Concorde fallacy. *Behav. Ecol. Sociobiol.*, **17**, 43–45.

Sih, A. (1987). Predator and prey lifestyles: an evolutionary and ecological overview. In: *Predation: Direct and Indirect Impacts on Aquatic Communities* (Ed. by W. C. Kerfoot and A. Sih), pp. 203–224. Hannover, London: University Press of New England.

Sih, A. (1989). The effects of predators on habitat use, activity and mating behaviour of a semi-aquatic bug. *Anim. Behav.*, **36**, 1846–1848.

Sih, A., Krupa, J. and Travers, S. (1990). An experimental study on the effects of predation risk and feeding regime on the mating behavior of the water strider. *Am. Nat.*, **135**, 284–290.

Svensson, I. (1988). Reproductive costs in two sex-role reversed pipefish species (Syngnathidae). *J. Anim. Ecol.*, **57**, 929–942.

Ukegbu, A. A. and Huntingford, F. A. (1988). Brood value and life expectancy as determinants of parental investment in male three-spined sticklebacks, *Gasterosteus aculeatus*. *Ethology*, **78**, 72–82.

Vaas, K. F., Vlasblom, A. G. and de Koeijer, P. (1975). Studies on the black goby (*Gobius niger*, Gobiidae, Pisces) in the Veerse meer, SW Netherlands. *Neth. J. Sea Res.*, **9**, 56–68.

Vestergaard, K. (1976). Nest building behaviour in the common goby *Pomatoschistus microps* (Krøyer) (Pisces, Gobiidae). *Vidensk. Medd. Dansk Naturh. Foren.*, **139**, 91–108.

Werner, E. E., Mittelbach, G. G., Hall D. J. and Gilliam, J. F. (1983). Experimental test of optimal habitat use in fish: the role of relative habitat profitability. *Ecology*, **64**, 1525–1539.

Wootton, R. J. (1984). *A functional biology of sticklebacks*. London: Croom Helm.

CHOOSING PREY SIZE:
A COMPARISON OF STATIC AND DYNAMIC
FORAGING MODELS FOR PREDICTING PREY
CHOICE BY FISH

PAUL J. B. HART and ANDREW B. GILL

Department of Zoology, University of Leicester, UK

INTRODUCTION

Foraging theory has developed extensively since the first papers by MacArthur and Pianka (1966) and Emlen (1966). An historical review of foraging theory is provided in Schoener (1987) which concentrates mostly on models that do not account for changes in motivation with time (static state-independent models). The introduction of tractable dynamic models by Mangel and Clark (1986, 1988) and by McNamara and Houston (1986, 1988) has made it possible to model decision making that depends on the internal state of the animal (state-dependent models). These models capture more of the behaviour, have more parameters but are less general than analytic static models.

In this paper we discuss first the general features of foraging that models try to capture. We then outline the differences between static and dynamic optimization models before discussing how successful static models have been in predicting the diets of bluegill sunfish (*Lepomis macrochirus*) and for threespine stickleback (*Gasterosteus aculeatus*). The simplest static model of diet choice (the Basic Prey Model or BPM) does not predict diet for stickleback feeding on *Asellus* spp., so we develop a stochastic dynamic programming model which we test against experimental data. Our final section discusses why the BPM works with bluegills eating *Daphnia* but not with sticklebacks eating *Asellus*, a difference we name Werner's Paradox after Earl Werner. He has this honour because his early test of the BPM predicted bluegill diets in nature despite his tests violating some of the model's assumptions. An analysis by Mangel (1992), allows an interesting resolution of the paradox and points up limitations to the application of the BPM.

A TAXONOMY OF OPTIMIZATION MODELS

Early foraging models were based on the assumption that each decision is made against a stable background of environmental influences (MacArthur, 1972). It was further hypothesized that over evolutionary time, natural selection would push phenotypes towards those that made the decisions yielding highest fitness. It was assumed further by these early models that optimal decisions were not changed by the internal state of the animal or by its experience of the foraging environment. The foraging cycle of a typical fish, as depicted by early models, is illustrated in Figure 1.

The assumption that natural selection leads to fitness maximization through optimal phenotypic design has been very fruitful. In foraging studies alone, theoretical work has lead to a wide variety of models that predict the behaviour of animals in nature (Stephens and Krebs, 1986). Other areas of behavioural ecology, such as mate choice and aggression are equally well endowed (Krebs and Davies, 1991).

Ethologists have recognised for a long time that decisions made by animals are state-dependent (Tinbergen, 1951; Hinde, 1966). Attempts to devise state-dependent models were developed by McFarland and coworkers (e.g. McFarland and Houston, 1981), using continuous mathematics. The difficulty in finding solutions to their equations made their efforts hard to apply to most behaviours. In the last five years a new paradigm, based on discrete mathematics has been developed by McNamara and Houston (1986, 1988) and Mangel and Clark (1986, 1988). The techniques of dynamic programming used in this 'new' departure were developed over thirty years ago (Bellman, 1957) and can be traced back to nineteenth century mathematicians such as Hamilton and Jacobi!

For the rest of this article we will concentrate on the BPM and on a stochastic dynamic programming (SDP) model of prey choice.

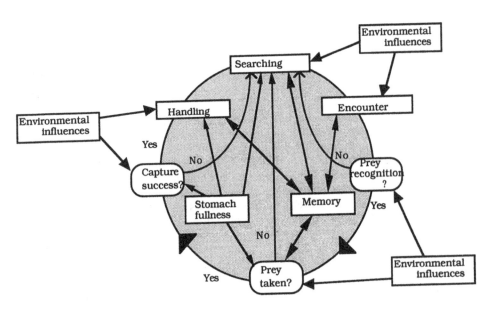

Figure 1 The foraging cycle. Rectangular boxes depict states while rounded boxes show decisions. The internal state of the animal is not considered important and environmental factors are assumed to be relatively constant from one turn of the cycle to the next.

THE BASIC PREY MODEL

Before discussing the application of the BPM it is important to be clear about it's structure, its assumptions and the predictions that can be derived from it. As with all optimization models, the BPM has three main components; a decision variable, a currency and a set of assumptions (Intriligator, 1971; Stephens and Krebs, 1986). Lifetime fitness is the ultimate currency, but as this is hard to measure, many theories use some other feature, such as long term rate of energy gain. This is justified when the characteristic is positively correlated with lifetime fitness. The BPM predicts the optimal diet assuming that prey are chosen to maximise the long term rate of energy gain (the currency). The equation for rate of gain is derived from Holling's (1959) disc equation (Stephens and Krebs, 1986) and is:

$$R = E/T = \frac{\sum_i P_i \lambda_i e_i}{1 + \sum_i P_i \lambda_i h_i}$$

where P_i is the probability that prey type i will be taken
λ_i is the encounter rate with prey type i
e_i is the energy gained per encounter with prey type i
and h_i is the handling time for prey type i.

To find the maximum of R we set $\partial R/\partial P_i = 0$ and solve for P_i ($0 \le P_i \le 1$), which is the decision variable. From this operation three predictions follow. These are:

1. Prey type i is either taken or ignored, there being no partial preferences. This is called the zero-one rule.

2. The predator ranks prey according to their profitability which is defined by e_i/h_i.

3. The inclusion of a prey type does not depend on its own encounter rate. Instead it is the encounter rate with more profitable prey types that is the deciding factor.

These predictions are valid when the following five biological assumptions are true (Stephens and Krebs, 1986).

1. The e_i's, h_i's and λ_i's are constant and independent of the decision variable P_i.
2. The forager does not encounter prey whilst handling another.
3. Prey are not encountered simultaneously and the interval between encounters is unpredictable (described by a Poisson process).
4. If prey are encountered but not attacked, the forager neither gains nor loses energy.
5. The forager is assumed to be built to make decisions in the way described by the model. This means it knows the values of the parameters and does not change its behaviour with time.

The assumption of most relevance to this article is that prey should be encountered sequentially and at random. Because of this assumption the BPM is a contingent model, meaning that the choice of prey is made on encounter.

OPTIMAL DIET THEORY AND BLUEGILL SUNFISH

The first ever test of the BPM, as well as an independent derivation of it, was by Werner and Hall (1974) using bluegill sunfish feeding on *Daphnia*. This study tested the BPM experimentally, controlling conditions stringently to nullify the effects of apparent size differences of prey and of hunger. The assumption of sequential encounter was violated as *Daphnia* were presented in swarms. Despite this, the fish behaved as predicted and dropped the less profitable prey sizes from the diet as the encounter rate with the most profitable type increased. Less profitable items were never entirely excluded so going against the zero-one rule. Many other tests have observed partial preferences (Stephens and Krebs, 1986).

For the decade after 1974, Werner and his coworkers used the BPM to understand better the feeding ecology of the bluegill and other sunfishes (Werner, 1984). The BPM was successful in predicting the optimal diet for bluegills in ponds and in predicting habitat shifts with season. The model was also used to predict the amount of niche overlap expected between bluegills, green sunfish (*Lepomis cyanellus*) and largemouth bass (*Micropterus salmoides*).

TESTING THE BPM WITH THREE-SPINE STICKLEBACK

The diet of three-spine stickleback feeding on a mixture of two *Daphnia* size classes was best predicted by the BPM in an experiment by Gibson (1980). Prey size classes were presented simultaneously and this violates the assumption of sequential encounter. In contrast, the BPM did not predict the diet of stickleback (ca 45 mm long) feeding on a sequence of *Asellus* of mixed sizes (Hart and Ison, 1991).

The apparatus used by Hart and Ison (1991) and their experimental protocol were an attempt to satisfy all the assumptions of the BPM. The fifth assumption could not be controlled for by good apparatus design. The experimental design attempted to control for experience. This was done by either presenting individual fish with an alternating mixture of the two treatments so that fish never habituated to one or the other, or by offering only one treatment to a fish so that it had the chance to learn the parameters and reach a stable performance. Details are in Hart and Ison (1991).

A preliminary experiment determined the handling and pursuit times for six, one millimetre, sizes classes of *Asellus* ranging between 3 and 10 mm. Energy values for the prey were obtained from Daoud (1984) allowing profitability to be calculated. Profitability was expressed as energy gained divided by pursuit plus handling time. The dome-shaped curve of profitability against prey size reached a maximum at *Asellus* sizes 5 and 6 mm. With a knowledge of the profitability of each size class two sequences of prey were devised that differed in rate of encounter with the most profitable prey sizes. Calling a sequence a Treatment, Treatment A had a high encounter rate with profitable prey while Treatment B had a low encounter rate. Using the BPM, it was then possible to calculate the optimal diet for the two Treatments.

Fish did not select the optimal diet (Figure 2). Prey size 3 mm should have been excluded under both Treatments, but were not. Prey size 7 mm should have been excluded from the diet under Treatment A, but was not. Under both Treatments 8 and

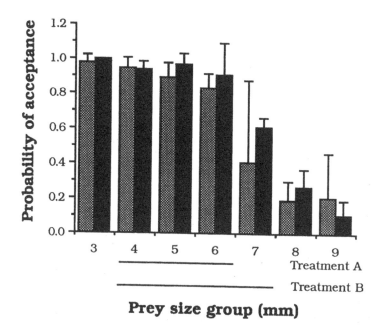

Figure 2 The probability of acceptance for each size class of *Asellus* (from Hart and Ison, 1991). Stippled columns represent the result from Treatment A and the black columns represent Treatment B (see text for details). Error bars are ± SD. The horizontal lines under the prey size axis show the optimal diet for the two treatments as calculated from the Basic Prey Model.

9 mm prey were eaten when they should have been ignored. For both Treatments, partial preferences were recorded.

Presenting prey in sequence meant that the state of the fish changed from one decision to the next. This change of state can be indexed by the notion of stomach fullness. Other physiological factors could also be important, but we assume that they would be linked to stomach fullness. To take account of changing fullness the captures for each prey size were divided into categories of energy already consumed. For example, all captures of 3 mm prey were divided up into those that were made when 0–99 J had already been consumed, 100–199, 200–299, and so on. This analysis showed that the amount of energy consumed had an influence on the acceptance of 7 mm or larger prey under both Treatments. As fish filled up they became more selective, particularly under Treatment A. Three millimetre prey were still readily consumed under all conditions.

The idea that the probability of a stickleback accepting large *Asellus* is influenced by stomach fullness was tested further by Hart and Gill (1992). Stickleback 45 mm in size were offered one of six different sequences of ten prey. Sequences differed in the number of 5 mm prey that filled the first five places. The number ranged between zero and five with the remaining positions being filled by 8 mm *Asellus*. The probability of taking the first 8 mm *Asellus* fell from about 0.9 when the fish had empty stomachs to less than 0.1 when five 5 mm prey had already been consumed. The experiment also showed that 45 mm stickleback continued to take prey up to a limit of about 450 J of energy consumed.

With stomach fullness a clear influence on prey choice, the state independent BPM is inadequate. We decided to develop a state dependent model to see whether it would provide a better prediction of diet when stickleback were eating *Asellus*.

A STOCHASTIC DYNAMIC PROGRAMMING MODEL OF STICKLEBACK PREY CHOICE

If all else fails, dynamic programming and a computer will usually produce results. (Stephens and Krebs, 1986)

The Model

Our SDP model will be described in detail in a separate publication. There is not space in this article to cover the details. Mangel and Clark (1988) present a state dependent optimal diet model and the model analysed here was derived from theirs. An SDP model dealing with prey choice has also been developed by Godin (1990).

In the experiments of Hart and Ison (1991) and Hart and Gill (1992), a trial lasted for about 10 minutes. For this reason we modelled fish feeding over a 10 minute interval dividing it up into one minute steps. The end of the feeding period then occurs at $T = 10$. In some SDP models T might coincide with a natural break in the animal's daily cycle, such as the fall of night, or life cycle, such as the onset of reproduction. Some authors, for example Huntingford and Metcalfe (1988) are worried by a T that does not correspond to some natural break. The argument is that the animal does not know when the period ends over which a policy is being optimized. In fact any time can be chosen for T, as pointed out by Houston and McNamara (1988), and by Clark (1991), because fitness at T is related to lifetime fitness, not just fitness at the end of the modelled interval.

There are two parts to SDP models; a fitness function and a difference equation describing state dynamics (Mangel and Clark, 1988). The fitness function, $F(x,t,T)$, gives the maximum value of the probability that the forager survives from period t to period T, given that at t the forager is alive and that the value of the state variable $X(t)$ is x. In our model we have assumed that at the end of the 10 minute period the fitness function, $F(x,T,T) = 1$, so long as the state variable is above a critical value x_c. When $x < x_c$, the fish dies from starvation.

At the start of each time period, the state (x) of the fish is calculated. During that period prey may or may not be encountered. If prey are not encountered, or they are encountered but rejected, then the fish gains no new energy but uses some in metabolism. As a result the change in state is described by

$$x(t+1) = x(t) - \alpha_0 = x_2$$

α_0 is the metabolic rate per unit time and $x(t)$ is stomach content in Joules, rescaled to a maximum of 10. To do this, the empirically determined maximum energy consumed in a feeding period, 450 J, is divided by itself and multiplied by 10. If x falls below x_c then the fish dies of starvation. Although fish do not starve to death over a short period without food, time spent below x_c is assumed to be related to the risk of starvation.

The second possibility is that a prey will be encountered during a time interval and consumed. If this is the case then the state is incremented by the energy gained from the food and reduced by the energy cost of maintenance and of prey handling. The difference equation is now

$$x(t+1) = x(t) - \tau_i \alpha_0 + Y_i = x_1$$

where τ_i is the handling time for prey size i and Y_i is the energy gained from prey size i. The model does not allow prey to be consumed if it would take the stomach contents beyond its limit of C=10. This takes account of the discreteness of the energy packet available in *Asellus* and the limits on the number of prey that can be packed into the stomach (Hart and Gill, 1992).

With a formulae derived for determining the fish's state it is now possible to calculate the fitness function. The two conditions of acceptance and rejection have to be accounted for separately. The probability of encountering a prey of size i is λ_i, so that the total probability of encountering prey is $\Sigma\lambda_i$. This means that λ_0, the probability of not encountering prey, is equal to $1 - \Sigma\lambda_i$. The fitness component resulting from encountering no prey in the interval is $\lambda_0 F(x_2,t+1,T)$. If prey are encountered then two outcomes are possible; either prey can be accepted, in which case fitness will be $F(x_1,t+\tau_i,T)$, or they can be rejected and fitness is expressed as $F(x_2,t+1,T)$. The model assesses the fitness gain from each of these options and takes the maximum. This comparison has to be done for all prey sizes available. The resulting equation is $\Sigma\lambda_i(\max[F(x_2,t+1,T), F(x_1,t+\tau_i,T)])$. This then has to be added to the fitness gained from not encountering prey. The complete equation for the fitness function at time t is then

$$F(x, t, T) = \lambda_0 F(x_2, t+1, T) + \sum \lambda_i (\max [F(x_2, t+1, T), F(x_1, t+\tau_i, T)])$$

Finding the Optimal Diet

Values for the variables, λ_0, λ_i, τ_i and Y_i were taken from Hart and Ison (1991 Tables IIb, IV and VIb), and those used are shown in Table 1. Median values for pursuit and handling times were used. Probabilities of encounter were recalculated from Table IIb in Hart and Ison (1991) so that $\Sigma\lambda_i + \lambda_0 = 1$. Energy values for each

Table 1 The values of the variables used in the SDP model of stickleback prey size selection. $\lambda_i =$ encounter rate with prey size i; $Y_i =$ energy gained from prey size i; $\tau_i =$ handling time for prey size i; $P_s(i) =$ probability that prey size i will be successfully captured once attacked

Prey size	Treatment A λ_i	Treatment B λ_1	Y_i	τ_i	$P_s(i)$
3	0.08	0.07	0.2	0.066	1.00
4	0.12	0.16	0.4	0.099	1.00
5	0.18	0.05	1.3	0.207	0.92
6	0.12	0.05	1.8	0.280	0.47
7	0.10	0.05	2.7	0.371	0.53
8	0.14	0.18	3.8	0.980	0.24
9	0.06	0.06	5.9	1.630	0.12

prey size were rescaled by dividing by the stomach capacity (450J) and multiplying by 10. This gives a range of stomach contents of from 1 to 10. The model was run on an IBM clone PC (a Tandon Plus).

The first run used the encounter probabilities from Treatment A and the second used those from Treatment B. The iteration required to calculate the optimal sequence of behaviours (designated by the matrix $A^*(x,t)$) starts at T and works backwards through time. The value of $A^*(x,t)$ changes as t approaches 1. A common property of SDP models is that the optimal policy converges on one pattern when the time to go $(T-t)$ becomes large, meaning that $A^*(x,t) = A^*(x)$ (Clark, 1991). An example of the optimal diet for $t = 3$ is shown in Figure 3a with $x_c = 0$ and $C = 10$. The predicted diet does not change with encounter rate. Instead, the main driving factor is the stomach capacity. When the fish's stomach fullness is close to 10, prey are rejected. Values of the fitness function $(F(x,t,T))$ did change with Treatment, being lower for Treatment B. This is a reflection of the lower probability of encountering prey.

Altering the SDP Model to Account for Prey Preferences

The failure of the SDP model to account for the observed change in diet with changed encounter rate means that some aspect of the system is still unaccounted for. The model assumes that if a prey is encountered in a time interval then it is either eaten or rejected. In the experiment reported by Hart and Ison (1991), the probability (P) that a prey of a particular size would be taken was a function of prey size (ps in mm), energy already consumed (stomach fullness) (x) and the encounter rate (λ). The multiple regression was

$$P = 1.52 - 0.16ps - 0.0009x + 0.11\lambda$$

Prey size accounted for 51.9% of the variance in P while energy consumed and encounter rate accounted for only 8.7 and 1.6% respectively. This shows that our very simple first SDP model, by failing to incorporate a response to prey size, was leaving out the most important factor. Prey acceptance was negatively related to prey size, so fish were increasingly reluctant to take larger prey. It was observed that small Asellus of 3 to 5 mm were eaten without hesitation, even as the stomach filled. Beyond 5mm, prey were approached with increasing hesitancy. When eating 5 mm prey Hart and Gill (1992) found that Asellus were eaten after only one manipulation, where manipulation is defined as spitting the prey out and recapturing it. For 8 mm prey, the majority were accepted after less than five manipulations and they rejected them after less than four.

To account for these observations in the SDP model we incorporated a new variable $(P_S(i))$, which is the probability of successfully ingesting a prey of size i once the fish had decided to attack. $P_S(i)$ decreases with increasing prey size. There would now be two outcomes that could follow encounter with a prey in a time interval. The prey can be rejected outright with no time spent handling, or it can be handled and then consumed.

The dynamic programming equation now becomes

$$F(x, t, T) = \lambda_0 F(x_2, t+1, T) + \sum_i \lambda_i \max \{F((x_2, t+1, T), P_s(i) F(x_1, t+\tau_i, T))\}$$

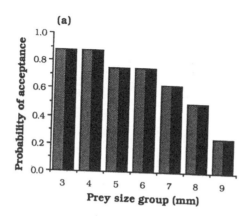

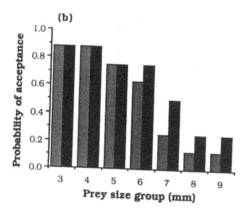

Figure 3 (a) The probability of acceptance for each size class of *Asellus* as predicted by the simplest SDP model. Stippled columns represent Treatment A and black columns Treatment B. For each prey size class the probability is calculated across all values of *x* between 1 and 10. At each *x* the model predicts that the prey should or should not be taken. (b) The predicted diets when the probability of successful attack is included in the model (see text for further details). This figure should be compared with Figure 2.

The optimal diet predicted by this equation for Treatment A and Treatment B of Hart and Ison's (1991) experiment is shown in Figure 3b. The diet shown is for $t=3$ and the $P_s(i)$ values given in Table 1. The predictions are in agreement with experimental results. As the internal state approaches 10, the fish become more selective, concentrating on the smaller prey items (Figure 4 a, b, and c). When the fish has an almost full stomach, space left in the stomach becomes the over-riding factor. This much is true for both Treatments, but we now also have a broader diet under Treatment B than under A.

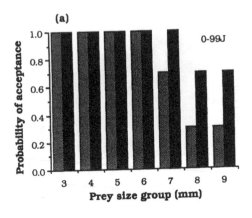

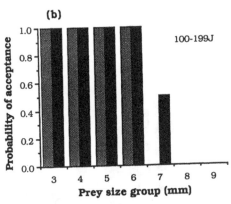

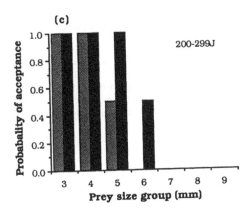

Figure 4 The probability of acceptance for each size class of *Asellus* as predicted by the SDP model incorporating the probability of successful attack. Stippled columns represent Treatment A and black bars Treatment B. (a) When the state variable *x* represents between 0–99 J, (b) when *x* is between 100–199 J and (c) when *x* is between 200–299 J. This figure should be compared with Figure 3.

Sensitivity Analysis

Gladstein *et al.* (1991) have pointed out that the sensitivity of SDP models to parameter changes always needs to be examined. A summary of a stability analysis is given in Table 2, which shows the general effect of increasing or decreasing the parameters or variables over those used in the 'standard' run of the model (see the Table's legend for these). To obtain the results shown in the Table, each parameter was either increased or decreased whilst all the other variables were held at their 'standard' values. Each λ_i, τ_i, and Y_i was increased or decreased by a constant amount. The effect of each change was greater when time to go $(T-t)$ was large and when x was close to C. The changes in diet are brought about by including or excluding more or less of the larger prey.

WERNER'S PARADOX AND ITS RESOLUTION

There is an interesting contrast between bluegills foraging on *Daphnia* and stickleback foraging on *Asellus*. For the bluegills the BPM makes predictions of diet that have been born out by experiment and field observation. For the sticklebacks feeding on *Asellus* the BPM does not predict prey choice but a state dependent model does. We call this interesting contrast, Werner's Paradox. What is it about a foraging environment that determines whether or not the BPM makes successful predictions?

The resolution of the paradox emerges from a consideration of a theoretical analysis of the problem by Mangel (1992). There is only space here to sketch his analysis. First an SDP model of a two prey system is established, using continuous mathematics for ease of comparison with the BPM. At the maximum of the SDP function, Mangel then looks for a rate maximizing solution. It should be remembered that SDP models usually take maximization of survival or reproduction as their

Table 2 The robustness of the second model to increases or decreases in the parameter values used in the calculations summarised in Figures 3b and 4; parameter values used in the 'standard run', $C = 10$, $x_c = 2$, $T = 10$, $\alpha_0 = 1$. Expanded diet means more prey sizes included; reduced means fewer included. All assessments of change made when the optimal strategy $A^*(x,t)$ has converged to a stationary strategy $A^*(x)$ (see text)

Parameter or variable	Increased from 'standard' value		Decreased from 'standard' value	
	Diet	Fitness	Diet	Fitness
Capacity (C)	expanded	higher	reduced	lower
Critical x (x_c)	reduced	lower	expanded	higher
Horizon (T)	expanded	lower	reduced	higher
α_0	expanded	lower	reduced	higher
τ_i	expanded	lower	reduced	higher
Y_i	reduced	higher	expanded	lower
λ_i	reduced	higher	expanded	lower

optimization criterion, whilst the BPM assumes that the forager is maximizing fitness through the maximization of the long term rate of energy gain. Finally, Mangel shows that the rate maximizing solution is the same as the state variable solution only when four conditions are the case. These are:

1. Metabolic rates are constant
2. Physiological constraints are unimportant, i.e. x_c and C do not play a role
3. Predation is unimportant
4. Expected future reproduction is equal to the value of the state variable.

For the resolution of Werner's Paradox, points one and two are probably the most important. The analysis by Mangel (1992) leads to a two stage prediction about the suitability of the BPM for predicting diet. According to this prediction, prey choice can be approximated by the BPM when prey items are small relative to the predator. The predictions of the BPM break down when each prey item contributes a significant proportion of the forager's energy requirement. When bluegills eat *Daphnia*, each prey item only changes the fish's state by a very small amount and search effort is small, as it will be for fish feeding on a swarm of prey. A 45mm stickleback eating even a 5mm *Asellus* is changing its internal state significantly. The contrast between the two forager/prey systems is illustrated in Figure 5.

Together with Hart and Ison's (1991) experiment, the experiment by Gibson (1980), mentioned earlier, provides a test of the prediction for stickleback. Gibson used fish of mean length 47mm, which is virtually the same length as those used by Hart and Ison (1991). Gibson's stickleback were fed *Daphnia* and the BPM best explained their prey choice when offered a mixture of two sizes.

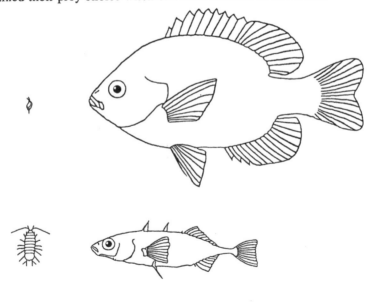

Figure 5 Relative sizes of predator and prey illustrating the resolution of Werner's Paradox (see text). The top picture shows a 75 mm bluegill sunfish and a 3 mm *Daphnia*, with the bottom pair showing a 45 mm stickleback and an 8 mm *Asellus*. The line represents 20 mm.

CONCLUSIONS

Where the underlying assumptions are valid, the BPM is a useful aid to understanding ecological processes at the population and community level (Werner, 1984; Persson and Diehl, 1990). It is simple and has a small number of parameters that can be easily estimated; the critical point being whether the BPM is valid for a given predator/prey system (Hart, 1989). SDP models are able to include more factors relevant to prey choice and we have shown in this article how an SDP model is a better predictor than the BPM of prey choice in the threespine stickleback feeding on *Asellus*. A significant conclusion of this article is that by analysing the conditions under which the two models make the same predictions, inspired by Mangel's analysis, it is now possible to see when the BPM might be valid and so can safely be used to analyse foraging relations in wild fish. The results will also probably translate to other taxonomic groups.

ACKNOWLEDGMENTS

The research on stickleback prey choice is supported by a grant from the UK Natural Environment Research Council (GR3/7293). We are grateful to Marc Mangel for sending us his unpublished manuscript, for help with its interpretation and for suggesting new interpretations of aspects of our SDP model. We thank also, Felicity Huntingford, Frietson Galis and Lennart Persson for comments on the manuscript.

References

Bellman, R. (1957). *Dynamic Programming*. Princeton: Princeton University Press.
Clark, C. W. (1991). Modeling behavioral adaptations. *Behav. Brain Sci.*, **14**, 85–117.
Daoud, Y. T. (1984). Ecology and bioenergetics of two species of *Asellus* in Rutland Water. Ph.D. thesis. University of Leicester, Leicester, UK.
Emlen, J. M. (1966). The role of time and energy in food preference. *Am. Nat.*, **100**, 611–617.
Gibson, R. M. (1980). Optimal prey-size selection by three-spined sticklebacks (*Gasterosteus aculeatus*): a test of the apparent-size hypothesis. *Zeitschrift fürTierpsychologie*, **52**, 291–307.
Gladstein, D. S., Carlin, N. F., Austad, S. N. and Bossert,W. H. (1991). The need for sensitivity analyses of dynamic optimization models. *Oikos*, **60**, 121–126.
Godin, J.-G. J. (1990). Diet selection under the risk of predation. In: *Behavioural Mechanisms of Food Selection* (Ed. by R. N. Hughes, NATO ASI Series Vol G20), pp. 739–770. Berlin: Springer-Verlag.
Hart, P. J. B. (1989). Predicting resource utilization: the utility of optimal foraging models. *J. Fish Biol.*, **35** (Supp. A), 271–277.
Hart, P. J. B. and Ison, S. (1991). The influence of prey size and abundance, and individual phenotype, on prey choice by three-spined stickleback, *Gasterosteus aculeatus* L. *J. Fish Biol.*, **38**, 359–372.
Hart, P. J. B. and Gill, A. B. (1992). Constraints on prey size selection by the three-spined stickleback: energy requirements and the capacity and fullness of the gut. *J. Fish Biol.*, **40**, 205–218.
Hinde, R. A. (1966). *Animal Behaviour. A Synthesis of Ethology and Comparative Psychology*. London: McGraw-Hill.

Holling, C. S. (1959). The components of predation as revealed by a study of small-mammal predation on the European sawfly. *Can. Entomol.*, **91**, 293–320.

Houston, A. I. and McNamara, J. M. (1988). A framework for the functional analysis of behavior. *Behav. Brain Sci.*, **11**, 117–163.

Huntingford, F. A. and Metcalfe, N. (1988). The functional analysis of behaviour: making room for Prufrock. *Behav. Brain Sci.*, **11**, 137–138.

Intriligator, M. D. (1971). *Mathematical Optimization and Economic Theory*. Englewood Cliffs: Prentice-Hall.

Krebs, J. R. and Davies, N. B. (1991). (editors) *Behavioural Ecology: An Evolutionary Approach*, 3rd edn. Oxford: Blackwell Scientific Publications.

MacArthur, R. H. (1972). *Geographical Ecology: Patterns in the Distribution of Species*. New York: Harper and Row.

MacArthur, R. H. and Pianka, E. R. (1966). On optimal use of a patchy environment. *Am. Nat.*, **100**, 603–609.

McFarland, D. J. and Houston, A. . (1981). *Quantitative Ethology: The State Space Approach*. London: Pitman.

McNamara, J. M. and Houston, A. I. (1986). The common currency for behavioural decisions. *Am. Nat.*, **127**, 358–378.

Mangel, M. (1992). Rate maximizing and state variable theories of diet selection. *Bull. Math. Biol.*, **54**, 413–422.

Mangel, M. and Clark, C. W. (1986). Towards a unified foraging theory. *Ecology*, **67**, 1127–1138.

Mangel, M. and Clark, C. W. (1988). *Dynamic Modeling in Behavioural Ecology*. Princeton: Princeton University Press.

Persson, L. and Diehl, S. (1990). Mechanistic individual-based approaches in the population/community ecology of fish. *Ann. Zool. Fennici*, **27**, 165–182.

Schoener, T. W. (1987). A brief history of optimal foraging theory. In: *Foraging Behaviour* (Ed. by A. C. Kamil, J. R. Krebs and H. R. Pulliam), pp. 5–67. New York: Plenum Press.

Stephens, D. W. and Krebs, J. R. (1986). *Foraging Theory*. Princeton: Princeton University Press.

Tinbergen, N. (1951). *The Study of Instinct*. Oxford: Oxford University Press.

Werner, E. E. (1984). The mechanisms of species interactions and community organization in fish. In: *Ecological Communities: Conceptual Issues and the Evidence* (Ed. by D. R. Strong Jr., D. Simberloff, L. G. Abele and A. B. Thistle), pp. 360–382. Princeton: Princeton University Press.

Werner, E. E. and Hall, D. J. (1974). Optimal foraging and the size selection of prey by the bluegill sunfish (*Lepomis macrochirus*). *Ecology*, **55**, 1042–1052.

FACTORS AFFECTING THE BEHAVIOURAL MECHANISMS OF DIET SELECTION IN FISHES

MICHEL J. KAISER[+#†] and ROGER N. HUGHES[+]

[+]*Functional and Evolutionary Biology Group, School of Biological Sciences, University of Wales, Bangor, Gwynedd, LL57 2UW, UK, and [#]Dunstaffnage Marine Laboratory, P.O. Box 3, Oban, Argyll, PA34 4AD, UK*

INTRODUCTION

Optimal Foraging Theory

Optimality models have proved to be useful to biologists interested in evolutionary aspects of behavioural ecology, particularly with regard to predator–prey interactions (Maynard Smith, 1978; Krebs and Davies, 1984; Stephens and Krebs, 1986; Hughes, 1990). The fundamental principle of any optimality model is that organisms act in a way that maximises fitness, defined in population-genetics terms as "probable contribution to the next generation, relative to that of individuals of other genotypes" (Dawkins, 1982). The successful acquisition of food is essential to maintain growth and reproduction, hence feeding behaviour has been the subject of intense research over the past 25 years (for reviews see Stephens and Krebs, 1986; Schoener, 1987; Hughes, 1990). The observed diets of predators in the field are rarely a random sample of the available prey (Stephens and Krebs, 1986; Schoener, 1987; Hughes, 1988). Selectivity could arise from mechanical relationships between predator feeding apparatus and prey morphology, or from complex behavioural responses to visual, mechanical and chemical characteristics of the prey (Case *et al.*, 1960; Curio, 1976; Moore and Moore, 1976; Hirtle, 1978; Liem, 1980; Holmes and Gibson, 1986; Hughes, 1988; Kaiser, 1991).

Economics models were first applied to predict optimal feeding strategies by MacArthur and Pianka (1966) and Emlen (1966), initiating what has become known as Optimal Foraging Theory (OFT). Although some have disputed the relevance of OFT (Pierce and Ollason, 1987), empirical studies, in general, have supported the predictions derived from the above energy maximisation premise (for reviews see Hughes, 1980; Pyke, 1984; Stephens and Krebs, 1986; Schoener, 1987). OFT, has been applied to animals in a wide range of trophic categories e.g. herbivores (Illius and Gordon, 1990), particle-feeders (Taghon *et al.*, 1978; Jørgensen, 1990), and carnivores (Kislalioglu and Gibson, 1976a; Elner and Hughes, 1978). In most of its forms, OFT has focused on patterns of diet choice and food-patch exploitation (Stephens and Krebs, 1986), also incorporating the effects on foraging behaviour of

---

†. Present address: MAFF, Fisheries Laboratory, Benarth Rd., Conwy, Gwynedd, LL32 8UB, UK

interspecific interactions (Huang and Sih, 1991) and time constraints (Real and Caraco, 1986).

Optimal Diet Theory

The sub-model of OFT applicable in diet choice, termed Optimal Diet Theory (ODT), predicts that animals acquire food in an economical fashion, i.e. that they maximise their net rate of energy gain, so increasing the energy available for reproduction and somatic growth (Charnov, 1976). When studying animals that live in highly dynamic habitats, additional factors such as tides, which limit foraging time and threaten survivorship, inter- and intraspecific competition and risk of predation, violate the simplifying assumptions of ODT (Hughes, 1990). However, these assumptions can be met in controlled laboratory studies, and the results taken as a guide to the underlying principles influencing selective feeding behaviour. Nevertheless few studies have attempted to elucidate the morphological and behavioural bases of diet selection, hence, as a step in this direction, we hope to bring together some of the relevant ideas.

FORAGING BEHAVIOUR OF FISH

Fish readily adapt to aquarium conditions and seem to exhibit natural behaviour in the laboratory. Consequently, they make excellent subjects for studies of mating (e.g. Sevenster, 1951; Losey and Sevenster, 1991; Rowland et al., 1991), agonistic (e.g. Huntingford and Turner, 1987) and foraging behaviour (e.g. Ibrahim and Huntingford, 1989; Wetterer, 1989; Gibson and Ezzi, 1990; Hart and Ison, 1991; Kaiser et al., 1992a).

Early studies of feeding behaviour centred on correlations between stomach contents and available prey, showing that fish were highly selective (e.g. Werner and Hall, 1974; Kislalioglu and Gibson, 1975). Further examination revealed that there was a close relationship between mouth dimensions and critical prey size (Kislalioglu and Gibson, 1975; Kaiser et al., 1992b), and more recently, dietary shifts have been related to ontogenetic changes in trophic apparatus (Galis, 1990). Werner and Hall (1974) proposed that observed diet selection was based on the rule of thumb, to eat the largest prey possible. This was observed in fish eating small prey, *Daphnia*, which were well within the limits of mouth constraints. Kislalioglu and Gibson (1976a), however, demonstrated that although fish were capable of eating larger prey, the preferred prey-size was significantly smaller than the critical prey-size based on maximum mouth diameter. Furthermore, when the net rate of energy intake (based on handling time per unit dry weight) was calculated, the predicted optimal prey-size correlated closely with the observed diet (Kislalioglu and Gibson, 1976a). These conclusions have been the basis of most studies of diet selection to the present date. Recently, however, attention has focused on the more complicated interactions of feeding behaviour with effects of experience (learning) (for reviews see Kieffer and Colgan, 1992; Hughes et al., 1992) and physiological state (hunger) of the fish (Croy and Hughes, 1990, 1991a,b) and on predatory capability relative to prey escape responses (Hart and Hamrin, 1990; Kaiser et al., 1992a). Predicted optimal diets, based on pre-digestive behaviour, generally have concurred with observed feeding behaviour (Kislalioglu and Gibson, 1976a; Wetterer, 1989; but see Hart and Ison, 1991).

Despite the wide application of the energy maximisation premise noted above, few studies have, until recently, attempted to quantify the net energetic effects of selective feeding. Predatory behaviour can account for a large proportion of daily energy expenditure. Lucas *et al.* (1991) followed the movements of individual pike using acoustic tags, measuring the heart-beat rate during various activities, and found that feeding behaviours accounted for up to 10% of daily activity metabolism. American eels, *Anguilla rostrata*, use three modes of feeding; suck, bite and spin feeding (Helfman, 1990; Helfman and Winkelman, 1991). Spin feeding, although energetically costly, allows eels to exploit carion that is otherwise too large to be swallowed. When eels were maintained on diets that required the use of one feeding mode exclusively, fish that had to use spin feeding lost more weight than those using suck feeding. Given a choice of prey, eels preferred to eat the food that required the least energetic mode of feeding (Helfman and Winkelman, 1991). Similarly, red-ear sunfish, *Lepomis microlophus*, feeding on different snail types chose those that required least energy to crush (Stein *et al.*, 1984). Planktivores, such as the herring, *Clupea harengus*, also use several modes of feeding (Gibson and Ezzi, 1985). When prey concentration was low fish used biting, switching to gulping and then filter-feeding as encounter rate increased (Gibson and Ezzi, 1992). Filter-feeding increases drag as fish move through the water (Crowder, 1985), hence this method was only used when the rate of energy intake exceeded the cost of swimming (Gibson and Ezzi, 1991).

Fish clearly have played a major role in experimental investigations of mechanisms underlying diet selection; past studies are well reviewed by Ringler (1983). In this paper we will examine more closely some recent studies on the feeding behaviour of the fifteen-spined stickleback, *Spinachia spinachia* (L.).

FEEDING BEHAVIOUR OF THE FIFTEEN-SPINED STICKLEBACK

Sticklebacks are amenable subjects for behavioural investigations; the foraging, reproductive and energetic biology of the three-spined stickleback, *Gasterosteus aculeatus* L., is especially well documented (for a review see Wootton, 1984). Conversely, the marine stickleback, *Spinachia spinachia*, until recently, has received little attention.

Spinachia is found in the subtidal zone of north-western Europe to a maximum depth of 20 m (Wheeler, 1969), but often moves into the intertidal zone to feed. *Spinachia* is an ambush predator (personal observation), hunting among stands of *Fucus spp.*, *Halydris siliquosa* and *Laminaria spp.*, where it is well camouflaged from predators (Croy, 1989). *Spinachia* is semelparous, but whereas early studies indicated an annual life cycle (Jones and Hynes, 1950), recent evidence suggests that some females may delay reproduction to a second year (Kaiser and Croy, 1991). The short life span of these fish makes them suitable for ontogenetic studies of diet selection, relating dietary changes to observed behaviour and morphology.

Constraints Associated with Feeding Apparatus

In a study of a Scottish population of *Spinachia,* stomach-contents analysis revealed that four types of prey were eaten; copepods, isopods, amphipods and mysids (Kaiser *et al.*, 1992b). A previous study had recorded a more diverse selection of prey in the diets of *Spinachia* from lochs on the west coast of Scotland (Kislalioglu and Gibson, 1977), but as beaches tend to have lower species diversity than sublittoral habitats, these results are not surprising. Copepods are available to the fish from March/April to September/October, but are excluded from the diet of large (length > 90 mm) fish, despite being abundant (Figure 1). In some instances gill-raker spacing is an important factor determining lower size limit of catchable prey (Wankowski, 1979; Dervo *et al.*, 1991), but cannot account for the exclusion of larger copepods, exceeding the gap between gill rakers, from the diet of large *Spinachia* (Figure 2). Copepods are probably unprofitable prey for fish > 90 mm in length, due to the large number that must be eaten to reach satiation, together with the associated increase in energy expended on searching and prey capture.

Therefore, large fish possibly ignore copepods. Isopods in the diet were exclusively *Idotea baltica*, these increase in importance in the diet of larger fish, but never assume a leading dietary role (Figure 1).

Mysids and amphipods appear to be the principal prey types in the diet, mysids becoming more important than amphipods as *Spinachia* increase in size (Kaiser *et al.*, 1992a). In the case of amphipods, irregular profile interferes with swallowing and increases ingestion time, so excluding larger items from the diet. As fish become larger, however, they are better able to handle the larger amphipods and therefore consume a much broader range of prey sizes. Conversely, mysids have a more streamlined profile than amphipods, hence all sizes of mysid were available to fish >50 mm in length. Therefore, although prey morphology explains the increase in range of amphipods eaten, it does not explain the greater importance of mysids in diets. We decided to examine the role of prey escape responses and fish predatory capabilities as a possible explanation of this observation.

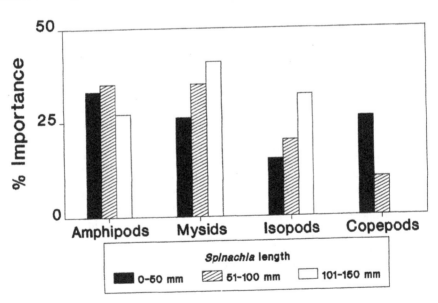

Figure 1 The change in importance of prey in the diet of *Spinachia* as fish increase in size. Data taken from Kaiser *et al.*, 1992a,b.

Predatory Capabilities

Spinachia feed on a diverse range of mainly crustacean prey, which have different methods of avoiding predators (Kislalioglu and Gibson, 1977; Kaiser, 1991; Kaiser *et al.*, 1992a). Mysids respond to fish attacks by using a fast-start tail flip (Figure 3), making them difficult to capture. The predatory behaviour used by sticklebacks was analysed from video recordings of prey capture (Kaiser *et al.*, 1992a). In the first 0.02 s of a strike at a mysid, *Spinachia* travelled the mean hover distance (Figure 3). During the following 0.02 s the fish reached their peak velocity, as did the mysid. The second 0.02 s of a strike is the critical stage when the mysid is actually captured. Mysids most commonly moved at an angle of approximately 90° to the line of fish attack (Kaiser *et al.*, 1992a). If *Spinachia*'s attack were to be in a straight line, mysids would evade the zone of interception, as described by Hart and Hamrin (1990). *Spinachia*, however, tended to angle their attack approximately 32° towards the anticipated direction of mysid escape (Figure 3), increasing capture rate. Kislalioglu and Gibson (1976b) suggested that *Spinachia* fixate on the mysid's head region, because of its prominence. During a mysid flip response, the head and antennae are posterior to the centre of mass and consequently closest to the predator's mouth. By fixating on the prey's head the probability of prey capture is increased. Analysis of the stomach contents of wild *Spinachia* revealed that the largest mysids eaten coincided with the upper limit of fish fast-start capability (Figure 4). Although mysid availability is not limited by mouth dimension in large fish, the fast-start capability of *Spinachia* is a critical constraint, possibly explaining why mysids increase in dietary importance as the fish become larger (Figure 1). Concomitantly, amphipods decrease in importance, so perhaps mysids are a more profitable food source for larger fish.

Amphipods most commonly escaped by simple swimming, with a mean velocity three times slower than the fast-start response of *Spinachia* (Figure 4). Moreover, the

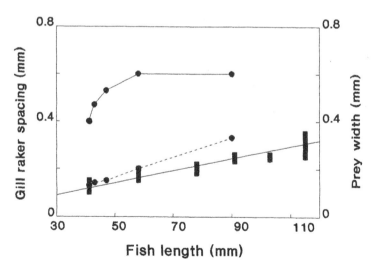

Figure 2   The greatest width (mm) of the largest (-●-)   and smallest (- ● -) copepods eaten by *Spinachia*, compared with gill-raker spacing (mm) (■) in the same fish. Taken from Kaiser *et al.*, 1992b.

mean velocity attained by *Spinachia*, striking at amphipods, was 1.1 times greater than the maximum amphipod swimming velocity. Thus, although amphipods can produce a tail-flip response, they tend not to use it when attacked by *Spinachia* (personal observation). Not only is their velocity much less than that achieved by mysids using the tail-flip response, but amphipods also may be unable to detect approaching fish in time to respond. This may explain why *Spinachia* succeed in hovering closer to amphipods than mysids (Kaiser *et al.*, 1992a).

Persistent predator strikes induced amphipods to adopt a motionless, curled-up, C-shape (thanatosis), increasing the height:length (H:L) ratio above 1:5, and causing *Spinachia* to lose interest in the attack. A motionless amphipod is well camouflaged against a coarse substratum (personal observation), furthermore visual predators, such as fish, find stationary prey less attractive than moving prey (Kislalioglu and Gibson, 1976b; Holmes and Gibson, 1986; Croy and Hughes, 1991c).

In a similar study the predatory behaviour of juvenile sticklebacks was compared with that of adults (Kaiser, 1992). When juveniles fed on nauplii of the brine shrimp, *Artemia* spp., they adopted the S-bend posture described for adults feeding on mysids (Kaiser *et al.*, 1992a). Fish hovered at a fixed distance, 1.0±0.1mm, from the prey before initiating an attack. The mean peak velocities attained by small and large juveniles were respectively 7.6 and 20.2 times greater than the mean swimming velocity of *Artemia*.

Unlike mysids or amphipods, *Artemia* used neither fast-start escape or thanatosis when attacked by fish. Escape was limited to an arbitrary change in direction, which occurred at a mean±SE rate of 54.3±2.9 turns/min (Kaiser, 1992). Linear swimming is the main escape mechanism used by *Artemia*, similar to gammarids. Juvenile fish were able to achieve much higher velocities than *Artemia*, and were expected to use a low-energy attack, similar to adults attacking gammarids (Figure 5). However, whereas mysids most frequently escaped at right angles to the line of attack (Kaiser *et al.*, 1992a), the direction of escape was unpredictable for *Artemia*. Consequently, juvenile *Spinachia* were unable to anticipate the direction of *Artemia* escape,

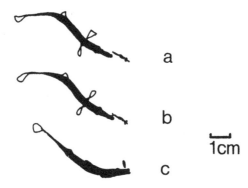

a

b

1cm

c

Figure 3 Silhouette images of *Spinachia* attacking a mysid, traced from a video recording (time interval 20 ms). Initially (a) the fish is stationary while fixating the mysid's head region. As the fish begins to move forward (b) the mysid does not respond, but as the fish reaches maximum velocity (c) the mysid tail-flips at right angles to the original line of attack. Redrawn from Kaiser *et al.*, 1992a.

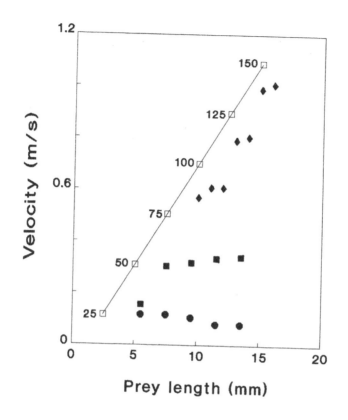

Figure 4 The mean velocities (m/s) achieved by *Spinachia* of varying lengths (mm) (next to points), predicted from: velocity = 7.8 fish length–0.0817 (Kaiser, 1992). The mean velocity during mysid (◆) and amphipod (■) tail-flip escapes, and amphipod swimming (●) (Kaiser *et al.*, 1992a).

Figure 5 The body form of *Spinachia* attacking different prey: a) high-energy attack on a mysid, note the pronounced S-bend producing maximum propulsion; b) low-energy attack on an amphipod with a much reduced S-bend; c) high-energy attack used by a juvenile attacking an *Artemia*. Taken from Kaiser *et al.*, 1992a; Kaiser, 1992.

continued to use the high-energy fast-start attack method (Kaiser, 1992). Fish thus seem to adjust their predatory tactics according to the prey-type encountered, using the most economical method whenever possible, and so maximising net rate of energy intake (Nyberg, 1971; Kaiser *et al.*, 1992a).

Behavioural Factors Affecting Diet Selection

Spinachia use a complex sequence of behaviours prior to the ingestion of prey (Figure 6; Croy and Hughes, 1990), but as experience with prey increases, fish retain fewer behavioural components in the pre-digestive repertoire; thus miss, spit, hold and chew occurred with decreasing frequency (Croy and Hughes, 1991a) (Figure 7). Also, hunger state, i.e. the motivation to feed, and learning, i.e. experience, affect foraging efficiency and hence the profitability of prey (Croy and Hughes, 1991a,b). Close to satiation, experienced sticklebacks use a complex array of behaviours when feeding on amphipods, attacking and then reorientating prey several times before finally swallowing them. Fish retain learned skills for up to eight days before relapsing to their original inexperienced state (Croy and Hughes, 1991a,b). This has important implications in heterogeneous environments where fish encounter many different prey types within a short space of time (Kaiser, 1991). Fish are able to improve their handling efficiency within a single foraging bout, rapidly increasing prey profitability. Moreover, if the same prey type is

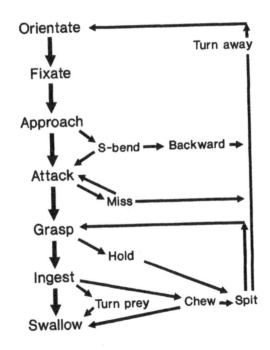

Figure 6 The complex sequence of behavioural components used by *Spinachia* prior to the ingestion of prey. The components used in the basic sequence are shown on the left (Croy and Hughes, 1990).

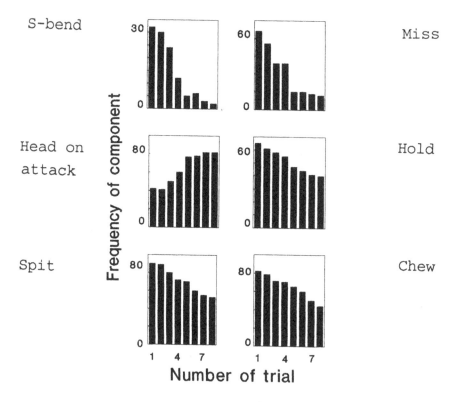

Figure 7 The change in frequency of each of the basic pre-ingestive behavioural components as experience with prey (amphipods) increased. Redrawn from Croy and Hughes, 1991a.

encountered on successive days, learnt foraging skills maintain a relatively high level of profitability (Croy and Hughes, 1991a,b).

Hunger affects the motivation to feed, and consequently as fish approach satiation they become more selective (Croy and Hughes, 1990, 1991b). As fish tend towards satiation, reaction distance and speed of attack decrease. Concomitantly prey are most profitable when fish are hungriest and least profitable as they near satiation. This is because of changes in handling behaviour and the likelihood of starvation. Thus as fish approach satiation, the band of profitable prey sizes narrows, leading to more selective feeding (Croy and Hughes, 1991a).

Croy and Hughes (1991b,d) suggested that learning could be a mechanism for switching. Clearly, fish improve foraging efficiency with experience, especially when prey are either difficult to ingest (amphipods) or have fast (mysids) or unpredictable (*Artemia*) escape responses. Frequent encounters with one prey type will increase its profitability, through learned handling skills, to the extent that other types then become relatively unprofitable and are ignored by the fish (Croy and Hughes, 1991d).

Prey Recognition

Optimal diet models assume that predators are able to rank prey in order of profitability (Charnov, 1976), leading to selective feeding. Presumably, then, predators respond to stimuli that indicate prey profitability.

In an early study Kislalioglu and Gibson (1976b) manipulated the dead bodies of mysids, altering shape, colour, size and movement. *Spinachia* responded to stimuli in the order:

$$\text{movement} \gg \text{size} > \text{colour} > \text{shape}$$

Manipulation of *Artemia* and amphipods yielded similar results (Croy and Hughes, 1991c). Fish responded more frequently to the patterned, contoured head region of mysids compared with the plainer abdomen (Kislalioglu and Gibson, 1976b) and used more subtle features, such as legs, to discriminate between different prey types (Kaiser *et al.*, 1992b). Holmes and Gibson (1986) found that turbot, *Scophthalmus maximus*, would not respond to model shrimps with an H:L ratio >1:5. In general, most of the prey eaten by predatory fish have H:L ratios less than this threshold (Holmes and Gibson, 1986). Movement is the strongest stimulus, initially drawing predators' attention to potential prey, thereafter other stimuli become more important in the decision to attack or ignore prey (Croy and Hughes, 1991c; Kaiser *et al.*, 1992b). Fish clearly respond to a variety of stimuli, enabling them to recognise prey and choose appropriate predatory behaviour (Kaiser *et al.*, 1992a).

Physiological Factors

On the basis of pre-digestive behaviours Kaiser *et al.* (1992a) predicted that mysids should be more profitable than amphipods, as was reflected in observed preference (Kaiser *et al.*, 1992b). However, although fish appear to behave according to ODT, this is based solely on the short term behavioural time-scale over periods of seconds or minutes. Although digestive kinetics in fish are well studied (e.g. Grove *et al.*, 1985; Dos Santos and Jobling, 1991) few applications of ODT have considered the time taken to extract energy from ingested prey. Prey differ in morphology and skeletal content, affecting stomach packing and the time required for enzymes to digest away a chitinous exoskeleton (Lindsay, 1984; Jackson *et al.*, 1987). Jaeger (1990) has shown this to be particularly important in salamanders, when confronted with a choice of ants and dipterans. Salamanders preferentially chose the less chitinous dipterans when these were available. Kaiser *et al.* (1992b) compared the absorption efficiencies (AE) and gut emptying times (GET) for amphipods and mysids when these were eaten by *Spinachia*. Fish had a high AE for both prey (97–98%) similar to estimates for minnows, *Phoxinus phoxinus* (Cui and Wootton, 1989). GET was much longer for mysids (30–35 h) than for amphipods (18–24 h), and varied according to the previous meal eaten. When the same prey were eaten to satiation on consecutive occasions, GET was much lower than when the alternative prey had been eaten previously. This indicates that there may be some effect of stomach pre-conditioning. When the effect of GET is included in the handling time equation, amphipods become the more profitable prey (Kaiser *et al.*, 1992b).

Although *Spinachia* maximise the net rate of energy intake estimated in terms of pre-digestive behaviours, diet selection appears to be suboptimal over the physiological time scale appropriate for digestive processing of the food. Possibly net rate of energy ingestion, as opposed to the net rate of energy absorption is more important in determining fitness. Minimising time exposed to risks such as predation (e.g. Croy and Hughes, 1991d) and stranding at high tide (e.g. Burrows and Hughes, 1991) may have a greater selective impact than optimizing energy intake over long-term, physiological time-scales, especially if the feeding opportunities are transitory.

CONCLUSIONS

Morphological constraints place upper and lower limits on available prey. Within these limits, behavioural mechanisms, which are affected by physiological state, fine tune pre-digestive predatory behaviour, leading to the selection of prey that are most profitable in the short term, although not necessarily in the long term. Time minimization therefore may be more important than energy maximization. Fish no doubt will continue to be fruitful subjects of future research on foraging behaviour, perhaps involving dynamic modelling to reveal the relative importance of, and interactions between, the different factors considered in this review.

ACKNOWLEDGMENTS

Frietson Galis and Andrew Gill made helpful comments on an earlier version of this manuscript. This work was carried out while M. J. K. was in receipt of an NERC/CASE studentship.

References

Burrows, M. T. and Hughes, R. N. (1991). Optimal foraging decisions by dogwhelks, *Nucella lapillus* L.: influences of mortality risk and rate-constrained digestion. *Funct. Ecol.*, 5, 461–475.

Case, J., Gwilliam G. F. and Hanson, F. (1960). Amino acid sensitivity of the dactyl chemoreceptors of *Carcinus maenas. Biol. Bull.*, Woods Hole, 121, 449–455.

Charnov, E. L. (1976). Optimal foraging: attack strategy of a mantid. *Am. Nat.*, 110, 141–151.

Crowder, L. B. (1985). Optimal foraging and feeding mode shifts in fishes. *Environ. Biol. Fishes*, 12, 57–62.

Croy, M. I. and Hughes, R. N. (1990). The combined effects of learning and hunger in the feeding behaviour of the fifteen-spined stickleback *Spinachia spinachia* L. Behavioural mechanisms of food selection (Ed. by R.N. Hughes). pp. 214–234. *NATO ASI Ser.*, Vol. G20. Berlin: Springer-Verlag.

Croy, M. I. and Hughes, R. N. (1991a). The role of learning and memory in the feeding behaviour of the fifteen-spined stickleback, *Spinachia spinachia* L. *Anim. Behav.*, 41, 149–160.

Croy, M. I. and Hughes, R. N. (1991b). The influence of hunger on feeding behaviour and on the aquisition of learned foraging skills by the fifteen-spined stickleback, *Spinachia spinachia* L. *Anim. Behav.*, 41, 161–170.

Croy, M. I. and Hughes, R. N. (1991c). Hierarchical response to prey stimuli and associated effects of hunger and foraging experience in the fifteen-spined stickleback, *Spinachia spinachia* L. *J. Fish Biol.*, 38, 599–607.

Croy, M. I. and Hughes, R. N. (1991d). Effects of food supply, hunger, danger and competition on choice of foraging location by the fifteen-spined stickleback, *Spinachia spinachia* L. *Anim. Behav.,* **42**, 131–140.

Cui, Y. and Wootton, R. J. (1988). Bioenergetics of growth of a cyprinid, *Phoxinus phoxinus*: the effect of ration, temperature and body size on food consumption, faecal production and nitrogenous excretion. *J. Fish Biol.,* **33**, 431–443.

Curio, E. (1976). *The Ethology of Predation.* Berlin: Springer-Verlag.

Dawkins, R. (1982). *The Extended Phenotype: The Gene as the Unit of Selection.* Oxford: W. H. Freeman.

Dervo, B. K., Hegge, O., Hessen, D. O. and Skurdal, J. (1991). Diel food selection of pelagic Arctic charr, *Salvelinus alpinus* L., and brown trout, *Salmo trutta* L., in Lake Atnsjø, SE Norway. *J. Fish Biol.,* **38**, 199–211.

Dos Santos, J. and Jobling, M. (1991). Factors affecting gastric evacuation in cod, *Gadus morhua* L., fed single-meals of natural prey. *J. Fish Biol.,* **38**, 697–714.

Elner, R. W. and Hughes, R. N. (1978). Energy maximisation in the diet of the shore crab, *Carcinus maenas* L. *J. Anim. Ecol.,* **47**, 103–116.

Emlen, J. M. (1966). The role of time and energy in food preference. *Am. Nat.,* **100**, 611–617.

Galis, F. (1990). Ecological and morphological aspects of changes in food uptake through the ontogeny of *Haplochromis piceatus.* In: *Behavioural mechanisms of food selection* (Ed. by R. N. Hughes). *NATO ASI Ser.* Vol. G20 pp. 281–302.

Gibson, R. N. and Ezzi, I. A. (1985). Effect of particle concentration on filter- and particulate-feeding in the herring, *Clupea harengus. Mar. Biol.,* **107**, 357–362.

Gibson, R. N. and Ezzi, I. A. (1990). Relative importance of prey size and concentration in determining the feeding behaviour of the herring, *Clupea harengus. Mar. Biol.,* **107**, 357–362.

Gibson, R. N. and Ezzi, I. A. (1992). The relative profitability of particulate- and filter-feeding in the herring, *Clupea harengus* L. *J. Fish Biol.* **40**, 577–590.

Grove, D. J. Moctezuma, M. A., Flett, H. R. J., Foott, J. S., Watson, T. and Flowerdew, M. W. (1985). Gastric emptying and the return of appetite in juvenile turbot, *Scophthalmus maximus* L., fed on artificial diets. *J. Fish Biol.,* **26**, 339–354.

Hart, P. J. B. and Hamrin, S. (1990). The role of behaviour and morphology in the selection of prey by pike. In: *Behavioural mechanisms of food selection* (Ed. by R.N. Hughes). *NATO ASI Ser.* Vol. G20 pp. 235–254.

Hart, P. J. B. and Ison, S. (1991). The influence of prey size and abundance, and individual phenotype on prey choice by the three-spined stickleback, *Gasterosteus aculeatus* L. *J. Fish Biol.,* **38**, 359–372.

Helfman, G. S. (1990). Mode selection and mode-switching in foraging animals. *Adv. Stud. Behav.,* **19**, 249–298.

Helfman, G. S. and Winkelman, D. L. (1991). Energy trade-offs and foraging mode choice in American eels. *Ecology,* **72**, 310–318.

Hirtle, R. W. M. (1978). Distance chemoreception and vision in the selection of prey by the American lobster (*Homarus americanus*). *J. Fish. Res. Bd. Can.,* **35**, 1006–1008.

Holmes, R. A. and Gibson, R. N. (1986). Visual cues determining prey selection by the turbot, *Scophthalmus maximus* L. *J. Fish. Biol.,* **29**, 49–58.

Huang, C. and Sih, A. (1991). Experimental studies on direct and indirect interactions in a three tropic-level stream system. *Oecologia,* **85**, 530–536.

Hughes, R. N. (1980). Optimal foraging theory in the marine context. *Oceanogr. Mar. Biol. Ann. Rev.,* **18**, 423–481.

Hughes, R. N. (1988). Optimal foraging in the intertidal environment: evidence and constraints. In: *Behavioural adaptation to the intertidal life* (Ed. by G. Chelazzi and M. Vannini) *NATO ASI Ser. A,* pp. 265–282.

Hughes, R. N. (Ed.) (1990). *Behavioural mechanisms of food selection. NATO ASI Ser.* Vol G20. Berlin: Springer-Verlag.

Hughes, R. N., Kaiser, M. J., Mackney, P. A. and Warburton, K. (1992). Optimizing foraging behaviour through learning. *J. Fish Biol.,* **41** (Suppl. B), 77–91.

Huntingford, F. A. and Turner, A. K. (1987). *Animal Conflict.* London: Chapman and Hall.

Ibrahim, A. A. and Huntingford, F. A. (1989). Laboratory and field studies on diet choice in three-spined sticklebacks, *Gasterosteus aculeatus* L., in relation to profitability and visual features of prey. *J. Fish Biol.,* **34,** 245–259.

Illius, A.W. and Gordon, I. J. (1990). Constraints on diet selection and foraging behaviour in mammalian herbivores. In: *Behavioural mechanisms of food selection* (Ed. by R. N. Hughes). *NATO ASI Ser.* Vol. G20. Berlin: Springer-Verlag. pp. 369–394.

Jaeger, R. G. (1990). Territorial salamanders evaluate size and chitinous content of arthropod prey. In: *Behavioural mechanisms of food selection* (Ed. by R. N. Hughes). *NATO ASI Ser.* Vol. G20. Berlin: Springer-Verlag. pp. 111–126.

Jackson, S., Duffy, D. C. and Jenkins, J. F. G. (1987). Gastric digestion in marine vertebrate predators: *In vitro* standards. *Funct. Ecol.,* **1,** 287–291.

Jones, J. W. and Hynes, H. B. N. (1950). The age and growth of *Gasterosteus aculeatus, Pygosteus pungitius,* and *Spinachia vulgaris,* as shown by their otoliths. *J. Anim. Ecol.,* **19,** 59–73.

Jørgensen, C. B. (1990). Water processing in filter-feeding bivalves. In: *Behavioural mechanisms of food selection* (Ed. by R. N. Hughes). *NATO ASI Ser.* G20. Berlin: Springer-Verlag. pp. 615–636.

Kaiser, M. J. (1991). *The behavioural and morphological basis of diet selection in fish and crabs.* Ph.D. thesis. University of Wales.

Kaiser, M. J. (1992). The ontogeny of predatory mechanisms in the fifteen-spined stickleback, *Spinachia spinachia* L. *J. Fish Biol.,* **40,** 485–487.

Kaiser, M. J. and Croy, M. I. (1991). Population structure of the fifteen-spined stickleback, *Spinachia spinachia* L. *J. Fish Biol.,* **39,** 129–131.

Kaiser, M. J., Gibson, R. N. and Hughes, R. N. (1992a). The effects of prey type on the predatory behaviour of the fifteen-spined stickleback, *Spinachia spinachia* L. *Anim. Behav.,* **43,** 147–156.

Kaiser, M. J., Westhead, A. P., Hughes, R. N. and Gibson, R. N. (1992b). Are digestive characteristics important contributors to the profitability of prey? A study of diet selection in the fifteen-spined stickleback, *Spinachia spinachia* L. *Oecologia.* **90,** 61–69.

Kieffer, J. D. and Colgan, P. W. (1992). The role of learning in fish behaviour. *Reviews in Fish Biology and Fisheries,* **2,** 125–143.

Kislalioglu, M. and Gibson, R. N. (1975). Field and laboratory observations on prey-size selection in *Spinachia spinachia* L. In: *Proc. 9th Europ. Mar. Biol. Symp.* pp. 29–41.

Kislalioglu, M. and Gibson, R. N. (1976a). Prey "Handling time" and its importance in food selection by the 15-spined stickleback, *Spinachia spinachia. J. Exp. Mar. Biol. Ecol.,* **25,** 151–158.

Kislalioglu, M. and Gibson, R. N. (1976b). Some factors governing prey selection by the 15-Spined Stickleback, *Spinachia spinachia* L. *J. Exp. Mar. Biol. Ecol.,* **25,** 159–169.

Kislalioglu, M. and Gibson, R. N. (1977). The feeding relationship of shallow water fishes in a Scottish sea loch. *J. Fish Biol.,* **11,** 257–266.

Krebs, J. R. and Davies, N. B. (1984). *An Introduction to Behavioural Ecology.* London: Blackwell Scientific.

Liem, K. F. (1980). Adaptive significance of intra- and interspecific differences in the feeding repertoires of cichlid fishes. *Am. Zool.,* **20,** 295–314.

Lindsay, G. J. H. (1984). Distribution and function of digestive tract chitinloytic enzymes in fish. *J. Fish Biol.,* **24,** 529–536.

Losey, G. S. and Sevenster, P. (1991). Can threespine sticklebacks learn when to display? 1. Punished displays. *Ethology,* **87,** 45–58.

Lucas, M. C., Priede, I. G., Armstrong, J. D., Gindy, A. N. Z. and De Vera, L. (1991). Direct measurements of metabolism, activity and feeding behaviour of pike, *Esox lucius* L., in the wild, by use of heart rate telemetry. *J. Fish Biol.,* **39,** 325–347.

MacArthur, R. H. and Pianka, E. R. (1966). On optimal use of a patchy environment. *Am. Nat.,* **100,** 603–609.

Maynard Smith, J. (1978). Optimisation theory in evolution. *Ann. Rev. Ecol. Syst.,* **9,** 31–56.

Moore, J. W. and Moore, I. A. (1976). The basis of food selection in flounder, *Platichthys flesus* L., in the Severn Estuary. *J. Fish Biol.,* **9,** 139–156.

Nyberg, D. W. (1971). Prey capture in the largemouth bass. *Am. Midl. Natur.,* **86,** 128–144.

Pierce, G. J. and Ollason, J. G. (1987). Eight reasons why optimal foraging theory is a complete waste of time. *Oikos,* **49,** 111–117.

Pyke, G.H. (1984). Optimal foraging theory: a critical review. *Ann. Rev. Ecol. Syst.,* **15,** 523–575.

Real, L. A. and Caraco, T. (1986). Risk and foraging in stochastic environments. *Ann. Rev. Ecol. Syst.,* **17,** 371–390.

Ringler, N. H. (1982). Variation in foraging tactics of fish. In: *Predators and Prey in Fishes* (Ed. by D. L. G. Noakes), pp. 159–171. The Hague: Dr W. Junk.

Rowland, W. J., Baube, C. L. and Horan, T. T. (1991). Signalling of sexual receptivity by pigmentation pattern in female sticklebacks. *Anim. Behav.,* **42,** 243–251.

Schoener, T. W. (1987). A brief history of optimal foraging theory. In: *Foraging Behaviour* (Ed. by A. C. Kamil, J. R. Krebs, and Pulliam), pp. 5–67. New York and London: Plenum Press.

Sevenster, P. (1951). The mating of the sea stickleback. *Discovery,* **12,** 52–56.

Stein, R. A., Goodman, C. G. and Marschall, A. (1984). Using time and energetic measures of cost in estimating prey value for fish predators. *Ecology,* **65,** 702–715.

Stephens, D. W. and Krebs, J. R. (1986). *Foraging Theory.* Princeton: Princeton Univ. Press.

Taghon, G. L., Self, R. F. L. and Jumars, P. A. (1978). Predicting particle selection by deposit feeders: a model and its implications. *Limnol. Oceanogr.,* **23,** 752–759.

Wankowski, J. W. J. (1979). Morphological limitations, prey size selectivity, and growth response of juvenile Atlantic salmon (*Salmo salar* L.). *J. Fish Biol.,* **14,** 89–100.

Werner, E. E. (1974). The fish size, prey size, handling time relation in several sunfishes and some implications. *J. Fish. Res. Bd. Can.,* **31,** 1531–1536.

Werner, E. E. and Hall, D. J. (1974). Optimal foraging and the size selection of prey by the bluegill sunfish, *Lepomis macrochirus. Ecology,* **55,** 1042–1052.

Wetterer, J. K. (1989). Mechanisms of prey choice by planktivorous fish: perceptual constraints and rules of thumb. *Anim. Behav.,* **37,** 955–967.

Wheeler, A. (1969). *Fishes of the British Isles and North Western Europe.* London: Macmillan.

Wootton, R. J. (1984). *A Functional Biology of Sticklebacks.* London: Croom Helm.

MORPHOLOGICAL CONSTRAINTS ON BEHAVIOUR THROUGH ONTOGENY: THE IMPORTANCE OF DEVELOPMENTAL CONSTRAINTS

FRIETSON GALIS

Department of Population Biology and Department of Organismal Zoology,
PO Box 9500, 2300RA Leiden, The Netherlands

Most fishes experience shifts in diet through ontogeny (Stoner and Livingston, 1984; Mittelbach, 1981; Holbrook *et al.*, 1985; Van Densen, 1985; Werner and Hall, 1988; Persson, 1990, Kaiser and Hughes this volume). The shifts in diet often coincide with shifts in habitat and with changes in the morphology of the feeding apparatus. Changes in dietary preference may induce habitat shifts, but habitat shifts (e.g. because of a changed predation risk) may also induce changes in food preference (Werner *et al.*, 1983; Mittelbach, 1984; Werner and Gilliam, 1984; Power, 1984). Morphological transformations of the feeding apparatus through ontogeny are likely to play an important role in dietary shifts (Liem and Osse, 1975; Barel, 1983; Wainwright, 1988; Motta, 1988). Changes in simple morphological parameters such as an increase in mouth gape or in inter-gill raker distance are often thought to bring about changes in food uptake (Lawrence, 1957; Hartman, 1958; Hansen and Wahl, 1981; Schmitt and Holbrook, 1984; Scheal *et al.*, 1991; see also Werner, 1974). However it is often questionable whether there is really a causal relation between the two phenomena or just a correlation. To understand the simultaneous changes of form and function (which mutually influence each other), careful analyses of the *relation* between form and function are necessary, especially when structures have several functions or when functions are carried out by several structures (Dullemeijer, 1974; Roth and Wake, 1989).

FUNCTION INFLUENCES FORM DURING ONTOGENY

While ontogeny is sometimes thought to be an interplay between genetic and epigenetic phenomena giving rise to the adult phenotype (Alberch, 1982), the interplay between morphology and environment through ontogeny is of crucial importance as shown in the following examples.

Young Pikeperch (*Stizostedion lucioperca*) are initially planktivorous and unable to eat fishes. When there is a sufficiently high availability of small prey fish juvenile Pikeperch of a minimum standard length (SL) of 2 cm switch to a piscivorous feeding mode and rapidly develop into specialized piscivores that can not return to a planktivorous feeding mode once larger than 10 cm SL (Van Densen, 1985). Pikeperch that do not manage to switch to piscivory in their first year grow more slowly and die at a much higher rate than conspecifics that become piscivores and possibly do not survive at all (Buijse and Houthuijzen, 1992). Although nutritional effects may play a role (see Wimberger, 1991), it seems unlikely that pikeperch will develop into specialized piscivores that are unable to feed on zooplankton if they are fed with fish meal. Apparently the food uptake (function) influences form parameters (including size) which in its turn influence the food uptake of the fish.

Spectacular examples of polymorphism in the pharyngeal jaw apparatus are found in the South American cichlids, *Cichlasoma minckleyi*, (Sage and Selander, 1975; Kornfield and Taylor, 1983; Liem and Kaufman, 1984), *C. citrinellum* (Meyer, 1990a and b) and *C. haitensis* (Meyer, 1991) and in several African cichlids including *Astatoreochromis alluaudi* (Greenwood, 1965; Hoogerhoud, 1986). These polymorphic fish have either strong, sturdy molariform pharyngeal jaws with hypertrophied branchial musculature or weak papilliform jaws with less well-developed muscles. The different morphs are correlated with differences in diet (Greenwood, 1965; Liem and Kaufman, 1984; Hoogerhoud, 1984, 1986; Meyer, 1990b). The molariform morphs are adapted to feeding on hard snails (Greenwood, 1965; Liem and Kaufman, 1984; Hoogerhoud, 1986; Meyer, 1989) and in *C. citrinellum* the papilliform morph was shown to be more efficient than the molariform one in feeding on soft snails (Meyer, 1989). It is not clear how much of the difference between the morphs can be attributed to genetical differences and how much to phenotypic plasticity. Nevertheless phenotypic plasticity certainly plays a role. For *C. citrinellum* (Meyer, 1990b) and for *Astatoreochromis alluaudi* (Hoogerhoud, 1986) it has been demonstrated that both morphs are papilliform first. Greenwood (1965) found that the differences between the two morphs of *A. alluaudi* are likely due to phenotypic plasticity and Hoogerhoud (1986) showed that most of the differences found in natural populations of *A. alluaudi* can be achieved by different feeding regimes within one generation (Figure 1). Meyer (1990b) also

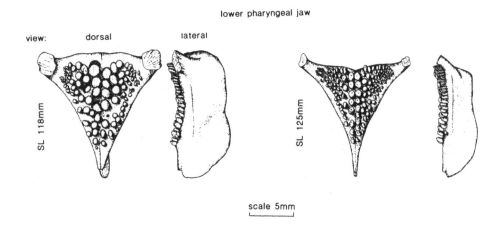

Figure 1 Variation in the lower pharyngeal jaws of wild-caught *Astatoreochromis alluaudi*. Hoogerhoud (1986) was able to induce these differences in young from the same brood that were fed on different diets. Young that were fed on hard-shelled food items developed the stout lower pharyngeal jaws with strongly molarized teeth of the left specimen, whereas the young that were fed on soft food items developed the more slender lower pharyngeal jaws with less molarized teeth of the right specimen. Illustration from Hoogerhoud (1984).

found indications of phenotypic plasticity in the pharyngeal jaws of *C. citrinellum* because two out of twelve wild-caught papilliform juveniles changed into molariform morphs when given hard food. Similarly, Wainwright *et al.* (1990 and Wainwright pers. comm.) found that phenotypic plasticity probably plays a role in the development of muscles and bones of the pharyngeal jaw apparatus of the moluscivorous centrarchid *Lepomis gibbosus*.

Thus far the relation between form and function of the different types of pharyngeal jaw apparatus seems straightforward; the molariform morphs with the hypertrophied pharyngeal jaw apparatus feed at least part of the time on hard snails that must be crushed and the papilliform morphs feed on soft prey. However many molluscivorous morphs and species with a strongly hypertrophied pharyngeal jaw apparatus prefer soft food items to hard snails when given the choice (molariform morphs: *C. minckleyi,* Liem and Kaufman, 1984; *C. citrinellum,* Meyer, 1991; *Astatoreochromis alluaudi,* Hoogerhoud, 1986; molariform species: *Haplochromis ishmaeli* Slootweg, unpublished data). When soft food is scarce, *C. minckleyi* switch to the less preferred hard snails (Liem and Kaufman, 1984). Thus the extremely specialized pharyngeal jaw apparatus of the molariform species and morphs does not necessarily increase the efficiency of the food uptake of the preferred prey, but enhances the use of a less preferred prey (Liem and Kaufman, 1984). In some molariform species and morphs (*A. alluaudi* and *H. ishmaeli*) the handling time for insects is shorter than that of the insectivorous *Haplochromis piceatus* (comparison of the data of Slootweg (unpublished data) and Galis and de Jong, 1988). The most likely explanation is that the handling time of the molluscivores is shorter because they ingest insects without much preparation as can be deduced from relatively intact insect remains in the fish faeces (Figure 2), whereas *H. piceatus* punctures the insects with its sharp pharyngeal teeth (Figure 3) in a time-consuming process (Galis and de Jong, 1988; Galis, 1990). These lesions allow the entrance of digestive fluids (see

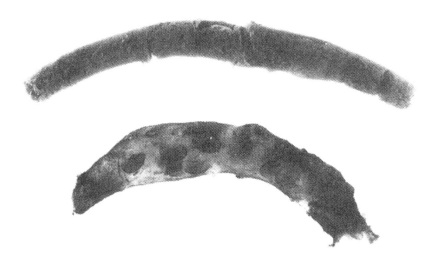

Figure 2 Fish faeces of *H. piceatus* (above) and *H. ishmaeli* (below) (same magn.). Both fish had been exclusively fed on insect larvae for three weeks. The faeces of the molluscivorous *H. ishmaeli* contain rather intact insect remains whereas those of the insectivorous *H. piceatus* have few intact remains.

Figure 3 Electroscan photograph of a *Chaoborus* larva from the stomach of an adult *H. piceatus* showing the crosswise damage that has been inflicted by the pharyngeal teeth.

Norton, 1988), resulting in faeces that are noticeably less recognizable (Figure 2). This may solve the problem of the shorter handling times observed in molluscivores, but does not clarify why molluscivores prefer insects that they can only inefficiently digest. Is it still more efficient for molluscivores to feed on insects than on molluscs?

FORM INFLUENCES FUNCTION DURING ONTOGENY

Here I want to discuss some of my work on how the form of the feeding apparatus influences function, i.e. prey selection, in *H. piceatus* throughout ontogeny. The diet of *H. piceatus* changes with size in that larger fishes eat on average larger prey that are more difficult to catch (Galis, 1990). Most prey items (cladocerans, copepods, insects and occasionally algae) are eaten by small and large fishes. The only absolute difference in diet between small and large fishes is that fish smaller than 4 cm SL do not eat insect pupae whereas larger fishes do (Figure 4). At a large range of food densities the most important factor limiting the efficiency of food uptake is the handling time within the pharyngeal jaws (Galis and de Jong, 1988). As small specimens don't eat pupae in the field (Figure 4) and in the laboratory (Galis, 1990), whereas they do eat the similarly sized, but softer, insect larvae, I decided to analyze the piercing of the prey within the pharyngeal jaws with a static biomechanical model of biting.

The movements of the lower and upper jaw are independent. When biting, the lower pharyngeal jaw presses against the prey and thereby pushes the upper jaw

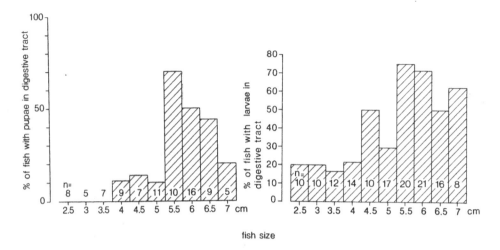

Figure 4 Frequency distributions of stomach contents. The left graph shows that small specimens of *H. piceatus* (below 4 cm SL) do not feed on *Chaoborus* pupae whereas they do feed on the softer *Chaoborus* larvae (right graph), including the fourth instar larvae that have the same size as pupae.

against the neurocranium (Figures 5 and 6). The upper jaw must exert a force on the prey that is equal in size, in line and opposite in direction. Part of this force is generated by the neurocranial reaction force. The position of the upper jaw is held constant (this usually requires muscle force). The rotation centre of the upper jaw is indicated in Figure 6D. The rotation centre of the lower jaw during biting is the point of contact with the prey (the lower pharyngeal jaw does not articulate with the shoulder girdle Galis, 1991, 1992). In a static analysis the sum of the forces in all directions and that of all torques equal zero. As there are more possible forces (muscle forces and the neurocranial reaction force) than equations, the system cannot be solved. As a result, in the analysis the combination of muscle forces was searched for that minimizes the total muscular force that generates a specified biting force. This optimization problem was solved with the SIMPLEX method for linear programming (Dantzig, 1963). The minimum total biting force was calculated for 19 different biting directions (Figure 6) at several prey positions and prey sizes. For each prey position and prey thickness the biting direction was chosen which requires the lowest total muscle force (thick arrows in Figure 7).

The force diagrams in Figure 7 show that the pharyngeal jaw apparatus can generate biting forces in a wide range of directions at specific biting points. The muscle combinations that must be active to generate the biting force differ depending on the size of the fish, size of the prey, position of the biting point and direction of the biting force. These predictions that are currently being tested using electromyography experiments strongly suggest that the muscle activity patterns must be flexible *during the biting phase*, both within an individual and throughout ontogeny.

Summation of the muscle forces on the lower and upper pharyngeal jaw leads to best biting directions that point posteriorly at rostral prey positions and point anteriorly at the remaining prey positions (Figure 8). The maximum possible biting

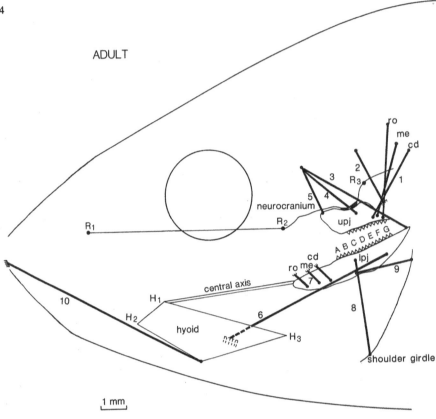

Figure 5 Projection of muscles and bony elements in the medial plane based on reconstructed points of an adult specimen (5 cm SL). R1, R2 and R3 are reference points. H1, H2 and H3 are points on the hyoid. A to G denote 7 chosen prey positions. ro, rostral part; cd, caudal part; me, middle part, apply to muscles 1 and 7. Numbers mark the lines of action of the following muscles: 1, m. retractor dorsalis; 2, m. levator posterior; 3, m. levator externus; 4, m. levator internus lateralis; 5, m. levator internus medialis; 6, m. pharyngohyoideus; 7, m. transversus ventralis; 8, m. pharyngocleithralis externus; 9, m. pharyngocleithralis internus; 10, m. geniohoideus. lpj, lower pharyngeal jaw; upj, upper pharyngeal jaw. From Galis (1992).

force (based on the cross-sectional area of the muscles) can be achieved at the central prey positions that require the least amount of muscle force for piercing the insects.

The directions of the teeth (Figure 9) coincide well with the best biting directions at the different prey positions (Figure 8). The directions of the teeth may also function to keep the prey within the dental area of the jaws.

The predictions of the maximum biting force were tested with a force gauge (Galis, 1992) to which a lower pharyngeal jaw was attached. The force gauge registered the peak force exerted as a *Chaoborus* larva or pupa was pierced by the teeth of a lower pharyngeal jaw. The results confirmed the hypothesis that adult *H. piceatus* are able to pierce *Chaoborus* larvae and pupae with their pharyngeal jaw

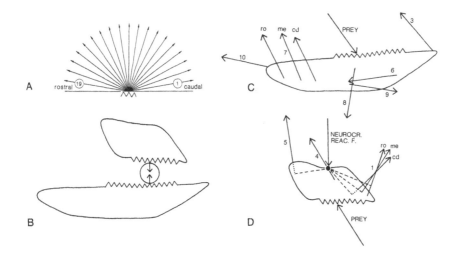

Figure 6 Simulation scheme (A) indicates the nineteen choices of biting directions. (B) is an example indicating forces that are exerted on the prey. The shape of the prey is arbitrarily chosen. (C) and (D) show a free-body diagram of the lower pharyngeal jaw and the upper pharyngeal jaw and the forces that are exerted on them. The length of the arrows is arbitrarily chosen. Numbers refer to the lines of action of the muscles, see Figure 5. The solid dot in (D) indicates the rotation centre. The broken lines indicate the lever arms of the lines of action.

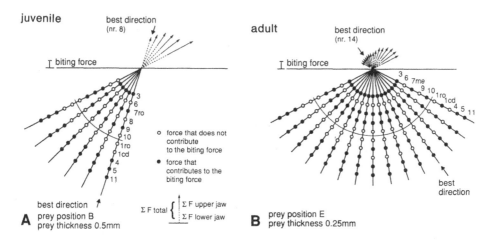

Figure 7 Force diagrams indicating (1) the contributing forces and (2) the feasible biting directions indicated by the arrows (of which the lines of action intersect the tooth surface of the upper jaw). The directions of the arrows refer to the directions indicated in Figure 6A. The length of the arrows indicates the muscular force that is exerted on the lower jaw (broken part of the arrow) and upper jaw (solid part) and the total muscular force (entire arrow). The best direction is the one of which the total muscular force is minimal. The size of the biting force (vertical bar) can be freely chosen because of the linear relationship between the forces. The solid points indicate active muscles, the open points indicate inactive muscles. Number 1 to 10 refer to muscles, see Figure 4; Nr. 11 is the neurocranial reaction force. Diagram A is an example for one prey position and thickness of a juvenile specimen of 2.5 cm SL and diagram B is an example for one prey position and thickness of an adult specimen of 2.5 cm SL.

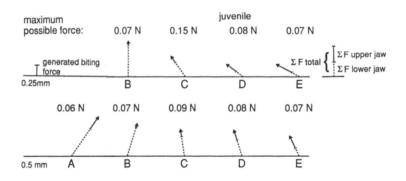

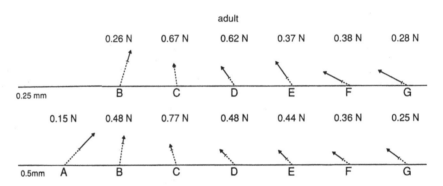

Figure 8   The best biting direction as indicated by the arrows (the direction for which the sum of the forces is minimal, see Figure 7) for prey thicknesses 0.25 mm and 0.5 mm at all prey positions (below for an adult specimen of 5 cm SL prey positions A to G and above for a juvenile specimen of 2.5 cm prey positions A to E). The directions of the arrows refer to the directions indicated in Figure 6A. The length of the arrows indicates the muscular force that is exerted (see Figure 7). The maximum possible biting forces, based on cross-sectional area of the muscles, are indicated above the arrows.

apparatus because the experimentally determined piercing forces (Table 1) fall within the range of possible biting forces that is predicted by the model (Figure 8). In addition the hypothesis that specimens smaller than 2.9 cm SL are able to pierce larvae but have problems or are unable to pierce pupae was confirmed for a specimen of 2.5 cm SL (compare Table 1 and Figure 8).

The model thus explains the absence of pupae in the diet of small *H. piceatus* (Figure 4) as due to morphological constraints. These constraints are mainly due to differences in size of the cross-sectional areas of the muscles. Predictions of the model for the adult based on measurements of the juvenile are similar as long as the values of the cross-sectional areas are taken from the adult. This is attributable to the almost isometric increase of most parameters (Galis, 1990). Therefore the morphological constraint is one imposed by the size of the organism. It is not a

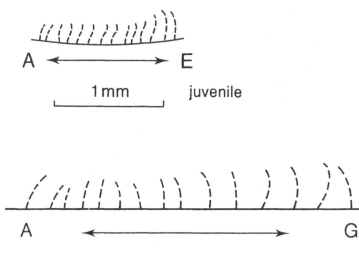

Figure 9   Longitudinal section through the medial plane showing the size and direction of teeth (from base to tip) on the lower pharyngeal jaw, below for an adult specimen of 5 cm SL and above for a juvenile specimen of 2.5 cm SL. A, E and G denote prey positions (see Figure 5B). These results correspond well with the results shown in Figure 7.

developmental constraint because it is not a constraint imposed by developmental changes. Yamaoka (1978) and Wainwright (1987, 1988) also found evidence that the size of the musculature is indicative of pharyngeal jaw crushing strength in the Labridae. Wainwright (1988) showed that ontogenetic stages of labrids switched from soft-bodied to hard-shelled prey at 3–5 N crushing strength, even though this crushing strength occurred at a different body size in each species. It seems that, as in *H. piceatus*, the absence in the diet of hard prey in certain ontogenetic stages is due to limitations of the crushing capacity of the pharyngeal jaw apparatus.

Table 1   Forces necessary for piercing *Chaoborus* larvae and pupae, measured with a lower jaw glued on a force gauge. The lower values of the piercing forces for the juvenile (2.5 cm SL) in comparison to the adult (5 cm SL) must be due to the finer points on the teeth of the juvenile. The forces necessary for piercing *Chaoborus* pupae with the juvenile pharyngeal jaw are higher than the maximum predicted biting forces (Figure 8, prey thickness 0.25 mm)

	Range of piercing forces in N juvenile adult	Average piercing force in N juvenile adult	n juvenile adult
Chaoborus	0.04–0.18	0.09	20
larvae	0.07–0.23	0.13	20
Chaoborus	0.11–0.20	0.15	20
pupae	0.15–0.31	0.23	20

DIVERSITY IN THE FOOD HABITS OF CICHLIDS

Liem has plausibly argued that the acquisition of an extra set of jaws in Cichlids and other Labroids has enabled a diversification of food preparation techniques and therefore of feeding habits (Liem, 1973; Liem and Osse, 1975; Liem, 1980; Liem and Greenwood, 1981). It is an example of a duplication of structures which allows a diversification of function (Lauder, 1981; Lauder and Liem, 1989; Roth and Wake, 1989; Bonner, 1988). Fishes of the cichlid family display a particularly wide array of feeding habits (e.g. Fryer and Iles, 1972; Greenwood, 1984; Witte, 1984; Lewis *et al.*, 1986). According to Liem and coauthors (Liem, 1973, 1979, 1980; Liem and Osse, 1975; Liem and Sanderson, 1986) the flexibility of the highly integrated pharyngeal jaw apparatus is the main factor that has enabled this diversity of feeding habits. They attribute much of the flexibility to the suspension of the lower pharyngeal jaw in a muscular sling and to the variability of muscle activity patterns (modulatory multiplicity) that Liem found in piscivorous cichlids (1978) and invertebrate pickers (1979). Studies on *H. piceatus* (Galis, 1991, 1992 and unpublished data) emphasize the importance of flexible muscle activity patterns and further suggest that the complex structure of the m. retractor dorsalis may play a role.

DEVELOPMENTAL CONSTRAINTS

At all stages of development the organism must function effectively (Bonner, 1988; Wake and Roth, 1989; Galis, 1990; Liem, 1991). How form changes during ontogeny (or during evolution) while preserving function is a fascinating problem. Many different aspects of an animal's relation to its physical environment are affected by change in any dimension (Horn *et al.*, 1982). This complicates the simultaneous changes of form and function and may constrain the performance level throughout any one stage (a developmental constraint).

During rapid transitions, when many form-function relations change in a rapid and more or less synchronous way, it is more likely that one or more form-function relations will be temporarily suboptimal than during gradual transitions. Therefore I want to contrast form-function changes during sudden transformations with more or less gradual ones.

Gradual Transitions

Most morphological parameters of the pharyngeal jaw apparatus of *H. piceatus* change isometrically (Galis, 1990). The increase in thickness of the muscles (and thus of their force generating capability) combined with the flexibility of the muscle activity patterns presumably causes a gradual increase in the capability of the pharyngeal jaw apparatus for biting (Galis, 1991, 1992). A gradual increase in the capability of the pharyngeal jaw apparatus can also be deduced from the increase in food uptake efficiency through ontogeny of *H. piceatus* (Figure 10).

The results of Heidweiller *et al.* (1992a and b) on drinking in chickens during development link up nicely with these results. They found that both the acquisition of novel motor activity patterns and flexibility of the motor activity patterns are necessary to maintain the ability to drink throughout ontogeny.

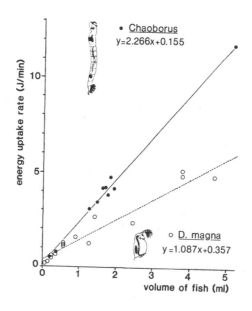

Figure 10 Relationship between fish volume and rate of energy intake from *Chaoborus* larvae (r=0.99, *P*<0.05) and *D. magna* (r=0.97, *P*<0.05). The energy uptake rate in these experiments is mainly determined by the processing time in the pharyngeal jaws. These data show that the processing efficiency increases in a gradual way.

Sudden Transitions

Metamorphosing animals are usually more vulnerable to predation than they are at other stages of their life (e.g. Arnold and Wasserzug, 1978; Stearns, 1982). For example garter snakes congregate in great numbers to feed on metamorphosing toads (Arnold and Wasserzug, 1978). Metamorphosing anurans are not well adapted to either the juvenile or the adult habitat (de Jongh, 1968; Wasserzug and Sperry, 1977). In particular the locomotor ability of metamorphosing anurans is limited, their swimming ability being inferior to that of pre-metamorphic stages and their jumping ability inferior to that of post-metamorphic stages (Wasserzug and Sperry, 1977). Temporarily the relation between form and function is less good and performance is hampered because many form-function relationships change at the same time. The vulnerability may be mitigated by cryptic coloration and behaviour, synchronisation and aggregation of metamorphosing animals and by rapid passage through the metamorphosis stage, but despite these compensatory traits vulnerability remains high (Wasserzug and Sperry, 1977; Arnold and Wasserzug, 1978).

In turbot (*Scophtalmus maximus*) there is a critical phase in tank-reared specimens right before the metamorphosis from symmetrical pelagic larva to asymmetrical benthic adult when large mortalities may occur for no apparent reason (Neave, 1984). Neave (1984) suggests that the low visual acuity at that stage may cause the vulnerability. It would be interesting to find out whether this high mortality also occurs in nature.

Not all sudden transitions are as drastic as metamorphoses. In cichlids (Otten, 1982; Liem, 1991) and pomacentrids (Liem, 1991) there is a shift in the mouth opening mechanism during early ontogeny. Early in the development the m. geniohyoideus plays an important role in the opening of the mouth (hyoid stage of jaw opening). Later after the ossification of the operculum and a shift in the line of action of the m. geniohyoideus, the levator operculi coupling becomes operational for mouth opening (opercular stage of jaw opening). During this fast transition there is shortly a higher risk for asphyxiation, i.e. a short vulnerable period (Liem, 1991).

Young Atlantic salmon (*Salmo salar*) undergo tremendous mortality right after the transition from endogenous feeding to exogenous feeding and this mortality is associated with the difficulties of first feeding (Coughlin, 1991). Lack of experience is important, but the morphology of the mouth also plays a role in the low catch success of the alavins (Coughlin, 1991). Initially there is a notch present in the mouth opening. The curvature of the mouth and the notch allow water to flow in from the sides during suction, reducing the directionality of suction and decreasing the volume of water sucked from directly ahead of the fish. Later suction is improved when the oral opening becomes flat and round. The reason for the presence of the notch and the different curvature is not clear.

A high mortality at first feeding is common in fishes (Blaxter, 1988). It is possible that in herring (*Clupea harengus*) and plaice (*Pleuronectus platessa*) the temporarily decreased resistance against hypoxia during the transition from cutaneous to branchial gas exchange may be responsible for part of the high mortality (De Silva and Tytler, 1973).

CONCLUSIONS

Apparently developmental constraints play a role in sudden transitions. Thus far no such developmental constraints have been found during more gradual transitions. The scanty evidence suggests that flexibility of muscle activity patterns provides organisms with a mechanism that buffers the effects of form changes and enables them to avoid developmental constraints that temporarily hamper their performance. Further research is needed and it seems that sudden transitions deserve our special attention. In addition studies are necessary in which muscle activity patterns are followed through ontogeny. In both cases it is crucial that the change in morphology is carefully studied *and* that performance tests (see Arnold, 1983; Wainwright, 1991) are carried out to measure the change in function.

The concept of developmental constraints is currently much in vogue. Not only developmental constraints on the performance level during any one stage but especially developmental constraints that limit evolutionary pathways. Attention has been paid to the constraining effect of genome organisation and the dynamics of epigenetic interactions (e.g. Oster and Alberch, 1982; Alberch, 1982; Alberch and Gale, 1985; Atchley and Hall, 1991). Relatively little attention has been paid to the constraining effect of the interactions between morphology and environment during the course of development (but see the work of de Jongh (1968) on *Rana temporaria*, of Wainwright (1988) on labridae, of Reilly and Lauder (1988) and Lauder and Reilly (1990) on salamandridae, of Galis (1990, 1991 and 1992) on *H. piceatus*, of Osse (1990) on fish larvae, of Liem (1991) on pomacentrid and embiotocid fishes, of Zweers (1991) on birds and of Heidweiller *et al.* a and b (1992) on *Gallus gallus*).

However to be able to predict alternative developmental pathways that are distinct from impossible alternatives (Alberch, 1982; Kauffman, 1983; Maderson *et al.*, 1982; Lauder and Liem, 1989; Lauder *et al.*, 1989; Dullemeijer, 1991) careful analyses of the relation between form and function through ontogeny are necessary. Thus studies of phylogenetic transformations should go hand in hand with studies of ontogenetic transformations.

The role of functional morphologists in enlarging our understanding of changes in feeding habits and of niche partitioning has been suggested above. In addition knowledge about the difference between potential and realized niches is necessary. For example *H. piceatus* is able to eat pupae at a size of 2.7 cm but only does so at a size of 4 cm under natural conditions (Galis, 1990). Competition by larger fish is probably responsible for this difference. Ecological field work and laboratory experiments should provide this kind of information. Therefore the joint effort of ecologists and functional morphologists is necessary to reach a better understanding of the varied feeding behaviour of organisms.

ACKNOWLEDGMENTS

I thank Jacques van Alphen, Gerrit Anker, Paul Hart, Mike Kaiser, Jaap de Visser, Frans Witte, and especially Ernie Wu, John Long and Peter Wainwright for comments on the manuscript. Peter Snelderwaard and Adrie 't Hooft made the photographs and Martin Brittijn made the drawings.

References

Atchley, W. R. and Hall, B. K. (1991). A model for development and evolution of complex morphological structures. *Biol. Rev.*, **66**, 101–157.

Alberch, P. (1982). Developmental constraints in evolutionary processes. In: *Evolution and Development*. Dahlem Conference Report no. 20. (Ed. by J. T. Bonner). Berlin, N.Y.: Springer-Verlag.

Alberch, P. and Gale, E. A. (1985). A developmental analysis of an evolutionary trend: digital reduction in amphibians. *Evolution*, **39**, 8–23.

Arnold, S. J. (1983). Morphology, performance and fitness. *Amer. Zool.*, **23**, 347–361.

Arnold, S. J. and Wasserzug, R. J. (1978). Differential predation on metamorphic anurans by garter snakes (*Thamnophis*): social behavior as a possible defense. *Ecology*, **59**, 1014–1022.

Barel, C. D. N. (1983). Towards a constructional morphology of cichlid fishes (Teleostei, Perciformes). *Neth. J. Zool.*, **33**, 357–424.

Blaxter, J. H. S. (1988). Pattern and variety in development. In: *Fish Physiology XI*, part a. (Ed. by W. S. Hoar and D. F. Randall). San Diego, Toronto: Academic Press, Inc. Harcourt Brace Jovanovich, Publishers.

Bonner, J. T. (1988). The evolution of complexity by means of natural selection. Princeton: Princeton University Press.

Buijse, A. D. and Houthuijzen, R. P. (1992). Piscivory, growth and size-selective mortality of age 0 pikeperch, *Stizostedion lucioperca* (L.). *Can. J. Fish. Aquat. Sci.*, **49**, 894–902.

Coughlin, D. J. (1991). Ontogeny of feeding behaviour of first-feeding Atlantic salmon (*Salmo salar*). *Can. J. Fish. Aquat. Sci.*, **48**, 1896–1904.

Dantzig, G. B. (1963). *Linear Programming and Extensions*. Princeton: Princeton University Press.

Densen, W. L. T. van (1985). Piscivory and the development of bimodality in the size distribution of 0+ pikeperch (*Stizostedion lucioperca* L.). *Z. angew. Ichthyol.*, **3**, 119–131.

De Silva, C. D. and Tytler, P. (1973). The influence of reduced environmental oxygen on the metabolism and survival of herring and plaice larvae. *Neth. J. Sea Res.*, **7**, 345–362.

Dullemeijer, P. (1974). *Concepts and Approaches in Animal Morphology*. Assen, The Netherlands: Van Gorcum and Comp. B.V.

Dullemeijer, P. (1991). Evolution of Biological Constructions: Concessions, Limitations, and Pathways. In: *Constructional Morphology and Evolution* (Ed. by N. Schmidt-Kittler and K. Vogel). Berlin: Springer-Verlag.

Fryer, G. and Iles, T. D. (1972). *The Cichlid Fishes of the Great Lakes of Africa*. Neptune City: T. F. H. Publications.

Galis, F. (1990). Ecological and morphological aspects of changes in food uptake through the ontogeny of *Haplochromis piceatus*. In: *Behavioural mechanisms of food selection* (Ed. by R. N. Hughes), NATO ASI Series, Vol.G 20. Berlin, Heidelberg: Springer-Verlag.

Galis, F. and de Jong, P. W. (1988). Optimal foraging and ontogeny; food selection by *Haplochromis piceatus*. *Oecologia (Berlin)*, **75**, 175–184.

Galis, F. (1992). A model for biting in the pharyngeal jaws of a cichlid fish: *Haplochromis piceatus*. *J. Theor. Biol.*, **155**, 343–368.

Galis, F. (1991). Interactions between the pharyngeal jaw apparatus, feeding behaviour and ontogeny in the cichlid fish, *Haplochromis piceatus*. A study of constraints in evolutionary ecology. Thesis, University of Leiden, The Netherlands.

Greenwood, P. H. (1965). Environmental effects on the pharyngeal mill of a cichlid fish, *Astatoreochromis alluaudi*, and their taxonomic implications. *Proc. Linn. Soc. Lond.*, **176**, 1–10.

Greenwood, P. H. (1974). The cichlid fishes of Lake Victoria, East Africa: the biology and evolution of a species flock. *Bull. Brit. Mus. Nat. Hist.*, Suppl. **6**, 1–134.

Hansen, M. J. and Wahl, D. H. (1981). Selection of small *Daphnia pulex* by Yellow Perch fry in Oneida Lake, New York. *Trans. Am. Fish. Soc.*, **110**, 64–71.

Hartman, G. F. (1958). Mouth size and food size in young Rainbow Trout, *Salmo gairdneri*. *Copeia*, 233–234.

Heidweiller, J. and Zweers, G. A. (1992). Development of drinking mechanisms in the chicken (*Gallus domesticus*; Aves, Galliformes). *Zoomorphology*, **111**, 217–228.

Heidweiller, J., Lendering, B. and Zweers, G. A. (1992). Motor patterns of cervical muscles during development of drinking in chicken. *Neth. J. Zool.*, **42**, 1–22.

Holbrook, S. J., Schmitt, J. R. and Coyer, J. A. (1985). Age-related dietary patterns of sympatric adult surfperch. *Copeia*, 986–994.

Hoogerhoud, R. J. C. (1984). A taxonomic reconsideration of the haplochromine genera *Gaurochromis* Greenwood, 1980 and *Labrochromis* Regan, 1920 (Pisces, Cichlidae). *Neth. J. Zool.*, **34**, 539–565.

Hoogerhoud, R. J. C. (1986). Ecological morphology of some cichlid fishes. Thesis, University of Leiden, The Netherlands.

Horn, H. S., Bonner, J. T., Dohle, W., Katz, M. J., Koehl, M. A. R. *et al.* (1982). Adaptive Aspects of development. Group report. In: *Evolution and Development*. Dahlem Conference Report no. 20 (Ed. by J. T. Bonner). Berlin, N.Y.: Springer-Verlag.

Jongh, H. J. de (1968). Functional morphology of the jaw apparatus of larval and metamorphosing *Rana temporaria* L. *Neth. J. Zool.*, **18**, 1–103.

Kauffman, S. A. and Hughes, R. N. (1983). Developmental constraints: internal factors in evolution. In: *Development and Evolution* (Ed. by B. Goodwin, N. Holder and C. C. Wylie). Cambridge, U.K.: Cambridge Univ. Press.

Kornfield, I. and Taylor, J. N. (1983). A new species of polymorphic fish, *Cichlasoma minckleyi*, from Cuatro Ciénegas, Mexico (Teleostei: Cichlidae). *Proc. Biol. Soc. Wash.*, **96**, 253–269.

Koutsikopoulos, C., Karakiri, M., Desaunay, Y. and Dorel, D. (1989). Response of juvenile sole (*Solea solea* L.) to environmental changes investigated by otolith microstructure analysis. *Rapp. P.-v. Réun. Cons. Int. Explor. Mer*, **191**, 281–286.

Lauder, G. V. (1981). Form and function: structural analysis in evolutionary morphology. *Paleobiology*, **7**, 430–442.

Lauder, G. V. and Liem, K. F. (1989). The role of historical factors in the evolution of complex organismal functions. In: *Complex Organismal Functions: Integration and Evolution in Vertebrates* (Ed. by D. B. Wake and G. Roth). New York: John Wiley and Sons Ltd.

Lauder, G. V., Crompton, A. W., Gans, C., Hanken, J., Liem, K. F. *et al.* (1989). Group report. How are feeding systems integrated and how have evolutionary innovations been introduced? In: *Complex Organismal Functions: Integration and Evolution in Vertebrates* (Ed. by D. B. Wake and G. Roth). New York: John Wiley and Sons Ltd.

Lauder, G. V. and Reilly, S. M. (1990). Metamorphosis of the feeding mechanism in tiger salamanders (*Ambystoma tigrinum*): the ontogeny of cranial muscle mass. *J. Zool. Lond.*, **222**, 59–74.

Lawrence, J. M. (1957). Estimated sizes of various forage fishes Largemouth bass can swallow. *Próc. S.E. Assoc. Game and Fish. Comm.*, **11**, 220–226.

Lewis, D. S. C., Reinthal, P. and Trendall, J. (1986). *A guide to the fishes of Lake Malawi National Park*. Gland, Switzerland: World Wildlife Fund.

Liem, K. F. (1973). Evolutionary strategies and morphological innovations: Cichlid pharyngeal jaws. *Syst. Zool.*, **22**, 425–441.

Liem, K. F. (1978). Modulatory multiplicity in the functional repertoire of the feeding mechanism in cichlid fishes. *J.Morph.*, **158**, 323–360.

Liem, K. F., (1979). Modulatory multiplicity in the feeding mechanism in cichlid fishes, as exemplified by the invertebrate pickers of Lake Tanganyika. *J. Zool., Lond.*, **189**, 93–125.

Liem, K. F. (1980). Adaptive significance of intra- and interspecific differences in the feeding repertoires of cichlid fishes. *Amer.Zool.*, **20**, 295–314.

Liem, K. F. (1991). A functional approach to the development of the head of teleosts: implications on constructional morphology and constraints. In: *Constructional Morphology and Evolution* (Ed. by N. Schmidt-Kittler and K. Vogel). Berlin: Springer-Verlag.

Liem, K. F. and Greenwood, P. H. (1981). A functional approach to the phylogeny of the pharyngognath teleosts. *Amer. Zool.*, **21**, 83–101.

Liem, K. F. and Kaufman, L. S. (1984). Intraspecific macroevolution: functional biology of the polymorphic cichlid species *Cichlasoma minckleyi*. In: *Evolution of Fish Species Flocks*. (Ed. by A. A. Echelle and I. Kornfield). Orono: University of Maine Press.

Liem, K. F. and Osse, J. W. M. (1975). Biological versatility, evolution and food resource exploitation in African cichlid fishes. *Amer. Zool.*, **15**, 427–454.

Liem, K. F. and Sanderson, S. L. (1986). The pharyngeal jaw apparatus of labrid fishes: A functional morphological perspective. *J. Morph.*, **187**, 143–158.

Maderson, P. F. A., Alberch, P., Goodwin, B. C., Gould, S. J., Hoffman, A. *et al.* (1982). The role of development in macroevolutionary change. In: *Evolution and Development*. Dahlem Conference Report no. 20. (Ed. by J. T. Bonner). Berlin, N.Y.: Springer-Verlag.

Meyer, A. (1989). Cost of morphological specialization: feeding performance of the two morphs in the trophically polymorphic cichlid fish, *Cichlasoma citrinellum*. *Oecologia*, **80**, 431–436.

Meyer, A. (1990a). Morphometrics and allometry in the trophically polymorphic cichlid fish, *Cichlasoma citrinellum*: Alternative adaptations and ontogenetic changes in shape. *J. Zool., Lond.*, **221**, 237–260.

Meyer, A. (1990b). Ecological and evolutionary consequences of the trophic polymorphism in *Cichlasoma citrinellum* (Pisces: Cichlidae). *Biol. J. Linn. Soc.*, **39**, 279–299.

Meyer, A. (1991). Trophic polymorphisms in cichlid fish: Do they represent intermediate steps during sympatric speciation and explain their rapid adaptive radiation? In: *New Trends in Ichthyology* (Ed. by J. H. Schroeder). Berlin: Paul Parey.

Mittelbach, G. G. (1981). Foraging efficiency and body size: a study of optimal diet and habitat use by bluegills. *Ecology*, **62**, 1370–1386.

Mittelbach, G. G. (1984). Predation and resource partitioning in two sunfishes (Caentrarchidae). *Ecology*, **65**, 499–513.

Motta, P. J. (1988). Functional morphology of the feeding apparatus of ten species of Pacific butterflyfishes (Perciformes, Chaetodontidae): an ecomorphological approach. *Env. Biol. Fish.*, **22**, 39–67.

Neave, D. A. (1984). The development of visual acuity in larval Plaice (*Pleuronectes platessa* L.) and Turbot (*Scophthalmus maximus* L.). *J. Exp. Mar. Biol. Ecol.*, **78**, 167–175.

Norton, S. (1988). Role of the gastropod shell and operculum in inhibiting predation by fishes. *Science*, **241**, 92–94.

Osse, J. W. M. (1990). Form changes in fish larvae in relation to changing demands of function. *Neth. J. Zool.*, **40**, 362–385.

Oster, G. and Alberch, P. (1982). Evolution and bifurcation of developmental programs. *Evolution*, **36**, 444–459.

Otten, E. (1982). The development of a mouth-opening mechanism in a generalized *Haplochromis* species: *H. elegans* Trewavas, 1933 (Pisces, Cichlidae). *Neth. J. Zool.*, **32**, 31–48.

Persson, L. (1990). In: *Behavioural mechanisms of food selection.* (Ed. by R. N. Hughes) NATO ASI Series, Vol.G 20. Berlin, Heidelberg: Springer-Verlag.

Power, V. E. (1984). Depth distributions of armored catfish: predator induced resource avoidance? *Ecology*, **65**, 523–528.

Reilly, S. M. and Lauder, G. V. (1988). Ontogeny of aquatic feeding performance in the Eastern Newt, *Notophthalmus viridescens* (Salamandridae). *Copeia*, 1988, 87–91.

Roth, G. and Wake, D. B. (1989). In: *Complex Organismal Functions: Integration and Evolution in Vertebrates.* (Ed. by D. B. Wake and G. Roth). New York: John Wiley and Sons Ltd.

Sage, R. D. and Selander, R. K. (1975). Trophic radiation through polymorphism in cichlid fishes. *Proc. Nat. Acad. Sci.*, **72**, 4669–4673.

Scheal, D. M., Rudstam, L. G. and Post, J. R. (1991). Gape limitation and prey selection in larval yellow perch (*Perca flavescens*), fresh water drum (*Aplodinotus grunniens*), and black crappie (*Pomoxis nigromaculatus*). *Can. J. Fish. Aquat. Sci.*, **48**, 1919–1925.

Schmitt, R. J. and Holbrook, S. F. (1984). Gape-limitation, foraging tactics and prey size selectivity of two microcarnivorous species of fish. *Oecologia*, **63**, 6–12.

Stearns, S. C. (1982). The role of development in the evolution of life histories. In: *Evolution and Development.* Dahlem Conference Report no. 20. (Ed. by J. T. Bonner). Berlin, N.Y.: Springer-Verlag.

Stoner, A. W. and Livingston, R. J. (1984). Ontogenetic patterns in diet and feeding morphology in sympatric Sparid fishes from seagrass meadows. *Copeia*, 1984, 174–187.

Wainwright, P. C. (1987). Biomechanical limits to ecological performance: mollusc-crushing by the Caribbean hogfish, *Lachnolaimus maximus* (Labridae). *J. Zool., Lond.*, **213**, 283–297.

Wainwright, P. C. (1988). Morphology and ecology: Functional basis of feeding constraints in Caribbean Labrid fishes. *Ecology*, **69**, 635–645.

Wainwright, P. C. (1991). Ecomorphology: Experimental functional anatomy for ecological problems. *Amer. Zool.*, **31**, 680–693.

Wainwright, P. C., Osenberg, C.W. and Mittelbach, G.G. (1990). Trophic polymorphism in the pumpkinseed sunfish (*Lepomis gibbosus* L.): effects of environment on ontogeny. *Func. Ecol.*, **5**, 40–55.

Wake, D. B. and Roth, G. (1989). The linkage between ontogeny and phylogeny in the evolution of complex systems. In: *Complex Organismal Functions: Integration and Evolution in Vertebrates* (Ed. by D. B. Wake and G. Roth). New York: John Wiley and Sons Ltd.

Wasserzug, R. J. and Sperry, D. G. (1977). The relationship of locomotion to differential predation on *Pseudacruis triseriata* (Anura: Hylidae). *Ecology*, **58**, 830–839.

Werner, E. E. (1974). The fish size, prey size, handling time relation in several sunfishes and some implications. *F. Fish. Res. Board Can.*, **31**, 1531–1536.

Werner, E. E., Gilliam, J. F., Hall, D. J. and Mittelbach, G. G. (1983). An experimental test of the effects of predation risk on habitat use in fish. *Ecology*, **64**, 1540–1548.

Werner, E. E. and Gilliam, J. F. (1984). The ontogenetic niche and species interactions in size structured populations. *Ann. Rev. Ecol. Syst.*, **15**, 393–425.

Werner, E. E. and Hall, D. J. (1988). Ontogenetic habitat shifts in the bluegill sunfish (*Lepomis macrochirus*): The foraging rate–predation risk tradeoff. *Ecology*, **69**, 1352–1366.

Wimberger, P. H. (1991). Plasticity of jaw and skull morphology in the neotropical cichlids *Geophagus brasiliensis* and *G. steindachneri*. *Evolution*, **45**, 1545–1556.

Yamaoka, K. (1978). Pharyngeal jaw structure in labrid fishes. *Publ. Seto Mar. Biol. Lab.*, **24**, 409–426.

Zweers, G. (1991). Transformation of avian feeding mechanisms: a deductive method. *Acta Biotheor.*, **39**, 15–36.

WHETHER OR NOT TO DEFEND?
THE INFLUENCE OF RESOURCE DISTRIBUTION

JAMES W. A. GRANT

Biology Department, Concordia University, 1455 blvd de Maisonneuve West, Montreal, Canada H3G 1M8

INTRODUCTION

Animals are often aggressive when competing for resources (for reviews see Huntingford and Turner, 1987; Archer, 1988). Our understanding of this phenomenon was revolutionized by Jerram Brown's (1964) concept of economic defendability. According to his model, animals will only defend resources when the benefits exceed the costs of defence or when the net benefits of defence exceed the net benefits of alternative tactics such as scrambling for the resource. The few studies that have measured both the benefits and costs of defence have supported the theory (e.g. Gill and Wolf, 1975; Carpenter and MacMillen, 1976; Davies and Houston, 1981; Ewald, 1985). The concept of economic defendability is now a unifying principle of behavioural ecology and a fundamental part of the theories of how spacing (e.g. Brown and Orians, 1970), mating (e.g. Emlen and Oring, 1977) and social (e.g. Lott, 1991) systems evolve.

Brown (1964) recognized that the decision of whether or not to defend is influenced by competitor density, resource density and resource distribution in space and time. The challenge for behavioural ecologists is to predict the ecological conditions in which aggression will be used to acquire resources. The purpose of this paper is to review the current theory and to provide a selective review of the evidence about how resource distribution influences the decision to defend or not. Whenever possible, I will use examples from the fish literature to explore this link.

DEFINITIONS

The dependent variable that I am interested in predicting is the occurrence of defence behaviour or, more generally, aggression in the context of competition for resources. Defence can be quantified in a number of ways including its occurrence (e.g. % of individuals defending territories), its intensity (e.g. % of conspecific intruders that are chased) or its exclusivity (e.g. % of range overlap with neighbours). Although I am primarily interested in aggressive behaviour in this paper, the same conditions that favour defence also favour sedentariness and resource monopolization (Wiens, 1976; Warner, 1980; Grant and Noakes, 1988). The prediction that mobility is inversely related to defence emerges from two lines of thought. The theory of optimal territory size assumes that larger areas cost more to defend (e.g. Schoener, 1983), so individuals will be less likely to defend large home ranges. Second, sit-and-wait foragers are predicted to be more aggressive than active searchers (Stamps, 1977). Resource monopolization, the uneven distribution of resources among individuals, is both a cause and a consequence of defence behaviour. Resources must be monopolizable before they are economically defendable (Emlen and Oring, 1977), but resource defence leads to more extreme resource monopolization.

Resources most commonly defended are mates and mating opportunities, food, and a place to live such as a shelter, burrow or space on the substrate on which to attach. Of the 33 taxa reviewed by Huntingford and Turner (1987), defence of mates, food and space/shelter are reported in 64, 33 and 24 percent of the taxa, respectively. However, the defence of food or space/shelter that is unrelated to reproduction occurs in only 15 and 12 percent of the taxa, respectively. This survey suggests that aggression occurs most frequently when animals compete for mating opportunities.

ENVIRONMENTAL PARAMETERS AND PREDICTIONS

The environmental parameters that are most often related to defence behaviour are population density and several aspects of resource distribution. Except where noted, I follow Warner's (1980) clear and concise definitions. The predictions are from Brown (1964), Emlen and Oring (1977), Warner (1980) and Wiens (1976).

Population Density

The density of competitors can be quantified as either number per area or as intruder pressure (e.g. intruders per time). When competitor density is extremely low, interactions with conspecifics over resources are rare so the benefits of defence are low. As competitor density increases, the costs of defence increase but the benefits of defence increase even faster until a lower threshold is reached where the benefits exceed the costs of defence. When competitor density is extremely high, defence becomes uneconomical because of the high costs. Therefore, defence is predicted to peak at an intermediate population density, between lower and upper thresholds (Figure 1a). Although I use the word threshold to define the point at which costs and benefits are equal, we should not necessarily expect to see a dramatic change from no defence to defence in all populations. Individuals might change their aggressive behaviour gradually in response to benefits and costs (see Craig and Douglas, 1986; Grant and Noakes, 1988). Alternatively, each individual might have a different threshold, so at the population level one might observe a gradual change in aggressive behaviour.

Resource Density

The abundance of a resource can be quantified as either a standing crop (mean amount per unit area) or as a renewal rate (production per unit area). When resource density is low, an individual must range over a large area to obtain sufficient resources, so the costs of defending such a range may be prohibitive. As resource density increases, the benefits of defence increase until a lower threshold

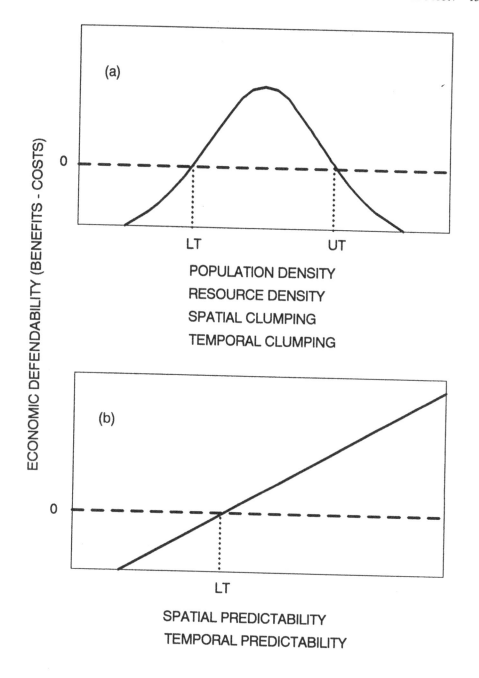

Figure 1   Economic defendability and hence the aggressiveness of animals are predicted to (a) peak at intermediate levels of population density, resource density and spatial or temporal clumping of resources or (b) increase as the spatial or temporal predictability of resources increase (LT=lower threshold for defence; UT=upper threshold for defence).

is reached where benefits exceed costs. When resources are extremely abundant an individual that devotes time to defence may waste energy and grow more slowly than an individual that ignores conspecifics. Doyle and Talbot (1986) produced a game-theory model that makes this verbal prediction more explicit. Defence is predicted to occur at intermediate resource densities between these lower and upper thresholds (Figure 1a).

Spatial Clumping

The patchiness or degree of aggregation of a resource can be quantified as the variance in resource density across space. When a resource is evenly dispersed, an individual must range over a large area to acquire sufficient resources, so the cost of defending such a range may be prohibitive. As spatial clumping increases, a resource becomes increasingly defendable because the individual can occupy a progressively smaller home range that is easier to defend. However, an extremely clumped resource that attracts many intruders may not be defendable. Defence is, therefore, predicted to occur at intermediate levels of spatial clumping (Figure 1a).

Temporal Clumping

The synchrony of a resource can be quantified as the variance in resource density at a particular place across time. A temporally clumped resource cannot be effectively monopolized because of the overlap in arrival times of the resource units. While the most competitive individual is occupied with one resource unit, less competitive individuals have an opportunity to obtain subsequent units. In contrast, when a resource arrives asynchronously the most competitive individual can potentially compete for each resource unit as it arrives. Blanckenhorn and Caraco (1992) recently developed an analytical version of the original verbal model (Trivers, 1972; Emlen and Oring, 1977; Wells, 1977).

Resource defence is predicted to increase as the temporal clumping of resource density decreases because the greater potential for monopolization increases the benefit of being the most successful individual and because individuals can use their 'free' time between resource arrivals to defend favourable sites (Trivers, 1972; Emlen and Oring, 1977; Wells, 1977). However, extremely asynchronous resources may not be defendable because of their low renewal rate (Emlen and Oring, 1977). Hence, defence is predicted to peak at intermediate levels of temporal clumping (Figure 1a).

Spatial Predictability

Spatial predictability is a measure of the dependability of good and poor sites over time (Warner, 1980). It can be quantified by ranking areas in terms of resource density at a particular time and then by measuring the correlation between these rankings over successive time intervals; a spatially predictable resource would have a high positive correlation coefficient. When resources are spatially unpredictable, individuals must range over large areas to acquire sufficient resources. As spatial predictability increases, an individual can restrict itself to a progressively smaller and effectively defended home range. Hence, sedentariness and defence are predicted to increase as spatial predictability increases (Figure 1b).

Temporal Predictability

Temporal predictability is a measure of the dependability of times of high and low resource density. It can be quantified by ranking times (e.g. months) in terms of resource density and then measuring the correlation among these rankings over years (Warner, 1980). Defence is predicted to increase as temporal predictability increases (Figure 1b).

REVIEW OF RESULTS

Competitor Density

Jones (1983) clearly demonstrated the influence of population density on the aggressive behaviour of juvenile *Pseudolabrus celiodotus* on reefs near New Zealand. These fish are not territorial, but feed in loose groups in overlapping home ranges. As population density increased, the frequency of agonistic interactions first increased then decreased (Figure 2a), suggesting lower and upper thresholds of population density for defence. However, the percentage of all encounters that resulted in aggression decreased monotonically with increasing density (Figure 2b). The apparent increase in aggressiveness as density initially increased was actually the result of greater opportunities to interact with competitors. At low densities, *P. celiodotus* interacted aggressively when feeding, but at high densities they resorted to scramble competition. Hence, Jones's study provides strong evidence of only an upper threshold of population density for defence.

Animals often cease defending territories entirely when intruder pressure is so high that defence is presumably uneconomical. Kawanabe's (1969) study of the ayu (*Plecoglossus altivelis*), a benthic algal grazer from Japan, is perhaps the best example of this in fishes. When population density was less than 2 m^{-2}, territorial fish occupied the best feeding sites and composed 30–50 % of the population. However, when densities exceeded 4 m^{-2} schooling fish predominated. Territorial fish initially try to defend against schools of intruders, but many give up and join schools, presumably because defence is uneconomical. Hence, the percent of fish defending territories decreases with increasing density (Figure 3), providing evidence for an upper threshold of population density for defence.

Population size and hence intruder pressure have a dramatic effect on the reproductive success of male blueheaded wrasse, *Thalassoma bifasciatum* (Warner and Hoffman, 1980a,b). In small populations (< 200 individuals), large territorial males successfully excluded small males from the best spawning sites and paired with most of the females. As population size increased, the mating success of territorial males declined because they spent an increasing amount of time chasing smaller males that intruded on their territories. In the largest populations (> 500 individuals), large males established breeding territories in peripheral habitats and left the best spawning sites to small males, presumably because of the high costs of defence. Their elegant field studies provide further evidence of an upper threshold of population density for defence.

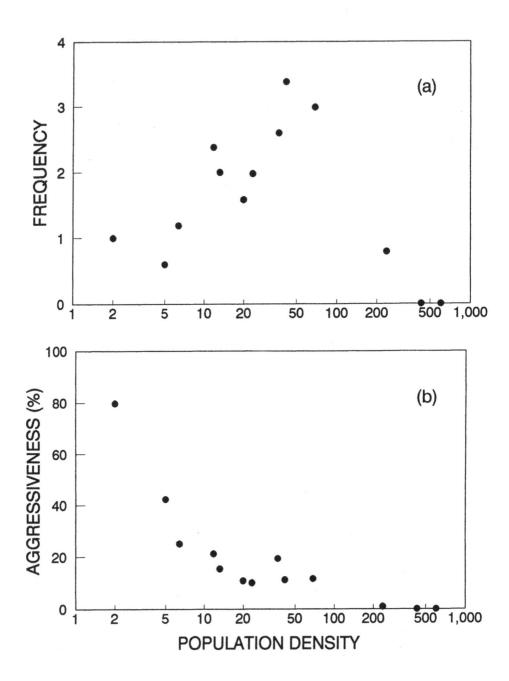

Figure 2   (a) Frequency of aggression (no./15 min) and (b) aggressiveness (% of encounters resulting in aggression) of juvenile *Pseudolabrus celiodotus* in relation to population density. Modified from Jones (1983).

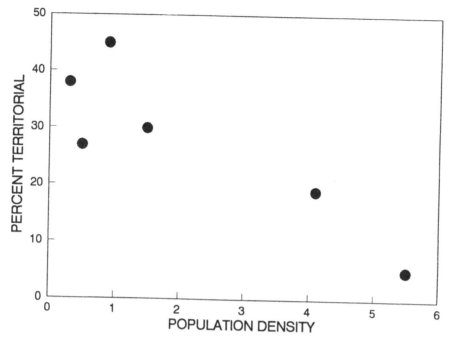

Figure 3   Percentage of ayu that defend feeding territories decreases with increasing population density (r=-0.88, P=0.021). Data are from Kawanabe (1969).

Reef fishes in particular appear to use schooling as a mechanism to circumvent the territorial behaviour of competitors. A school of fish can temporarily overwhelm the defence capabilities of the resident and thereby feed in an area that a solitary fish would normally be excluded from (Barlow, 1974; Robertson *et al.*, 1976). While the school is present, the resident does not cease defending, but its aggressive behaviour is diluted among the members of the school.

Aquaculturists are well aware of the upper threshold of population density for defence and try to maintain densities of salmonid fishes above this threshold to minimize the deleterious effects of aggression (Refstie and Kittleson, 1976; Wallace *et al.*, 1988; Noakes and Grant, 1992).

I am not aware of any study that has clearly shown evidence of a lower threshold of population density for defence.

Resource Density

The best evidence of lower and upper thresholds of food density for defence come from studies of nectar-feeding birds. Both Hawaiian honeycreepers, *Vestiaria coccinea*, (Carpenter and MacMillen, 1976) and golden-winged sunbirds, *Nectarina reichenowi*, (Gill and Wolf, 1975) defend territories at intermediate levels of food density and cease defending when food density is extremely high.

No such elegant studies exist for fish in the wild. However, intraspecific variation in aggressiveness of stream-dwelling salmonid fishes has been related to the renewal rate of food in their environment. These fish feed primarily on stream drift, aquatic invertebrates that are carried by water currents, and the renewal rate of drift increases linearly with current velocity (Grant and Noakes, 1987). The aggressiveness of brook charr, *Salvelinus fontinalis*, increased significantly as current velocity initially increased (Figure 4a), providing evidence of a lower threshold of defence. Aggressiveness also declined significantly at velocities greater than 6 cm/s, not because of an excess of food but presumably because the costs of defence increased and foraging efficiency decreased in these fast velocities (Grant and Noakes, 1988). The mobility of brook charr was also influenced by current velocity: in zero-current habitats they were primarily active searchers, whereas in running water they were sit-and-wait foragers (Figure 4b). As predicted, aggressiveness and mobility decreased as the renewal rate of food increased.

Magnuson's (1962) study of medaka (*Oryzias latipes*) provides the best example for a fish of an upper threshold of food density for defence. He paired large and small fish in aquaria under three food regimes: none, limited and excess. The chasing rate of the larger fish increased, but not significantly, between the none and limited treatments, providing only weak evidence of a lower threshold (Figure 5). However, when food was provided in excess, the chase rate of the dominant decreased significantly compared to the limited treatment and was not different from the smaller fish's chase rate (Figure 5). The aggressive behaviour of the dominant in the limited treatment was economical because dominants grew faster than subordinates in this treatment; there were no differences in the growth rates of dominants and subordinates in the none or excess treatments.

To my knowledge, nectar-feeding birds provide the only examples of upper thresholds for defence under natural conditions. All other examples of the cessation of defence because of high food density occur in laboratory experiments (e.g. Magnuson, 1962; Wyman and Hotaling, 1988) or in provisioned field populations (e.g. Wilcox and Ruckdeschel, 1982). Upper thresholds may be common in the wild but remain undetected because of the difficulty of measuring food abundance. Hence, nectar-feeding birds may be notable exceptions, simply because nectar abundance is relatively easy to measure. But, even some nectar-feeding birds do not cease defence when food is unlimited; hummingbirds continue to defend (for a review, see Carpenter, 1987). In addition, the defence of territories that are exclusively for feeding is rare in most fishes (Barlow, this volume) and in animals in general (see Huntingford and Turner, 1987). This suggests that food is usually found at levels below a lower threshold for defence. Hence, while upper thresholds of food density for defence are theoretically possible, they may be rare in wild populations.

Spatial Clumping

Despite the large literature on the effects of spatial clumping of resources on behaviour and ecology (e.g. Wiens, 1976), only a handful of experimental studies have measured aggressive behaviour and/or resource monopolization in response to spatial clumping. Monaghan and Metcalfe's (1985) study is perhaps the simplest and clearest example. Wild brown hares, *Lepus americanus*, were allowed to forage on pieces of apple that were either gathered in clumps of 0.2 m^2 or dispersed over a grid of 50 m^2. When the apples were clumped, the frequency of aggression increased significantly because the dominant hare attempted to defend the food patch. Dominants also spent more time feeding than subordinates, at least in small

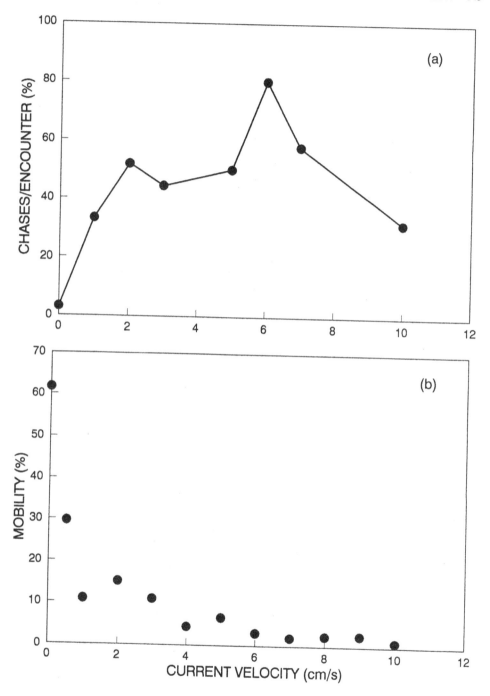

Figure 4   (a) Aggressiveness and (b) mobility (% of search time as active searcher) of 20–29 mm brook charr in relation to current velocity. Modified from Grant and Noakes (1988).

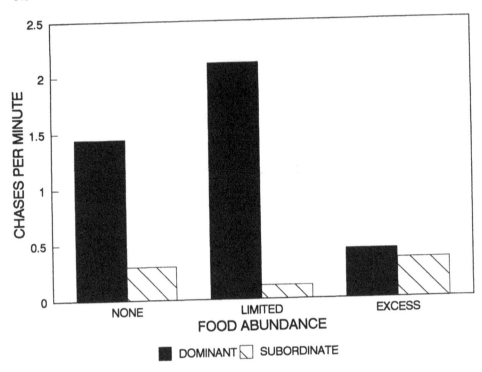

Figure 5 Frequency of aggression by dominant and subordinate medaka in relation to food abundance. Data are from Magnuson (1962).

group sizes. Hence, increasing the spatial clumping of the resource appeared to increase both the aggressiveness and foraging success of dominant hares.

Fish seem to respond in a similar way to the spatial clumping of resources. Large, dominant medaka defended feeding territories and had high rates of aggression when food was clumped in patches of 19.6 or 50 cm^2, but did not defend space and had low rates of aggression when food was dispersed in patches of 100 or 400 cm^2 (Magnuson, 1962). Similarly, male pygmy sunfish, *Elassoma evergladei*, defended territories when food was clumped in a patch of 26.5 cm^2 but did not defend territories when food was randomly dispersed over the aquarium (1250 cm^2)(Rubenstein, 1981).

These studies and other similar ones with pied wagtails, *Motacilla alba alba*, (Zahavi, 1971), rhesus monkeys, *Macaca mulatta*, (Southwick, 1967) and dark-eyed juncos, *Junco hyemalis*, (Theimer, 1987) are summarized in Figure 6. Together these results suggest that either aggressiveness or monopolization or both increase as the spatial clumping of food increases. I am not aware of any studies showing that extremely clumped resources are indefensible because of high intruder pressure. This interaction between spatial clumping and intruder pressure needs investigation.

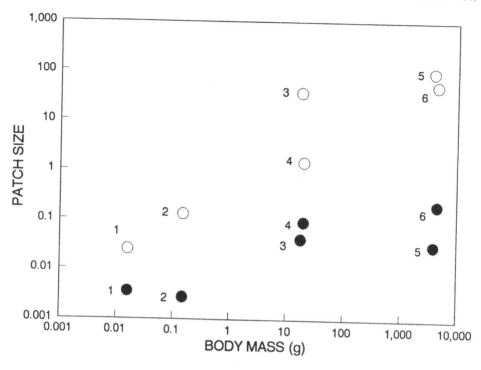

Figure 6 Size of food patches (m²) that were either defended and monopolized (solid circles) or not defended and shared with conspecifics (open circles). Numbers refer to data for: 1=medaka (Magnuson, 1962); 2=pygmy sunfish (Rubenstein, 1981); 3=pied wagtails (Zahavi, 1971); 4=dark-eyed juncos (Theimer, 1987); 5=rhesus monkeys (Southwick, 1967) and 6=brown hares (Monaghan and Metcalfe, 1985).

Temporal Clumping

The temporal clumping hypothesis was originally proposed for females as the limiting resource for competing males (Trivers, 1972; Emlen and Oring, 1977; Wells, 1977). With the exception of a study by Grant and Kramer (1992), the few tests of the hypothesis focus only on measures of monopolization, primarily the variance of male mating success, and the results have either been equivocal (Gatz, 1981; Ridley, 1986, but see Altmann, 1990) or opposite to the predicted direction (Ryan, 1985). However, Blanckenhorn (1991) has recently shown that monopolization of food by water striders, *Gerris remigis*, decreased when temporal clumping increased, as predicted by the theory.

Grant and Kramer (1992) tested the temporal clumping hypothesis using groups of six, individually tagged zebrafish, *Brachydanio rerio*, competing for 300 *Daphnia pulex* prey. Temporal clumping of prey arrival was manipulated by varying the duration of the period (3, 10, 30, 100 or 300 min) over which they arrived through a single feeding tube. Resource monopolization increased as trial duration increased (Figure 7a). The effectiveness with which dominants defended the feeding tube appeared to increase with increasing trial duration, because the number of fish over the feeding platform decreased significantly (Figure 7b). However, there was only equivocal evidence that the aggressiveness of dominants actually changed with trial duration. Chase rate by dominants initially increased with trial duration (Figure 7c) despite fewer fish over the feeding platform (Figure 7b), suggesting that dominants were chasing a higher proportion of intruders. However, dominant fish may have been using a single behavioural rule throughout; e.g. 'eat a prey if one is in sight, otherwise chase competitors'. The apparent change in defence may have arisen simply as a consequence of the conflict in time between eating and chasing during short trials. This study underscores the difficulty in documenting real changes in the aggressiveness of individuals when environmental conditions change. In summary, Grant and Kramer provide evidence of an upper threshold of temporal clumping for monopolization and perhaps defence in zebrafish.

Spatial Predictability

I know of only two studies that have investigated the influence of spatial predictability of resources on aggression. Grand and Grant (in press) used groups of six individually tagged convict cichlids, *Cichlasoma nigrofasciatum*, competing for food in aquaria. The six fish were fed 72 *Daphnia pulex* per trial for six trials over three days; one prey arrived every 15 s in one of four feeding patches. The patches differed in the probability that they would receive the food item (either 0.67, 0.167, 0.083 or 0.083). Spatial predictability was manipulated by the number of different spatial arrangements of the patch probabilities that the fish were exposed to over the three days: 1 (predictable), 6 (intermediate) and 36 (unpredictable). There were 11 replicate groups in each treatment.

The most striking result was that fish were more sedentary (i.e. changed patches less often) as spatial predictability increased ($P<0.001$). In addition, dominant fish appeared to become more aggressive with increasing spatial predictability ($P=0.007$); they chased 15% of intruders in the unpredictable treatment and 30% in the predictable treatment. Dominants also tended to eat a greater percentage of the food as spatial predictability increased ($P<0.001$), about 20% in the unpredictable treatment and 35% in the predictable treatment. Taken together, these results suggest real behavioural change by dominant fish in response to changes in spatial predictability. It was encouraging that the results were consistent with the theory even though the manipulations were made on such a small spatial scale.

Pimm (1978) manipulated the spatial predictability of artificial feeders and measured the aggressiveness of hummingbirds, *Lampornis clememciae*, while using the feeders. Because his experimental design manipulated both spatial predictability and clumping at the same time, it is difficult to interpret the results. Nevertheless, as spatial predictability increased (and spatial clumping decreased) the amount of time spent defending feeders increased.

Temporal Predictability

I am not aware of any studies that have tested the hypothesis that aggressiveness increases as the temporal predictability of a resource increases.

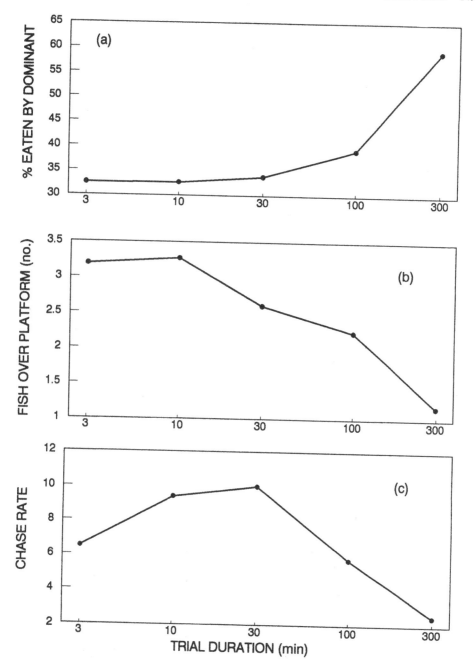

Figure 7   (a) Monopolization of food by the dominant zebrafish, (b) number of zebrafish over the feeding platform and (c) chase rate (no./min) by the dominant zebrafish in relation to the duration of the period over which 300 *Daphnia pulex* prey appeared in a single feeding patch. Modified from Grant and Kramer (1992).

DISCUSSION

The available evidence suggests that population density and resource distribution exert a strong influence on the decision of whether or not to defend a resource. Furthermore, the qualitative changes in aggressiveness and/or monopolization are generally consistent with the predictions of the theory of economic defendability. However, many of the hypothesized effects, such as an upper threshold of spatial clumping and lower thresholds of population density, temporal clumping or temporal predictability for defence, remain untested. In addition, the upper threshold of temporal clumping and the lower threshold of spatial predictability for defence need further testing.

If the theory of economic defendability is to have more than heuristic value, it should be able to predict the direction of changes in aggressiveness in response to changes in environmental variables. This will be difficult for the independent variables depicted in Figure 1a because, unless the peak of the curve can be specified a priori, almost any data will be consistent with the predictions. The only way to clearly falsify the hypothesis is to show that aggressiveness is lowest at intermediate levels of the independent variables. However, our empirical observations can help. For most of the variables, one of the two thresholds is relatively unlikely, at least in natural conditions. As I have already suggested, an upper threshold of food abundance seems to be rarely encountered in wild populations, so for most practical purposes Figure 1a can be redrawn as a straight line increasing with resource density. Similarly, the available evidence suggests that lower thresholds are more important than upper thresholds for spatial clumping (Figure 6) and that upper thresholds are more important than lower thresholds for temporal clumping. Hence, for practical purposes, these curves may also be redrawn as straight lines, increasing and decreasing with spatial and temporal clumping, respectively. Therefore, we may be able to make predictions about the change of behaviour for all independent variables except population density.

Few studies have attempted to model or manipulate more than one environmental parameter at a time. For example, resource density and spatial clumping are probably positively correlated in the wild (e.g. Gillis et al., 1986). Experiments on the interactions of more than one parameter at a time would be valuable.

Ideally, the theory should make quantitative predictions about when defence will occur. I do not think this is currently feasible, but data in the literature may be sufficient to begin the process. For example, imagine an extensive data set like that shown in Figure 6 for a group of animals, say birds or fishes. Such a data set could be used to predict the likelihood that food distributed in a patch of a given size will be defended or not. I suggest that integrating empirical data with theory will provide a more powerful method for predicting when animals will or will not defend resources.

Fish behavioural ecologists are in a good position to continue their strong contribution to this developing area of research. While feeding territories are relatively rare in most animals, a notable exception are coral-reef fishes (Barlow, this volume). They provide important and tractable natural systems for testing the theory of economic defendability. Second, many freshwater fishes adapt readily to the lab and defend feeding territories, even though they do not do so in the wild (Barlow, this volume). Such species provide ideal model systems for experimental studies of defence. Third, reproductive success should be more strongly related to body size for fishes, at least for females, than for endothermic vertebrates. Hence, net energy gain or growth rate may be more reasonable estimates of the net benefits of defence for fishes than for most other animals.

ACKNOWLEDGMENTS

I thank G. W. Barlow, T. C. Grand and J. A. Hutchings for helpful comments on an earlier version of this paper. I am supported by operating grants from NSERC (Canada) and FCAR (Québec).

References

Altmann, J. (1990). Primate males go where the females are. *Anim. Behav.*, **39**, 193–195.
Archer, J. (1988). *The behavioural biology of aggression*. Cambridge: Cambridge University Press.
Barlow, G. W. (1974). Extraspecific imposition of social grouping among surgeonfishes (Pisces: Acanthuridae). *J. Zool., Lond.*, **174**, 333–340.
Blanckenhorn, W. U. (1991). Foraging in groups of water striders (*Gerris remigis*): effects of variability in prey arrivals and handling times. *Behav. Ecol. and Sociobiol.*, **28**, 221–226.
Blanckenhorn, W. U. and Caraco, T. (1992). Social subordinance and a resource queue. *Am. Nat.*, **139**, 442–449.
Brown, J. L. (1964). The evolution of diversity in avian territorial systems. *Wilson Bull.*, **76**, 160–169.
Brown, J. L. and Orians, G. H. (1970). Spacing patterns in mobile animals. *Ann. Rev. Ecol. Syst.*, **1**, 239–262.
Carpenter, F. L. (1987). Food abundance and territoriality: to defend or not to defend? *Am. Zool.*, **27**, 387–399.
Carpenter, F. L. and MacMillen, R. E. (1976). Threshold model of feeding territoriality and test with a Hawaiian honeycreeper. *Science*, **194**, 639–642.
Craig, J. L. and Douglas, M. E. (1986). Resource distribution, aggressive asymmetries and variable access to resources in the nectar feeding bellbird. *Behav. Ecol. Sociobiol.*, **18**, 231–240.
Davies, N. B. and Houston, A. I. (1981). Owners and satellites: the economics of territory defence in the pied wagtail, *Motacilla alba*. *J. Anim. Ecol.*, **50**, 157–180.
Doyle, R. W. and Talbot, A. J. (1986). Artificial selection on growth and correlated selection on competitive behaviour in fish. *Can. J. Fish. Aquat. Sciences*, **43**, 1059–1064.
Emlen, S. T. and Oring, L. W. (1977). Ecology, sexual selection, and the evolution of mating systems. *Science*, **197**, 215–223.
Ewald, P. W. (1985). Influence of asymmetries in resource quality and age on aggression and dominance in black-chinned hummingbirds. *Anim. Behav.*, **33**, 705–719.
Gatz, A. J. Jr. (1981). Non-random mating by size in American toads, *Bufo americanus*. *Anim. Behav.*, **29**, 1004–1012.
Gill, F. B. and Wolf, L. L. (1975). Economics of feeding territoriality in the golden-winged sunbird. *Ecology*, **56**, 333–345.
Gillis, D. M., Kramer, D. L. and Bell, G. (1986). Taylor's power law as a consequence of Fretwell's ideal free distribution. *J. Theor. Biol.*, **123**, 281–287.
Grand, T. C. and Grant, J. W. A. (In press). Spatial predictability of food influences its monopolization and defense by juvenile convict cichlids. *Anim. Behav.*
Grant, J. W. A. and Kramer, D. L. (1992). Temporal clumping of food arrival reduces its monopolization and defence by zebrafish, *Brachydanio rerio*. *Anim. Behav.*, **44**, 101–110.

Grant, J. W. A. and Noakes, D. L. G. (1987). Movers and stayers: foraging tactics of young-of-the-year brook charr, *Salvelinus fontinalis*. *J. Anim. Ecol.*, **56**, 1001–1013.

Grant, J. W. A. and Noakes, D. L. G. (1988). Aggressiveness and foraging mode of young-of-the-year brook charr, *Salvelinus fontinalis* (Pisces, Salmonidae). *Behav. Ecol. Sociobiol.*, **22**, 435–445.

Huntingford, F. A. and Turner, A. (1987). *Animal Conflict*. London: Chapman and Hall.

Jones, G. P. (1983). Relationship between density and behaviour in juvenile *Pseudolabrus celidotus* (Pisces: Labridae). *Anim. Behav.*, **31**, 729–735.

Kawanabe, H. (1969). The significance of social structure in production of the "ayu", *Plecoglossus altivelis*. In: *Symposium on Salmon and Trout in Streams* (Ed. by T. G. Northcote), pp. 243–251. Vancouver: Institute of Fisheries, The University of British Columbia.

Lott, D. F. (1991). *Intraspecific Variation in the Social Systems of Wild Vertebrates*. Cambridge: Cambridge University Press.

Magnuson, J. J. (1962). An analysis of aggressive behavior, growth, and competition for food and space in medaka (*Oryzias latipes* (Pisces, Cyprinodontidae)). *Can. J. Zool.*, **40**, 313–363.

Monaghan, P. and Metcalfe, N. B. (1985). Group foraging in wild brown hares: effects of resource distribution and social status. *Anim. Behav.*, **33**, 993–999.

Noakes, D. L. G. and Grant, J. W. A. (1992). Feeding and social behaviour of brook and lake charr. In: *The Importance of Feeding Behaviour for the Efficient Culture of Salmonid Fishes* (Ed. by J. E. Thorpe and F. A. Huntingford), pp. 13–20. Baton Rouge: World Aquaculture Society.

Pimm, S. L. (1978). An experimental approach to the effects of predictability on community structure. *Am. Zool.*, **18**, 797–808.

Refstie, T. and Kittleson, A. (1976). Effects of density on growth and survival of artificially reared Atlantic salmon. *Aquacult.*, **8**, 319–326.

Ridley, M. (1986). The number of males in a primate troop. *Anim. Behav.*, **34**, 1848–1858.

Robertson, D. R., Sweatman, H. P. A., Fletcher, E. A. and Cleland, M. G. (1976). Schooling as a mechanism for circumventing the territoriality of competitors. *Ecology*, **57**, 1208–1220.

Rubenstein, D. I. (1981). Population density, resource patterning, and territoriality in the everglades pygmy sunfish. *Anim. Behav.*, **29**, 155–172.

Ryan, M. J. (1985). *The Tungara Frog, a Study in Sexual Selection and Communication*. Chicago: University of Chicago Press.

Schoener, T. W. (1983). Simple models of optimal territory size: a reconciliation. *Am. Nat.*, **121**, 608–629.

Southwick, C. H. (1967). An experimental study of intragroup agonistic behavior in rhesus monkeys (*Macaca mulatta*). *Anim. Behav.*, **28**, 182–209.

Stamps, J. A. (1977). Social behavior and spacing patterns in lizards. In: *Biology of the Reptilia*, Vol. 7 (Ed. by C. Gans and D. W. Tinkle), pp. 265–334. London: Academic Press.

Theimer, T. C. (1987). The effect of seed dispersion on the foraging success of dominant and subordinate dark-eyed juncos, *Junco hyemalis*. *Anim. Behav.*, **35**, 1883–1890

Trivers, R. L. (1972). Parental investment and sexual selection. In: *Sexual Selection and the Descent of Man* (Ed. by B. G. Campbell), pp. 136–179. Chicago: Aldine.

Wallace, J. C., Kolberinshavn, A. G. and Reinsnes, T. G. (1988). The effects of stocking density on early growth in Arctic charr, *Salvelinus alpinus* L. *Aquacult.*, **73**, 101–110.

Warner, R. R. (1980). The coevolution of behavioral and life-history characteristics. In: *Sociobiology: Beyond Nature/Nurture?* (Ed. by G. W. Barlow and J. Silverberg), pp. 151–188. Boulder: Westview.

Warner, R. R. and Hoffman, S. G. (1980a). Population density and the economics of territorial defense in a coral reef fish. *Ecology*, **61**, 772–780.

Warner, R. R. and Hoffman, S. G. (1980b). Local population size as a determinant of mating system and sexual composition in two tropical marine fishes (*Thalassoma spp.*). *Evolution*, **34**, *508–518*.

Wells, K. D. (1977). The social behaviour of Anuran amphibians. *Anim. Behav.*, **25**, 666–693.

Wiens, J. A. (1976). Population responses to patchy environments. *Ann. Rev. Ecol. Syst.*, **7**, 81–120.

Wilcox, R. S. and Ruckdeschel, T. (1982). Food threshold territoriality in a water strider (*Gerris remigis*). *Behav. Ecol. Sociobiol.*, **11**, 85–90.

Wyman, R. L. and Hotaling, L. (1988). A test of the model of the economic defendability of a resource and territoriality using young *Etroplus maculatus* and *Pelmatochromis subocellatus kribensis*. *Env. Biol. Fishes*, **21**, 69–76.

Zahavi, A. (1971). The social behaviour of the white wagtail *Motacilla alba alba* wintering in Israel. *Ibis*, **113**, 203–211.

THE PUZZLING PAUCITY OF FEEDING TERRITORIES AMONG FRESHWATER FISHES

GEORGE W. BARLOW

Department of Integrative Biology, and Museum of Vertebrate Zoology
University of California, Berkeley, California 94720, USA

INTRODUCTION

As a beginning graduate student I read two papers by Gerking (1950, 1953) that left an impression on me. His thesis was that fishes living in lotic freshwaters have restricted home ranges. However, Gerking (1959: 221) referred to home range and territory collectively. He assumed that patrilocal species are usually territorial, as Lowe-McConnell (1987) apparently does. That assumption obscures the paucity of nonbreeding territories among freshwater fishes.

When treating home range and territoriality, Gerking (1959) also did not distinguish between freshwater and marine fishes. That perspective persists, evidently because most who investigate fish behavioral ecology tend to specialize either on freshwater or on marine species (but see Barlow, 1981; Baylis, 1981; Sazima, 1986). When feeding territories are compared across those environments, however, marine reef fishes commonly have them but freshwater species do not. Why? Whatever the reasons, my experiences and the literature have led me to a new perspective on the difference. In keeping with theory, mating systems among marine reef fishes have often been derived from feeding territories. The reverse is true among freshwater fishes, insofar as is known: Except for salmonids and some exceptional characiforms, feeding territories appear to have evolved from mating territories.

BEHAVIORAL ECOLOGY AND SOCIAL SYSTEMS

I focus on a single basic precept of behavioral ecology, that social or mating systems are tailored to the ecology of the organism. The complexities inherent in relating mating systems to ecology, however, are daunting (Lack, 1954, 1968; Crook, 1964; Jarman, 1974; Pitelka *et al.*, 1974; Clutton-Brock, 1989). The biology of the organism, such as mode of reproduction and presence or absence of parental care, strongly influences social organization; seasonality, and nocturnal versus diurnal activity, also play a role (e.g. Hobson, 1975; Helfman, 1981). Small wonder that so few comprehensive attempts have been made to explain mating systems among fishes in terms of their ecological substrate (Barlow, 1974, 1991; Reese, 1975; Thresher, 1977; Hourigan, 1989), and that they remain rudimentary.

Despite the complexity inherent in such analyses, the foremost influences narrow down to a few. The main ones are the nature of the physical habitat (Crook, 1964), the risk of predation (Crook, 1964; Farr, 1975; Gross and MacMillan, 1981) and feeding behavior (Lack, 1954; Crook, 1964; Pitelka *et al.*, 1974). This approach was crystallized by Emlen and Oring (1977) (see also Davies, 1991) in their seminal paper relating mating systems to ecology in birds. Their theoretical framework has been applied liberally to many other kinds of animals. In their scheme, females space themselves in relation to some resource such as food or a place of refuge, which may be defended as a territory. Central to this line of reasoning about ultimate cause is that the spatial configuration is primary, especially that of females.

Males then map onto the distribution of females. Ideally, from a male's perspective, several females reside within the boundaries of his territory. Here the essential aspect of territoriality is exclusive access to the resource (Pitelka, 1959), and the result is polygyny. This mating system is favored when the distribution of resources is patchy in space, or in time, allowing the male to defend a number of females economically from other males. If the resources are uniformly and relatively thinly distributed, the females become too widely spaced for a male to sequester more than one at a time, favoring monogamy. However, if the individuals live in undefended home ranges the consequences can differ for social systems.

I leave aside thornier issues such as lekking, polyandry, and too many females available simultaneously at the same place. However, the effect of parental care needs mentioning. In mammals and insects the female is the principal care giver. In birds biparental care prevails. Theory has been shaped by this perception. Among fishes with parental care, however, the male is almost always the caretaker (Blumer, 1979) and thus a resource for the female. Should females then map onto the distribution of males? In most cases, probably not.

Caretaking males typically receive the eggs of more than one female, so males still compete for the eggs of females. Thus females remain the limiting sex (*sensu* Trivers, 1972; Baylis, 1981). Selection favors males who situate their nest sites in accordance with the distribution of females. When nest sites are not occupied continuously, food need not limit choice of nest sites. Male fishes typically do not feed while parenting. When care extends to swimming young, an unusual situation, food for the fry may play a role (McKaye and Barlow, 1976; Perrone, 1978). In this respect, the mating systems of freshwater fishes that care for eggs are generally unlike those of birds and mammals who regularly feed themselves and provision their young from the resources within their breeding territories.

This didactic prelude makes an essential point: A fertile starting place in the study of social systems is an examination of the pattern of spacing that exists *prior* to mating (Pitelka *et al.*, 1974). If the fish is associated with a complex substrate, that facilitates defending a feeding territory or a place of refuge (Lowe-McConnell, 1987). This is a profitable approach to understanding the social organization of coral-reef fishes that are closely associated with the reef and bottom-dwelling riverine species; I will focus mainly on these two fish assemblages, and emphasize the effect of feeding territories.

TERRITORY

Noble (1939) provided the original, terse, and often quoted definition of territory: any defended area. With the ascendancy of behavioral ecology, territory took on another definition: exclusive access to resources within a given space (Pitelka, 1959). Either meaning applies, depending on whether the focus is on proximate or ultimate causation. The proximate explanation turns on the behavioral mechanisms required for the appearance of a territory: first, the animal must stay within an area and, second, it must aggressively defend it. The ultimate definition depends on the consequences of the first: by defending the territory the owner benefits from proprietary access to the resources contained therein, be it food, refuge, breeding site or potential mates.

For territoriality to arise, the benefits of holding a territory must outweigh the costs (Brown, 1964; Brown and Orians, 1970; Davies and Houston, 1984). Thus in the case of a trout holding station in a stream, the current might be too strong and thus too energetically costly despite a rich supply of food. Presence or absence of territory, and its size, should therefore be set by the optimal trade-off between costs and benefits.

MARINE REEF FISHES

Feeding Territories and Mating Systems

Feeding territories are well known among marine fishes that are associated with the substrate, such as herbivores and coralivores (e.g. Ebersole, 1980; Hixon, 1980; Nursall, 1981; Fricke, 1986; Robertson and Gaines, 1986). Although less immediately apparent, feeding territories also occur among marine carnivores. For instance, in sandperches (Pinguipedidae) the territory of one male embraces the feeding territories of two or three females (Clark et al., 1991), as is the case in a reef grouper (Shpigel, 1985). Two species of the omnivorous sharpnose puffers, one in the Gulf of California (Kobayashi, 1986) and one in the Caribbean Sea (Sikkel, 1990), are territorial and haremic. Because feeding territories are so well known and accepted among benthic reef fishes, I elaborate on them in relation to mating systems in only a few cases.

Damselfish territoriality has probably received the most attention among reef fishes (Clarke, 1970; Ebersole, 1980; Hixon, 1980). Pomacentrids commonly farm algae to eat (Brawley and Adey, 1981) and live alone. However, some regularly form colonies of tightly packed feeding territories (pers. observ.). Females have their own territories, whether spaced or colonial, and spawn in the males' territories. Males care for more than one clutch of eggs, so the system is polygynous. If males could care for only one clutch of eggs, the predicted mating system would be monogamy or polyandry, depending on the time needed for the eggs to hatch and for the female to produce another clutch of eggs. The existing system raises the intriguing possibility that females might map onto the distribution of caretaking males, feeding territories permitting.

The social systems of butterflyfishes are best understood through their relationship to the reef (Reese, 1975). Although much remains to be explained, one monogamous species has been studied experimentally (Hourigan, 1989). Pairs of *Chaetodon multicinctus* occupy a territory, feeding on the coral polyps there and spawning planktonic eggs. The female cannot hold the territory alone, but the male can. The female benefits from the defense of the territory by the male, and the male is assured a fecund mate for spawning.

Life, however, is never simple. Another species, the chevron butterflyfish, *Chaetodon trifascialis*, has as its territory a patch of coral, of a different type, on which it feeds. In most respects, its feeding ecology and mode of reproduction are like that of the monogamous *C. multicinctus* that feeds on table *Acropora* (Reese, 1975). Nevertheless, the chevron butterflyfish is solitary, much like herbivorous damselfishes. One wonders whether the key lies in the difference in guarded coral species, such as the rate of renewal of polyps.

The longnose filefish, *Oxymonacanthus longirostris*, parallels the monogamous butterflyfishes but with instructive differences (Barlow *et al.*, unpublished data). The male and female jointly hold and defend a territory where they ingest the polyps of a branching coral. When they spawn they commonly wander out of the territory in search of a suitable tuft of algae to receive their eggs (Barlow, 1987). Thus the feeding territory may bind the pair but is not a necessity for spawning.

A study that nicely illustrates the Emlen-Oring model (1977) of males mapping onto territories of females has been provided by Gladstone (1987). Mature females of the small sharpnose puffer *Canthigaster valentini* occupy enduring exclusive territories on the coral reef where they feed on a variety of benthic organisms and aufwuchs. They spawn demersal eggs within their territories but provide no care. Defense of the food resource seems the critical factor driving the system. One dominant male holds a territory that encompasses those of one to five females.

That the male is drawn to the females, not the other way round, was revealed experimentally. When Gladstone (1987) removed one female puffer, the male stopped defending the vacant female's former territory. When Gladstone took away all the females, most of the effected territorial males left the area and were not seen again; apparently they went off in search of females.

Social systems within a given species of reef fish vary geographically, as Reese (1975) has shown for butterflyfishes. You have just read a synopsis of the haremic system of *C. valentini* on the Great Barrier Reef. However, *C. valentini* is monogamous on the island of Moorea in French Polynesia (Barlow and Barlow, unpublished data) The distribution of food probably clusters the females of this species on the Great Barrier Reef but spaces them on Moorea.

Some triggerfishes (Balistidae) and wrasses (Labridae), are patrilocal, apparently territorial, and live in harems (Fricke, 1980; Neudecker and Lobel, 1982; Tribble, 1982; Hoffman, 1983). They move about within a large territory in search of food. One male territory encompasses that of a number of females

Refuging Among Reef Fishes

Some of the better examples of the how mating systems relate to the spatial distribution of fishes come from the use of defended refuges rather than feeding territories. This is most clearly seen in plankton-feeding species, who form a nice parallel with drift-feeding salmonids (see below). Anemonefishes (*Amphiprion*) fit this pattern. Typically, a social group consists of an enduring monogamous pair that apparently remains with their anemone for most of their lives (Allen, 1972; Ochi, 1989). They defend and take cover in the anemone. The pair spawns under the umbrella of the anemone and the male cares for the eggs. Space is limiting, so other breeding adults are excluded.

The plankton-feeding damselfishes in *Dascyllus* hover in groups over branching coral, to which they flee for protection, and where males care for eggs. Depending on the size of the coral head and the number of fish occupying it, the system within a single species can vary from apparent monogamy to polygyny and most likely polygynandry (Fricke, 1980). Thus in these species, refuging sites, hence breeding sites are defended. Defending the feeding area, in open water, is not economically feasible.

Local Spawning Migrations

I have not mentioned one of the more common and notably different patterns of spawning among reef fishes. Indeed, it characterizes the conspicuous larger and mobile reef species: They migrate to the edge of the reef and spawn in groups (e.g. Randall, 1961; Johannes, 1978; Robertson *et al.*, 1979; Myrberg *et al.*, 1988; Colin and Bell, 1991). Some species set up temporary territories at the reef's margin for the sole purpose of spawning, a situation that resembles lekking (Loiselle and Barlow, 1978; Warner, 1988; Colin and Bell, 1991).

Most of these species are wandering feeders, not tied to a small area on the reef. Some, however, are territorial herbivores. Many are to some degree patrilocal (Robertson *et al.*, 1979; Myrberg *et al.*, 1988). Nonetheless, they depart to the reef's edge to spawn.

These examples demonstrate that the mating systems of a large segment of the species found on coral reefs do not derive directly from feeding territories.

Dispersal and the General Absence of Care of Fry Among Reef Fishes

Reef fishes and freshwater fishes differ in two fundamental ways. First, freshwater fishes, but not reef fishes, are able to disperse at all stages of their life history. Second, reef fishes almost never provide care for free-swimming young; several freshwater species in different families do. Why don't reef fish care for free-swimming offspring? After all, one species of damselfish does and it manages well (Robertson, 1973).

I have argued (Barlow, 1981) that coral-reef fishes, especially small species, cannot disperse as adults and must do so as numerous, tiny, pelagic propagules. Caring for fry results in offspring too large, and too adapted to reef life, to disperse across open water. Further, reproduction among coral-reef fishes is almost continuous. Individuals often spawn on consecutive days, when tides and currents are right, and almost year round (Colin and Bell, 1991). Apparently selection favors extraordinary fecundity to overcome the great loss of eggs and larvae during dispersal. This applies to marine fishes in temperate and boreal regions as well, though their reproduction is seasonally constrained.

FRESHWATER FISHES

Dispersal and Mobility

Freshwater fishes of the tropics diverge from the pattern that characterizes coral-reef fishes. Their dispersal is not limited to an early planktonic stage because of their riverine environment, the primary condition of freshwater fishes. Even fishes that spend most of their lives in lakes or marshes commonly return to rivers to reproduce (Whitehead, 1959; Corbet, 1961; Fryer and Iles, 1972; Goulding, 1980). Rivers and streams are continuous and do not usually present barriers to movement by fishes at any stage in their life history. Of course, waterfalls and rapids limit some species in certain rivers. Small species are deterred by major rivers or by deep water and open bottom in lakes. Freshwater fishes are often patrilocal (see below). However, in general, their dispersal is not constrained to any stage of development, with notable exceptions. That means they are relatively free to move about, sometimes over long distances. Spectacular spawning migrations range from the ocean-basin and up-river journeys of anadromous salmonids (Jonsson, 1991) to movements up or down-river of up to thousands of kilometers by catfishes and characoids (Lopez, 1978; Goulding, 1980; Lowe-McConnell, 1987). Nothing comparable is seen among coral-reef fishes, not even the spawning migrations of some of the large reef groupers (Shapiro, 1987). Moreover, the young stages make the return journey to the parental feeding grounds.

Unlike typical coral reefs, fresh waters experience pronounced seasonality, even in the tropics. Seasonal changes in rivers are characterized by rainy versus dry periods. Reproduction is characteristically confined to the most favorable season. Depending on the kind of fish and its mode of reproduction, that can mean the rainy season for some species and the dry season for others (Lowe-McConnell, 1987). In nonmigratory species, parents may increase their fitness by investing in care of swimming young with no cost to dispersal of those young. As the young become larger and behaviorally more competent they become ever better equipped to move through the environment. Prolonged parental investment, including biparental care of swimming young, is well known in teleosts such as the Cichlidae and also occurs in a number of 'lower' and subteleostan fishes (Barlow, 1984, 1986).

Feeding Territories

I have emphasized the mobility of riverine fishes, in contrast to marine reef inhabitants, to highlight the question of feeding territories among freshwater fishes. You may therefore be surprised to learn that territories have been reported for many freshwater fishes. Several of those reports, however, are of questionable value. In some instances, feeding territories are mentioned without explaining how the investigator came to that conclusion (e.g. MacLean, 1980).

Most often, claims of feeding territories are based on aquarium observations (Morris, 1958; Gerking, 1959; Lowe-McConnell, 1987). Food is predictably localized in aquaria. In my experience, fishes that are associated with the substrate regularly defend either the entire aquarium or a specific location where food is introduced. These fishes include many kinds of cichlids, characins, and cyprinids. When I observed some of the same cichlids in the field, they never held feeding territories. Loricariid catfishes appear to be territorial in captivity (Barlow, pers. obs.), but are not when studied in the field. However, they do space out in response to aggression (Power, 1984a). Thus observations of territorial behavior in captivity indicate only that the species has the capacity to be territorial.

In nature, fishes that feed off or around the bottom in complex environments often express the behavioral mechanisms needed for territoriality. For instance, when Midas cichlids in Nicaraguan lakes find a rich source of food, they stay with it and defend it. However, the Midas cichlids soon abandon their feeding places and move on, rejoining loose schools of conspecific fish (Barlow, 1976). That makes the question even more intriguing. If these fishes have the proximate mechanisms needed for territoriality, why do they seldom express it when feeding territories in the field are well known in one family? That family is the Salmonidae.

Territories among the salmonids are so common that the very knowledge of them may have led to the false impression that territoriality is a regular feature of the behavior of freshwater fishes. In the following, I document their territoriality, and also recount a rare report of feeding territories in two characins in Brazil. Then I describe the situation in a temperate-zone minnow and among some algal-scraping loricariid catfishes in the tropics; I do that to illustrate how territories fail to materialize where one would expect them to occur. I close by reporting on cichlid feeding territories which appear to have evolved from breeding territories; this is the reverse of what I seek in trying to predict mating systems from feeding ecology.

Salmonids

Trout fishermen have known for generations that trout and charr in streams hold exclusive stations from which they feed. Pioneering papers by Kalleberg (1958) and Keenleyside and Yamamoto (1962) made the scientific community aware of feeding territories among juvenile salmon, even though Kalleberg's studies were of captive fish. Investigations of salmonid feeding territories have increased so much in the last few decades that their territoriality has become almost paradigmatic. Recent essays (Noakes and Grant, 1986; Grant and Kramer, 1990) cover much of the relevant literature and center on the behavior of juveniles.

In general, salmonids hold tear-drop shaped territories (Grant *et al.*, 1989) with the blunt end facing up stream. They feed on invertebrate prey, called drift, wafting downstream. The territory is defended against conspecific and heterospecific salmonids. Across seven species, size of territory was proportional to size of fish, and the fitted regressions for the species had similar slopes and y-intercepts. That suggests an optimal trade-off between size of territory and cost of defense. That is also evidence for the minimum size of territory limiting population density. Grant and Kramer (1990: 1736) wrote, "We suspect that territoriality is not the mechanism that actually limits density, but rather that territory size predicts the spatial requirements of stream-dwelling salmonids, whether the space is defended or not."

Compared to terrestrial animals, the territories of juvenile salmonids are remarkably small (Grant and Kramer, 1990), probably because they stay put where the current brings food to them. Compared with lizards and birds, their territories are about four and five orders of magnitude smaller, respectively, on a scale of $\log_{10}$ (Grant and Kramer, 1990).

Not all individuals are territorial, however. They grade continuously, as assessed by aggressive behavior, from subordinate nonterritorial to dominant territorial (Grant, 1990). Puckett and Dill (1985) analyzed the energetics of holding feeding territories in juvenile coho salmon (*Oncorhynchus kisutch*). They categorized the juveniles into three classes, territorial, nonterritorial but localized, and floaters. Territorial and floater fish had similar feeding efficiencies, but that of the nonterritorial fish was less than one-half that of the others. Noakes and Grant (1986) listed a number of species and their social organization, demonstrating the general occurrence of alternative strategies, such as territoriality or no territory or schooling, depending on the economics of the situation.

The presence of alternative social systems within species provides an opportunity to relate the systems to ecological variables. Thus Grant and Noakes (1987) found that juvenile brook charr (*Salvelinus fontinalis*) that fed on drift were the fish that held feeding territories in flowing water. Those that fed more on benthos, or prey on the water surface, were fish that roamed about, often where the current was slow. Dominant fish held territory where the current was optimal and subordinate fish were relegated to roaming (Puckett and Dill, 1985; Noakes and Grant, 1986; Grant, 1990).

The most striking alternative social systems are territoriality and schooling, as seen in the ayu, *Plecoglossus altivelis* (Kawanabe, 1969). This species has noteworthy parallels with coral-reef fishes because it grazes algae from the substrate. At high population densities, many fish were excluded from feeding territories associated with the bottom; they then formed schools (for analogs in the marine environment, see Barlow, 1974; Foster, 1985). Nonetheless, schooling ayu grew as well as territorial ones.

Salmonids, then, seem a good family for relating mating systems to spatial patterns imposed by feeding behavior. Unfortunately, salmonids, do not spawn where they feed. They move to areas with suitable substrate for burying their eggs in highly oxygenated water. In the extreme case of the salmons, *sensu stricto*, feeding territories are held by juveniles in streams, but the adults mature in the open ocean, returning to natal streams to spawn (Jonsson, 1991). Thus the best known and well documented case of feeding territoriality among freshwater fishes is of no help relating mating system to feeding ecology.

Territorial characins in the Pantanal of Brazil

The Pantanal is a vast area in Western Brazil that floods during the rainy season. Sazima (1988) observed feeding territories in two species of characiform fishes in clear pools there that contained up to 60 species of fish. One, the characid *Catoprion mento*, is a lurking predator that darts out from cover to pluck scales from its victims. Individuals establish feeding territories centered on clumps of vegetation, the ambush sites. Territories persisted throughout the day, and were confirmed to endure at least a few days.

Another characiform fish, the curimatid *Curimata spilura*, is a microphagous herbivore (Sazima, 1988). It has two modes of feeding, in loose schools and as individual territory holders. The territorial fish focussed their feeding on a cluster of submerged, grass-like vegetation. Defense was more overtly aggressive than in *C. mento*, with fierce fighting that even featured jaw locking.

One would like to know how the mating systems of these characins relate to their feeding territories, especially when compared to near relatives that do not have feeding territories. Further, do feeding territories occur in both sexes? Do the fish spawn where they hold territories in the flooded Panatanal or in the streams and rivers during the dry season? If the latter, do they have feeding territories in the streams? If not, then feeding territories in the Pantanal would be irrelevant to the question of mating systems.

A temperate-zone minnow

At first glance, drawing a simplistic parallel from algal and coral-grazing reef fishes, candidates for feeding territoriality in temperate-zone streams ought to be species that graze algae. One such fish, the minnow *Campostoma anomalum*, has been studied in some detail (Power *et al.*, 1985; Power, 1987). It is not territorial over food, however, and its biology reveals why.

In Oklahoma this minnow and largemouth bass (*Micropterus salmoides*) and spotted bass (*M. punctulatus*) inhabit the same streams. The bass prey on the minnows. Minnows avoid bass by avoiding deeper pools where the bass dwell; this was confirmed experimentally. But when the minnows move into water less than about 15 cm deep they are susceptible to avian predators. In consequence, minnow schools have to stay on the move. Predators may prevent them from holding territories

Minnows, however, are not territorial in bass-free pools. This may be because they deplete the algae too quickly, or because their behavioral mechanisms are too attuned to the risk of predation. Further, freshets change the streams, e.g. the depth of pools, and both minnows and bass sometimes redistribute themselves as a consequence. That also works against territoriality.

Loricariid catfishes in Panama

The loricariids are another candidate for territoriality. They feed on periphyton (attached algae and aufwuchs) and can be observed in clear streams. They are more abundant in sunny pools because productivity is highest there. The catfishes remove the obvious epiphyton at depths greater than about 20 cm, though a thin but productive layer of diatoms remains to sustain them. The abundance of catfishes in relation to the meager food supply appears to make territorial defense uneconomic, but other explanations are possible, such as the effect of predators, especially in the shallows.

The epiphyton becomes more abundant in inverse proportion to water depth. However, the large catfishes do not venture into the shallows during the day, although a few do so at night. Even during the dry season, when the large catfishes stop growing and loose weight because food is scarce, they avoid the shallows. Small catfishes penetrate the shallows but they are too few to severely reduce the epiphyton there (Power, 1983, 1984a).

Large loricariid catfishes thus shun the shallows and consequently are concentrated in deeper water. When algal-covered rocks were moved from the shallows to deeper water, the encrusting algae was quickly consumed by the catfishes (Power, 1989). And when catfishes were removed from a pool, periphyton on deeper substrates increased. Some catfishes were experimentally starved but they still avoided the shallows. Finally, large catfishes were exposed to avian predators in enclosures of various depths. The two shallower pens were relentlessly depredated, but the two deeper pens experienced no losses. Predation in the shallows is so severe that large catfishes attempting to feed there are selected against. Even at night, nocturnal avian waders and mammals are deterrents to feeding in the shallows.

Small catfishes in the shallows are kept at low densities by small piscine predators, mostly characins. Consequently, the lush algae in shallow water remains an under-utilized resource, as in the streams in Oklahoma (Power, 1987, 1989).

Marking catfishes revealed that they are patrilocal (Power, 1984a). The larger individuals, however, sometimes shifted to other pools, their size giving them relative immunity from predacious fishes. Movement was evident when the pools changed, as when the falling of a tree opened the canopy to sun, or when a freshet covered the substrate with sediment.

The picture that emerges disfavors territories, despite aggressive sparring among the catfishes when feeding (Power, 1984a). Territories in deeper water are not economic because of severe competition for scarce food. The availability of food, further, is unpredictable in time and place, favoring exploratory behavior and mobility. A link between feeding territories and mating system is therefore out of the question.

On the other hand, refuges might be important. When Power (1987) added small logs to a bare pool, within two weeks the number of *Ancistrus spinosus* increased but the *Rineloricaria* moved out. *Rineloricaria* is more cryptic and therefore better adapted to living in exposed areas. *Ancistrus* needs hollow logs, or the likes, in which to nest. Such sites are in short supply and are guarded by males as breeding territories; the search for them results in further movement among pools (Moodie and Power, 1982).

When plastic tubes of the proper size and orientation were placed in the stream, breeding males, guarders of the eggs, soon took up residence in them and emitted stridulatory calls. Males accepted a number of clutches of eggs. Thus the distribution of breeding sites might be important to the mating system. A scarcity of sites favors polygyny while an abundance could promote monogamy if females prefer males without clutches, which is unlikely (Knapp and Sargent, 1989; Sikkel, 1989).

Cichlid fishes in African Rift Lakes

I mentioned above that cichlid fishes I observed in Central America do not have enduring feeding territories. This applies both to riverine and to lake dwelling species, and appears to be general among cichlids (e.g. Sjölander, 1972). I reasoned, nonetheless, that if feeding territories occur among cichlids, they would most likely arise in a geologically old lake and among herbivores. The obvious place to look was among the cichlids in the rift lakes of Africa.

Territory-based harems are now known for substrate-breeding cichlids in Lake Tanganyika (Kuwamura, 1986; Yanagisawa, 1987). In haremic lamprologines, a male has a large territory embracing those of a few females. Females of at least one species have been observed to breed continuously, producing another brood of young as soon as the last one reached independence. The females remained on territory (Yanagisawa, 1987), so they must feed there as well. However, feeding appears to have arisen as a secondary consequence of attachment to a breeding territory. The same conclusion appears to apply to the mouthbrooding species *Lamprologus moorii* in that lake, though the mating system is different (Yanagisawa and Mutsumi, 1991).

Ribbink *et al.* (1983), in their extensive monograph of the reef dwelling mbuna of Lake Malawi, described territoriality among many species. I focussed on the *Pseudotropheus* group of rock dwelling cichlids in their study because the information for them was the most complete.

Male breeding territories characterize haplochromines such as mbuna and is the ancestral condition. A displaying territorial male is approached by a female who spawns with him and then swims off with the fertilized eggs in her mouth.

In Lake Malawi (Ribbink *et al.*, 1983) males of several species also maintain feeding territories. From the pattern of occurrence among them, feeding territories must have evolved from breeding territories, not the other way round. In support of this, feeding territories are not found in the absence of breeding territories but breeding territories occur without feeding territories.

Further evidence comes from the distribution of breeding and feeding territories within mbuna in the genus *Pseudotropheus*. The typical pattern is seen in several species: Males maintain breeding territories but not feeding territories. In many species, however, male breeding territories are also feeding territories; this is especially likely when breeding occurs year round. Algal gardens develop, from which the males feed and which the males defend.

The degree to which feeding territoriality is expressed varies across species of mbuna (in some of these, the species status is unclear pending evaluation by systematists). Unusual species such as those in the group *P. williamsi* have only weakly defended breeding territories. More common are species that regularly hold breeding territories but manifest no algal gardens, as in several forms of *P. zebra*. In the next step, males of species such as *P. tropheops* 'chinyankwazi' have breeding territories, but only a few of them have algal gardens. "Progressing" further, in species such as *P. elongatus* 'bar' males regularly possess feeding territories. Another "advance" is represented by *P. tropheops* 'broadmouth' whose males have algal gardens; a few females do as well. Apparently many but not all females have feeding territories in *P. tursiops*. Finally, in *P. elongatus* 'aggressive' algal gardens are held by males, females, and by juveniles longer than 60mm SL.

Thus, starting with the condition of males having breeding territories but no feeding territories, territoriality has apparently gone in two directions. Some species of *Pseudotropheus* have given up territoriality or have it only weakly expressed; this characterizes the entire genus *Melanochromis* and its species complex and the aufwuchs feeding *Cyrtocara*. In the other direction, and apparently together with prolonged holding of breeding territories, feeding territories have arisen among males. Somehow, feeding territoriality then passed on to females in some species and even to larger juveniles in at least one.

Ribbink *et al.* (1983) and Ribbink (1991) even observed some territorial females to return to their feeding territories while carrying eggs in their mouths, even though females do little or no feeding while mouthbrooding. Ribbink *et al.* (1983) also commented that increasing aggressiveness appears to correlate with holding breeding territories that become feeding territories.

In this example, I make no inference about phylogeny, except to hold that the stem condition is male breeding, but not feeding, territoriality. A continuum exists across species, from nonfeeding to feeding territories, including females and juveniles. In each of these species, therefore, the behavior has current utility and is a plausible intermediate step in an evolutionary progression.

Across the many mbuna on the reefs, another relationship emerges (Ribbink *et al.*, 1983). The food habits of species affects their territoriality. Mobile species, e.g. those that roam about feeding on invertebrates (*Labidochromis*) or taking scales from other fishes (*Genyochromis*), are typically not territorial at all. The possibility therefore exists of relating social system to patterns of spacing as a result of feeding habits.

As Ribbink (1991: 52) concluded from a comparison of the cichlids in Lake Tanganyika with those in Lake Malawi, "Although the available data preclude generalizations, it appears that those fishes with large territories such as the maternal mouthbrooders are primarily herbivores, whereas those which depend upon small territories for an extended period, during which they raise several families, are primarily carnivores. If this is correct, then the greater abundance of invertebrate prey in Lake Tanganyika may have been instrumental in the development of communities of substratum spawners."

So, Why are Territories Infrequent Among Freshwater Fishes?

A source of motivation for writing this essay was my frustration with the literature on this issue. Many observers have failed to distinguish between feeding and breeding territories, or have assumed that the presence of feeding territories in aquaria means the same applies in the field. I want to stimulate field biologists to observe more critically, to try to disprove my claim that territoriality over food among freshwater fishes is generally selected against. I propose some hypotheses to falsify.

Freshwater fishes are territorial but unobserved

This is a real possibility. Most species are found in the tropics, where few behaviorally minded ichthyologists reside. Nevertheless, even the numerous ichthyologists who work on freshwater fishes in the temperate zones have been slow to recognize the value of entering the aquatic realm to observe fishes directly. "Until recently, underwater observations of lotic fishes were uncommon, although snorkeling and SCUBA diving have been routinely used to study coral reef and lentic fishes." (Greenberg and Holtzman, 1987: 22). Recall, however, that Sazima (1988) was able to watch 60 species in clear freshwater pools and saw feeding territories in only two species. And in one of these territoriality was an alternative strategy. Another possibility is that territories are common but impossible to see because fresh waters are often too turbid.

One of the strongest arguments for the possibility of nonbreeding territories among freshwater fishes is that they often have small home ranges. This was the main point of Gerking's (1953) review. He also mentioned the tendency of stream fishes to return to their home sites when displaced. Tenacity to site is to be expected especially among fishes closely associated with the substrate. Candidate fishes in North America include the darters (Percidae), sculpins (Cottidae) and some sunfishes (Centrarchidae) that take cover among rocks, debris, and undercut banks. However, even mobile species within the Cyprinidae and Catostomidae that are patrilocal might be territorial.

Greenberg and Holtzman (1987) marked banded sculpin (*Cottus carolinae*) in a variety of ways and then repeatedly entered the stream to observe them, even at night. They concluded that a home range was characteristically about 47 m², or a stream length of six m. They attributed missing fish to tag loss but those data could also be interpreted to mean that some sculpin were leaving the area.

Hill and Grossman (1987) did a similar study, tagging fish by injecting acrylic paint. They saw marked fish for up to 18 months. Two of the species were closely associated with the bottom, a sculpin (*Cottus bairdi*) and a dace (*Rhinichthys cataractae*); the third, a minnow (*Clinostomus funduloides*), swam in midwater. Their estimates of home range, which they suspect where high, were on the order of 10 to 20 m of stream for all three species. After a rain storm, only the range of the minnows increased, and by 50%.

Other studies have produced conflicting conclusions about the effect of storms, some showing no effect and others claiming heightened movements of fish (see Hill and Grossman, 1987). The analysis of the activity of loricariid catfishes in Panama (Power, 1984a) offers a partial resolution. The catfish were indeed patrilocal, but they explored up and downstream. As the stream changed, the catfish moved to optimize their intake of food. Power's (1984a) analysis of these movements by loricariids, in fact, may be the only persuasive demonstration in a natural system of the ideal-free distribution (Fretwell, 1972). Even when the catfish remained in a given pool they did not develop feeding territories. Staying put does not mean that individual necessarily hold territories.

It therefore seems likely that freshwater fishes in general, and especially those living in lotic environments, are seldom territorial. That forces us to ask why that should be.

Differences between freshwater and marine fishes are phylogenetically explained

Communities of freshwater fishes, especially in streams and rivers, are dominated by the Ostariophysi. Coral-reef fishes, in contrast, are predominantly percomorphs. Perhaps the bauplan of ostariophysan fishes does not suit them to holding feeding territories but that of the acanthopterygiian percomorphs does, irrespective of ecological differences. In support of this, Lowe-McConnell (1987: 284) wrote when comparing ostariophysan with acanthopterygian fishes that "Full advantage of cover can only be taken by fishes that have reached a certain stage of morphological evolution; in many acanthopterygians the pelvic fins have moved forward, making the fish maneuverable, and their dorsal fin spines, important fin strengtheners, contributing to maneuverability, also deter predators." Thus ostariophysan fishes, despite their prevalence in fresh waters, especially in lentic habitats, would seem morphologically unfit for the demands of holding a territory in a complex substrate.

The evidence speaks against the argument that ostariophysans are unable to hold territories. First, salmonids have much the same bauplan as cyprinids yet they defend feeding territories in complex stream situations. Admittedly, maintaining a territory to feed on drift presents different demands compared with, say, owning a patch of algal turf.

Second, many ostariophysan fishes are admirably adapted to living in intimate association with the substrate. Good examples are provided by numerous species of catfishes and gymnotoids, but also by some characoids, catastomids and even cyprinids (e.g. *Gyrnocheilus*). Gymnotoids probably defend refuge sites in nature but not feeding territories.

Third, even generalized ostariophysans are capable of holding territories associated with complex substrate. Cyprinids of the genus *Rhodeus* (e.g. Wunder, 1931) have breeding territories. And some minnows and characins regularly defend feeding territories in captivity, such as African cyprinids of the genus *Labio*.

Fourth, acanthopterygians that dwell in freshwater have rarely been reported to possess feeding territories, though that conclusion bears close examination, yet they often hold breeding territories. Cichlids with feeding territories, described in the foregoing, are exceptional because most cichlids, so far as known, do not hold such territories.

The freshwater fishes of Australia present a singular opportunity. The fauna lacks ostariophysans, save for a few catfishes, but is remarkably rich in atherinomorphs, percomorphs and a variety of other orders of fishes (Merrick and Schmida, 1984). I suspect that ecology will prove the important factor there, and that feeding territories in the streams and rivers will be few, despite their phylogenetic affinities with the reef assemblage of fishes.

Feeding territories in freshwaters are not economically defensible

The examples of salmonids in high-latitude streams, characins in pools and cichlids in stable tropical lakes proves that territoriality over food is possible in both lotic and lentic environments. The advances made on salmonid territoriality show that its expression is energetically based and serve as a prototype for the study of feeding territories in other fishes. To possess a territory, a salmonid must get enough food to make its defense worthwhile. That depends on current, abundance of food and number of intruders (Puckett and Dill, 1985; Noakes and Grant, 1986). In addition, those factors that make territoriality feasible must have reasonable stability in space and time for the territory to persist. Finally, even when territoriality is economically viable, the risk of predation can be so high that territoriality is precluded (Power, 1987).

Using one taxon of fishes to make the general case is fraught with danger. Despite this, I turn to the loricariid catfishes in Panama (Power, 1983, 1984a) to set up a scenario for others to refute. Territoriality in those fishes seems precluded by several factors. First, the population density is high relative to the resource, epiphyton; that makes defense too costly. Second, and in conjunction with the high pressure on the food base, abundance of food shifts from pool to pool. Not only is food removed by competitors, it may be covered with sediment after a freshet (Power, 1984b). That has more than one consequence.

Sediment blanketing the bottom may cause the catfishes to move to another pool. Large catfish sometimes clear an area of sediment, in which case small catfishes move close to them to feed (Power, 1984b). This is a special case of a locally rich patch attracting competitors, a subtle effect on social organization.

Power (1987, pers. comm.) is especially impressed by the impact of predators on the distribution of grazing stream fishes. But it remains unclear whether predators are important in the lack of territoriality. If all predators were removed from a section of stream in Panama, or for that matter in Oklahoma, I doubt the resident herbivores would become territorial. Rather, I speculate that the local population would simply concentrate in the shallows and strip the lush algae there. Territories should be economically too costly because of excess competitors. The territory would soon be depleted and not worth defending, or the food contained in it might be eliminated by physical disturbance.

CLOSING COMMENTS

Territoriality probably seldom occurs among freshwater fishes because it is not economic. This may be exacerbated by the unstable distribution of food in space and time. In comparison, coral reefs are more stable, though at first glance they might seem not. Cyclonic storms devastate large areas of reef (Harris *et al.,* 1984). Storms can damage reef structure inside an atoll lagoon at depths to 30 meters (pers. observ. at Enewetok Atoll; see also Bak and Luckhurst, 1980). I have argued (Barlow, 1981) that selection has favored dispersal among coral-reef fishes because of the threat of local extinction from such disturbances.

The temporal and spatial scales of those disturbances, nevertheless, differ from rivers and streams. A given place on a coral reef may not experience the life threatening degradation of a cyclonic storm or collapsing reef face within a span of time corresponding to many generations of piscine inhabitants. Other disturbances on reefs, such as habitat alteration by algae, corals, urchins or echinoderms, are relatively slow and local albeit sometimes profound (Bak and Luckhurst, 1980; Kaufman, 1981; Harris et al., 1984). Individual fish, including the smallest species, can usually adjust to such changes.

In contrast, severe disturbance of habitat is a constant fact of life for lotic fishes and in some water systems is regularly extreme (e.g. Hall et al., 1991). An individual may be subjected to freshets and floods many times within its life-span. Further, the physical aspects of freshwater habitats, unlike marine ones, are profoundly influenced by local conditions because of the relatively small volume of water. Consequently, "freshwater bodies are likely to be more heterogenous with regard to purely physical and chemical factors" (Baylis, 1981: 233).

One reservation in this comparison is that the shallower the reef the more subject it is to disturbance (Bak and Luckhurst, 1980; Baylis, 1981). Fishes living in littoral and sublittoral zones are regularly subjected to pounding surf whose waves may have originated thousands of kilometers away. For that reason, fishes adapted to those zones, such as blennies, could provide fruitful comparison with stream fishes. They should be less territorial than sublittoral reef fishes. However, they are so small that they utilize the substrate differently than do large reef fishes; in particular, they brace themselves in the interstices of the rocky substrate to withstand the turbulence.

If lakes differ from streams and rivers, and in the direction of coral reefs, feeding territories should be more common in them. Lakes, it might be argued, have a short life and they do not persist long enough in the history of a species to allow sufficient specialization for the development of territoriality. Against that, fishes have radiated in lakes and manifest profound morphological adaptations (Tchernavin, 1944; Fryer and Iles, 1972), and not all of these lakes are geologically old (Myers, 1960). In addition, many lotic fishes have the behavioral mechanisms needed for territoriality. Therefore, I doubt the short life of lakes prevents feeding territoriality.

To return to the starting point, the paucity of feeding territories among freshwater fishes results in a parallel paucity of mating systems based on feeding territories. Lekking systems are found on both reefs and in fresh water, but they are independent of feeding territories (Loiselle and Barlow, 1978; McKaye, 1983). Monogamous pairs and harems are relatively common on the reef, tied to feeding territories and refuges. Monogamous freshwater species, in so far as known, become territorial only to breed, and then to protect swimming young (Barlow, 1984, 1986). Although the breeding territory may be important for the feeding by the young, it plays no role for the parents (Perrone, 1978).

Harems have been suggested for South American cichlids (Loiselle, 1985; Barlow, 1991), based on aquarium observations. Those harems most likely arose in response to female breeding, not feeding, territories because nonbreeding males and females in captivity are not territorial. Thus mating systems among these fishes may result from females distributing themselves in relation to favorable breeding sites, and males mapping onto that distribution. The haremic systems in some lamprologine cichlids in Lake Tanganyika (Kuwamura, 1986; Yanagisawa, 1987) are good examples. The absence of mating systems derived from feeding territories among freshwater fishes is a notable difference from marine fishes. Why this should be so remains a mystery, given that freshwater cichlids can evolve feeding territories from breeding territories. The mystery might disappear if we knew more about feeding territories among freshwater fishes and how they relate to mating systems. Closer examination of freshwater species is urgently needed, especially in the tropics. With a robust empirical base, more fruitful comparisons could be made and general relationships deduced. I would be delighted if this essay inspires investigators to prove that feeding territories are abundant in tropical fresh waters and play an important role in the mating systems of those fishes.

ACKNOWLEDGMENTS

I am grateful to W. A. Grant, D. L. G. Noakes, F. A. Pitelka, and M. E. Power for their helpful suggestions for improving the essay. I also thank Felicity Huntingford and Patrizia Torrecilli for inviting me to Erice.

References

Allen, G. R. (1972). *Anemonefishes: Their Classification and Biology.* Neptune City: TFH Publications.
Bak, R. P. M. and Luckhurst, E. (1980). Constancy and change in coral reef habitats along depth gradients at Curacao. *Oecologia (Berlin)*, **47**, 145–155.
Barlow, G. W. (1974). Contrasts in social behavior between Central American cichlid fishes and coral-reef surgeon fishes. *Amer. Zool.*, **14**, 9–34.
Barlow, G. W. (1976). The Midas cichlid in Nicaragua. In: *Investigations of the Ichthyofauna of Nicaraguan Lakes* (Ed. by T. B. Thorson), pp. 333–358. Lincoln: School of Life Sciences, University of Nebraska, Lincoln.
Barlow, G. W. (1981). Patterns of parental investment, dispersal and size among coral-reef fishes. *Env. Biol. Fish.*, **6**, 65–85.
Barlow, G. W. (1984). Patterns of monogamy among teleost fishes. *Arch. FischWiss.*, **35** (Beiheft 1), 75–123.
Barlow, G. W. (1986). A comparison of monogamy among freshwater and coral-reef fishes. *Proc. Second Intern. Conf. Indo-Pac. Fish.*, pp. 767–775.
Barlow, G. W. (1987). Spawning, eggs and larvae of the longnose filefish *Oxymonacanthus longirostris*, a monogamous coralivore. *Env. Biol. Fish.*, **20**, 183–194.
Barlow, G. W. (1991). Mating systems among cichlid fishes. In: *Cichlid Fishes. Behaviour, Ecology and Evolution* (Ed. by M. H. A. Keenleyside), pp. 173–190. New York: Chapman and Hall.
Baylis, J. R. (1981). The evolution of parental care in fishes, with reference to Darwin's rule of male sexual selection. *Env. Biol. Fish.*, **6**, 223–251.
Blumer, L. S. (1979). Male parental care in the bony fishes. *Quart. Rev. Biol.*, **54**, 149–161.
Brawley, S. H. and Adey, W. H. (1981). The effect of micrograzers on algal community structure in a coral reef microcosm. *Mar. Biol.*, **61**, 167–177.
Brown, J. L. (1964). The evolution of diversity in avian territorial systems. *Wilson Bull.*, **76**, 160–169.
Brown, J. L. and Orians, G. H. (1970). Spacing patterns in mobile animals. *Ann. Rev. Ecol. Syst.*, **1**, 239–262.
Clark, E., Pohle, M. and Rabin, J. (1991). Stability and flexibility through community dynamics of the spotted sandperch. *Nat. Geogr. Res. Explor.*, **7**, 138–155.
Clarke, T. A. (1970). Territorial behavior and population dynamics of a pomacentrid fish, the garibaldi, *Hypsypops rubicunda*. *Ecol. Monogr.*, **40**, 189–212.

Colin, P. L. and Bell, L. J. (1991). Aspects of the spawning of labrid and scarid fishes (Pisces: Labroidei) at Enewetak Atoll, Marshall Islands with notes on other families. *Env. Biol. Fish.*, **31**, 229–260.

Corbet, P. S. (1961). The food of non-cichlid fishes in the Lake Victoria basin, with remarks on their evolution and adaptation to lacustrine conditions. *Proc. Zool. Soc. Lond.*, **136**, 1–101.

Crook, J. H. (1964). The evolution of social organisation and visual communication in the weaver birds (Ploceinae). *Behaviour Suppl.*, **10**, 1–178.

Davies, N. B. (1991). Mating systems. In: *Behavioural Ecology, An Evolutionary Approach*, 3rd edn (Ed. by J. R. Krebs and N. B. Davies), pp. 263–294. Boston: Blackwell.

Davies, N. B. and Houston, A. I. (1984). Territory economics. In: *Behavioural Ecology: An Evolutionary Approach* (Ed. by J. R. Krebs and N. B. Davies), pp. 148–169. Oxford: Blackwell.

Ebersole, J. P. (1980). Food density and territory size: An alternative model and a test on the reef fish *Eupomacentrus leucostictus*. *Amer. Nat.*, **115**, 492–509.

Emlen, S. T. and Oring, L. W. (1977). Ecology: Sexual selection and the evolution of mating systems. *Science*, **197**, 215–223.

Farr, J. A. (1975). The role of predation in the evolution of social behavior of natural populations of the guppy, *Poecilia reticulata* (Pisces: Poeciliidae). *Evolution*, **29**, 151–158.

Foster, S. A. (1985). Size-dependent territory defense by a damselfish. A determinant of resource use by group-foraging surgeonfishes. *Oecologia (Berlin)*, **67**, 499–505.

Fretwell, S. D. (1972). *Populations in a Seasonal Environment*. Princeton: Princeton University Press.

Fricke, H. W. (1980). Control of different mating systems in a coral reef fish by one environmental factor. *Anim. Behav.*, **28**, 561–569.

Fricke, H. W. (1986). Individuelles Erkennen bei dem Anemonenfisch *Amphiprion bicinctus*. *Publ. Wissensch. Film. Biol.*, **18** (10), 3–7.

Fryer, G. and Iles, T. D. (1972). *The Cichlid Fishes of the Great Lakes of Africa. Their Biology and Evolution*. Edinburgh: Oliver and Boyd.

Gerking, S. D. (1950). Stability of a stream fish population. *J. Wildl. Mgmt*, **14**, 194–202.

Gerking, S. D. (1953). Evidence for the concepts of home range and territory in stream fishes. *Ecology*, **34**, 347–365.

Gerking, S. D. (1959). The restricted movement of fish populations. *Biol. Rev.*, **34**, 221–242.

Gladstone, W. (1987). Role of female territoriality in social and mating systems of *Canthigaster valentini* (Pisces: Tetraodontidae): Evidence from field experiments. *Mar. Biol.*, **96**, 185–191.

Goulding, M. (1980). *The Fishes and the Forest. Explorations in Amazonian Ecology*. Berkeley: University of California Press.

Grant, J. W. A. (1990). Aggressiveness and the foraging behaviour of young-of-the-year brook charr (*Salvelinus fontinalis*). *Can. J. Fish. Aquat. Sci.*, **47**, 915–920.

Grant, J. W. A. and Kramer, D. L. (1990). Territory size as a predictor of the upper limit to population density of juvenile salmonids in streams. *Can. J. Fish. Aquat. Sci.*, **47**, 1724–1737.

Grant, J. W. A. and Noakes, D. L. G. (1987). Movers and stayers: Foraging tactics of young-of-the year brook charr, *Salvelinus fontinalis*. *J. Anim. Ecol.*, **56**, 1001–1013.

Grant, J. W. A., Noakes, D. L. G. and Jones, K. M. (1989). Spatial distribution of defence and foraging in young-of-the-year brook charr, *Salvelinus fontinalis*. *J. Anim. Ecol.*, **58**, 773–784.

Greenberg, L. A. and Holtzman, D. A. (1987). Microhabitat utilization, feeding periodicity, home range and population size of the banded sculpin, *Cottus carolinae*. *Copeia*, 19–24.

Gross, M. R. (1981). Predation and the evolution of colonial nesting in bluegill sunfish (*Lepomis macrochirus*). *Beh. Ecol. Sociobiol.*, **8**, 163–174.

Gross, M. R. and MacMillan, A. M. (1981). Predation and the evolution of colonial nesting in bluegill sunfish (*Lepomis macrochirus*). *Beh. Ecol. Sociobiol.*, **8**, 167–174.

Hall, J. W., Smith, T. I. J. and Lamprecht, S. D. (1991). Movements and habitats of shortnose sturgeon, *Acipenser brevirostrum* in the Savannah River. *Copeia*, 695–702.

Harris, L. G., Ebeling, A. W., Laur, D. R. and Rowley, R. J. (1984). Community recovery after storm damage: A case of facilitation in primary succession. *Science*, **224**, 1336–1338.

Helfman, G. S. (1981). Twilight activities and temporal structure in a freshwater fish community. *Can. J. Fish. Aquat. Sci.*, **38**, 1405–1420.

Hill, J. and Grossman, G. D. (1987). Home range estimates for three North American stream fishes. *Copeia*, 376–379.

Hixon, M. A. (1980). Food production and competitor density as the determinants of feeding territory size. *Amer. Nat.*, **115**, 510–530.

Hobson, E. S. (1975). Feeding patterns among tropical reef fishes. *Amer. Sci.*, **63**, 382–392.

Hoffman, S. G. (1983). Sex-related foraging behavior in sequentially hermaphroditic hogfishes (*Bodianus* spp.). *Ecology*, **64**, 798–808.

Hourigan, T. F. (1989). Environmental determinants of butterflyfish social systems. *Env. Biol. Fish.*, **25**, 61–78.

Jarman, P. J. (1974). The social organisation of antelope in relation to their ecology. *Behaviour*, **48**, 215–267.

Johannes, R. E. (1978). Reproductive strategies of coastal marine fishes in the tropics. *Env. Biol. Fish.*, **3**, 65–84.

Jonsson, N. (1991). Influence of water flow, water temperature and light on fish migration in rivers. *Nordic. J. Freshw. Res.*, **66**, 20–35.

Kalleberg, H. (1958). Observations in a stream tank of territoriality and competition in juvenile salmon and trout (*Salmo salar* L. and *S. trutta* L.). *Inst. Freshw. Res. Drottningholm*, **(39)**, 1–98.

Kaufman, L. (1981). There was biological disturbance on Pleistocene coral reefs. *Paleobiology*, **7**, 527–532.

Kawanabe, H. (1969). The significance of social structure in the production of the "ayu", *Plecoglossus altivelis*, In: *Symposium on Salmon and Trout in Streams* (Ed. by T. G. Northcote), pp. 243–251. Vancouver: University of British Columbia, Fish. Res. Inst.

Keenleyside, M. H. A. and Yamamoto, F. T. (1962). Territorial behavior of juvenile Atlantic salmon (*Salmo salar* L.). *Behaviour*, **19**, 138–169.

Knapp, R. A. and Sargent, R. C. (1989). Egg-mimicry as a mating strategy in the fantail darter, *Etheostoma flabellare*: Females prefer males with eggs. *Beh. Ecol. Sociobiol.*, **25**, 321–326.

Kobayashi, D. R. (1986). Social organization of the spotted sharpnose puffer, *Canthigaster punctissima* (Tetraodontidae). *Env. Biol. Fish.*, **15**, 141–145.

Kuwamura, T. (1986). Parental care and mating systems of cichlid fishes in Lake Tanganyika: A preliminary field survey. *J. Ethol.*, **4**, 129–146.

Lack, D. (1954). *The Natural Regulation of Animal Numbers*. Oxford: Oxford University Press.

Lack, D. (1968). *Ecological Adaptations for Breeding in Birds*. London: Methuen.

Loiselle, P. V. (1985). *The Cichlid Aquarium*. Melle: Tetra-Press.

Loiselle, P. V. and Barlow, G. W. (1978). Do fishes lek like birds? In: *Contrasts in Behavior* (Ed. by E. S. Reese and F. J. Lighter), pp. 33–75. New York: Wiley.

Lopez, S. M. I. (1978). Migracion de la sardina *Astyanax fasciatus* (Characidae) en el rio Tempisque, Guanacaste, Costa Rica. *Rev. Biol. Trop.*, **26**, 261–275.

Lowe-McConnell, R. H. (1987). *Ecological Studies in Tropical Fish Communities*. New York: Cambridge University Press.

MacLean, J. (1980). Ecological genetics of threespine sticklebacks in Heisholt Lake. *Can. J. Zool.*, **58**, 2026–2039.

McKaye, K. R. (1983). Ecology and breeding behavior of a cichlid fish, *Cyrtocara eucinostomus*, on a large lek in Lake Malawi. *Env. Biol. Fish.*, **8**, 81–96.

McKaye, K. R. and Barlow, G. W. (1976). Competition between color morphs of the Midas cichlid, *Cichlasoma citrinellum*, in Lake Jiloà, Nicaragua. In: *Investigations of the*

Ichthyofauna of Nicaraguan Lakes (Ed. by T. B. Thorson), pp. 465–475. Lincoln: School of Life Sciences, University of Nebraska.

Merrick, J. R. and Schmida, G. E. (1984). *Australian Freshwater Fishes: Biology and Management*. Netley: Griffen.

Moodie, G. E. E. and Power, M. (1982). The reproductive biology of an armoured catfish, *Loricaria uracantha*, from Central America. *Env. Biol. Fish.*, **7**, 143–148.

Morris, D. (1958). The reproductive behaviour of the ten-spined stickleback (*Pygosteus pungitius* L). *Behaviour Supp.*, **6**, 1–154.

Myers, G. S. (1960). The endemic fish fauna of Lake Lanao, and the evolution of higher taxonomic categories. *Evolution*, **14**, 323–333.

Myrberg, A. A. Jr., Montgomery, W. L. and Fishelson, L. (1988). The reproductive behavior of *Acanthurus nigrofuscus* (Forskal) and other surgeonfishes (Fam. Acanthuridae) off Eilat, Israel (Gulf of Aqaba, Red Sea). *Ethology*, **79**, 31–61.

Neudecker, S. and Lobel, P. S. (1982). Mating systems of chaetodontid and pomacanthid fishes at St. Croix. *Z. Tierpsychol.*, **59**, 299–318.

Noakes, D. L. G. and Grant, J. W. A. (1986). Behavioural ecology and production of riverine fishes. *Pol. Arch. Hydrobiol.*, **33**, 249–262.

Nursall, J. R. (1981). The activity budget and use of territory by a tropical blenniid fish. *Zool. J. Linn. Soc.*, **72**, 69–92.

Ochi, H. (1989). Mating behavior and sex change of the anemonefish, *Amphiprion clarkii*, in the temperate waters of southern Japan. *Env. Biol. Fish.*, **26**, 257–275.

Perrone, M. (1978). Mate size and breeding success in a monogamous cichlid fish. *Env. Biol. Fish.*, **3**, 193–201.

Pitelka, F. A. (1959). Numbers, breeding schedule, and territoriality in pectoral sandpipers of northern Alaska. *Condor*, **61**, 233–264.

Pitelka, F. A., Holmes, R. T. and MacLean, S. F. Jr. (1974). Ecology and evolution of social organization in Arctic sandpipers. *Amer. Zool.*, **14**, 185–204.

Power, M. E. (1984a). Depth distributions of armored catfish: Predator-induced resource avoidance? *Ecology*, **65**, 523–528.

Power, M. E. (1984b). The importance of sediment in the grazing ecology and size class interactions of an armored catfish, *Ancistrus spinosus*. *Env. Biol. Fish.*, **10**, 173–181.

Power, M. E. (1987). Predator avoidance by grazing fishes in temperate and tropical streams: Importance of stream depth and prey size. In: *Predation* (Ed. by W. C. Kerfoot and A. Sih), pp. 333–351. Hanover: University Press of New England.

Power, M. E. and Matthews, W. J. (1983). Algae-grazing minnows (*Campostoma anomalum*), piscivorous bass (*Micropterus* spp.), and the distribution of attached algae in a small prairie-margin stream. *Oecologia (Berlin)*, **60**, 328–332.

Power, M. E., Dudley, T. L. and Cooper, S. D. (1989). Grazing catfish, fishing birds, and attached algae in a Panamanian stream. *Env. Biol. Fish.*, **26**, 285–294.

Power, M. E., Matthews, W. J. and Stewart, A. J. (1985). Grazing minnows, piscivorous bass, and stream algae: Dynamics of a strong interaction. *Ecology*, **66**, 1448–1456.

Puckett, K. J. and Dill, L. M. (1985). The energetics of feeding territoriality of juvenile coho salmon (*Oncorhynchus kisutch*). *Behaviour*, **42**, 97–111.

Randall, J. E. (1961). A contribution to the biology of the convict surgeonfish of the Hawaiian Islands, *Acanthurus triostegus sandvicensis*. *Pacif. Sci.*, **15**, 215–272.

Reese, E. S. (1975). A comparative field study of the social behavior and related ecology of reef fishes of the family Chaetodontidae. *Z. Tierpsychol.*, **37**, 37–61.

Ribbink, A. J. (1991). Distribution and ecology of the cichlids of the African Great Lakes. In: *Cichlid Fishes: Behaviour, Ecology and Evolution* (Ed. by M. H. A. Keenleyside), pp. 36–59. New York: Chapman and Hall.

Ribbink, A. J., Marsh, B. A., Marsh, A. C., Ribbink, A. C. and Sharp, B. J. (1983). A preliminary survey of the cichlid fishes of rocky habitats in Lake Malawi. *S. Afr. J. Zool.*, **18**, 149–310.

Robertson, D. R. (1973). Field observations on the reproductive behaviour of a pomacentrid fish, *Acanthochromis polyacanthus*. *Z. Tierpsychol.*, **32**, 319–324.

Robertson, D. R. and Gaines, S. D. (1986). Interference competition structures habitat use in a local assemblage of coral reef surgeonfishes. *Ecology*, **67**, 1372–1383.

Robertson, D. R., Polunin, N. V. C. and Leighton, K. (1979). The behavioral ecology of three Indian Ocean surgeonfishes (*Acanthurus lineatus, A. leucosternon* and *Zebrasoma scopas*): Their feeding strategies, and social and mating systems. *Env. Biol. Fish.*, **4**, 125–170.

Sazima, I. (1986). Similarities in feeding behaviour between some marine and freshwater fishes in two tropical communities. *J. Fish. Biol.*, **29**, 53–65.

Sazima, I. (1988). Territorial behaviour in a scale-eating and a herbivorous Neotropical characiform fish. *Rev. Brasil. Biol.*, **48**, 189–194.

Shapiro, D. Y. (1987). Reproduction in groupers. In: *Tropical Snappers and Groupers* (Ed. by J. J. Polvina and S. Ralston), pp. 295–327. Boulder: Westview Press.

Shpigel, M. (1985). Aspects of the biology and ecology of Red Sea grouper *Cephalopholis* (Serranidae), Teleostei. Ph.D. dissertation. Tel-Aviv University.

Sikkel, P. C. (1989). Egg presence and developmental stage influence spawning-site choice by female garibaldi. *Anim. Behav.*, **38**, 447–456.

Sikkel, P. C. (1990). Social organization and spawning in the Atlantic sharpnose puffer, *Canthigaster rostrata* (Tetraodontidae). *Env. Biol. Fish.*, **27**, 243–254.

Sjölander, S. (1972). Feldbeogachtungen an einigen westafrikanischen Cichliden. *Aquar. Terr.*, **19** (2), 42–45, **19** (3), 86–88, **19** (4), 116–118.

Tchernavin, V. (1944). A revision of the subfamily Orestiinae. *Proc. Zool. Soc. Lond.*, **114**, 140–233.

Thresher, R. E. (1977). Ecological determinants of social organization of reef fishes. *Proc. Third Intern. Coral Reef Symp.*, pp. 551–559.

Tribble, G. W. (1982). Social organization, patterns of sexuality, and behavior of the wrasse *Coris dorsomaculata* at Miyake-jima, Japan. *Env. Biol. Fish.*, **7**, 29–38.

Trivers, R. L. (1972). Parental investment and sexual selection In: *Sexual Selection and the Descent of Man* (Ed. by B. Campbell), pp. 136–179. Chicago: Aldine.

Warner, R. R. (1988). Traditionality of mating-site preferences in a coral reef fish. *Nature*, **335**, 719–721.

Whitehead, P. J. P. (1959). The anadromous fishes of Lake Victoria. *Rev. Zool. Bot. Afr.*, **59**, 329–363.

Wunder, W. (1931). Brutplfege und Nest bau bei Fischen. *Ergeb. Biol.*, **7**, 118–192.

Yanagisawa, Y. (1987). Social organization of a polygynous cichlid *Lamprologus furcifer* in Lake Tanganyika. *Jap. J. Ichthyol.*, **34**, 82–90.

Yanagisawa, Y. and Nishida, M. (1991). The social and mating system of the maternal mouthbrooder *Tropheus moorii* (Cichlidae) in Lake Tanganyika. *Jap. J. Ichthyol.*, **38**, 271–282.

KNOWLEDGE OF PROXIMATE CAUSES AIDS OUR UNDERSTANDING OF FUNCTION AND EVOLUTIONARY HISTORY

GEORGE S. LOSEY, JR.

Department of Zoology and Hawaii Institute of Marine Biology, University of Hawaii, Honolulu, HI 96822, USA

It is important for animal behaviorists to keep distinct ideas about proximate causation and the ultimate causes or adaptive significance of actions. However, study of proximate causes can offer clues as to the evolutionary history and adaptive nature of behavior. Three case histories are presented: (i) Do individuals learn how and when to display? Evidence suggests that many species do have "open" developmental programs that allow individuals to mold their aggressive actions to best fit their environment. But work with the head down display of the threespined stickleback indicates strong constraints on any changes in its use. These constraints give clues as to the multiple functions of this aggressive act. (ii) Do interspecifically aggressive species learn "who" to attack? The Hawaiian damselfish uses ecological cues to learn new species of apparent food competitors and excludes them selectively from its territory. This demonstration is strong evidence that such interspecific aggression has been selected for and is not merely a case of mistaken identity. (iii) Do host fishes for cleaning symbiosis have special responses for responding to cleaners? Hosts in Hawaii lack any causal systems that would be expected if their behavior had been selected for in a co-evolved mutualism. Hosts probably learn to respond to cleaners as sources of tactile reward. Cleaning has likely evolved through specialization of cleaners to "parasitize" existing tactile reward systems in their hosts.

LEVELS OF STUDY IN ANIMAL BEHAVIOR

In our study of animal behavior, it is a well known lesson that we must keep hypotheses and explanations separate for proximate and ultimate causation, or "How vs. Why" questions about behavior (Alcock, 1984). For example, just because it may be adaptive for a fighter to submit *because* its opponent has superior fighting ability does not necessarily mean that the fighter will base its decision on an accurate assessment of the opponent's prowess. The ultimate causes of submission must be studied separate from its physiological basis.

Stamps (1991) discusses the use of indirect cues in various assessment processes: For optimal foraging, individuals should attend to various cues that indicate the value of a food patch relative to others. However, simple hedonistic perception (Bindra, 1978) might suffice to govern intake of valuable but (at least historically) scarce food commodities such as sugar and fat (Drewnowski *et al.*, 1989). The hedonistic option was painfully evident during the Erice conference as we responded in a less-than-fit manner to the gustatory delights of the food and wine of Sicily.

It is just as critical to keep ideas regarding the current function of a behavior distinct from hypotheses as to its evolutionary history. Kroodsma and Byers (1991) wisely caution that "Moving too easily between levels of explanation, often as if they were equivalent, confuses both the investigator and reader." For example, to claim that fishes visit cleaning organisms "to" have their parasites removed as a co-evolved reciprocal altruism confuses all three levels of explanation: the proximate causes, current functions and evolutionary history. This example will be examined in more detail later.

Unfortunately, the separation of levels of research can result in our having a thorough analysis of the ultimate causes of behavior, its adaptive significance, with little information regarding its proximate causes. Others may have much knowledge of the physiological causes of behavior with only guesses as to its adaptive value. More dangerously, the proximate causal systems that produce a behavior might be inferred and assumed to exist because of functional arguments or *vice-versa*. All of these conditions impede our ability to develop rigorous evolutionary inferences.

Stamps (1991) gives a broad review of cases where evolutionary problems have led to questions about proximate causation. She sees a growing coalition between students of proximate and ultimate causation. In some cases, such as optimal foraging, study of behavioral mechanisms can help to explain why some of the predictions of optimality models are not upheld in practice. In other cases, knowledge of behavior was previously thought to have little relevance. But now even students of the evolution of morphological traits must attend to the effects of maternal and paternal care on development and survival of their young.

Barlow (1989) gave a persuasive argument that the recent surge in interest in sociobiology should serve to revitalize ethological studies of proximate causation. He pictured proximate mechanisms as setting the boundaries for ultimate explanations. An individual simply cannot realize the ultimate advantages of a behavior if it lacks the proximate systems required for its implementation. He also gave several examples of how physiological and behavioral findings can open doors to new areas of research and give meaning to formerly obscure behavior.

In this paper, I will review how study of proximate causation for its own sake has led to a better understanding of the adaptive nature of behavior. Study of the mechanisms of behavior in fishes has frequently given me new information of value in interpreting their behavioral ecology and formulating hypotheses as to their evolution. I echo the views of Barlow (1989), Halpin (1991) and Stamps (1991): We must foster an increase in the study of proximate causes from its current state of neglect if we have hopes of truly understanding behavior. Huntingford, Csani and Galis in this volume give examples of how experience can shape predator responses and even the morphology of the feeding apparatus in fishes.

EXPERIENCE AND INTRASPECIFIC AGGRESSIVE BEHAVIOR

Display behavior has long fascinated ethologists. Social displays provide us with a modal unit of behavior that we can measure, that is more or less species specific and constant, and that appears to have heritable components that have evolved for the purpose of communication (Barlow, 1989). My first exposure to theories of the proximate and ultimate causes of displays in reading Tinbergen (1950) led to a lasting fascination with animal communication and adaptations that revealed an individual's motivational state. Such concepts were central to the birth of ethology in general (Barlow, 1989).

Later, the new applications of game theory and the "selfish gene" led us to question whether displays are accurate indicators of mood or future behavior of the individual (Dawkins and Krebs, 1978). Hawks, doves, retaliators, etc. might all exist depending on various costs and benefits of aggressive display (Maynard Smith, 1982). They might exist as separate genotypes in the populations or form a conditional strategy based upon individual experiences. Since the introduction of these ideas, a lively interchange has ensued as to the conditions that would favor truthful vs. deceptive communication (e.g. see Bond, 1989a,b). Barlow (1989) depicts this as an area of sociobiological theory that has moved furthest ahead of data available to support it.

Classical ethological theory is easily brought up to date with these ideas. Naive individuals of a single genotype might all use display in a similar and predictable manner. But, depending on feedback from their experiences, some might modify their naive pattern and vary in features of their display behavior (Schleidt, 1962; Baerends, 1985). They might learn how or when to display based on the consequences of showing the behavior. Due to chance and variability in size, fighting prowess, etc., display could indicate a somewhat different mood in each individual. But do animals actually do this? Can individuals learn to use an aggressive activity to their own best advantage?

This is far from being a simple question. The consequences of aggression for any individual may vary widely depending on the opponent. In addition, as stressed by Baerends (this volume), the same display will often by used to communicate with very different kinds of individuals, even different species. At the very least, the result will be an extremely complex learning paradigm. Never the less, aggressive displays have been a popular subject for the study of the effects of experience. These diverse studies can be organized into three major areas:

1. Operant Conditioning

First is the study of the reinforcing properties of aggressive behavior. Fish have figured centrally in showing that animals will learn to perform tasks that result in the presentation of a conspecific that is attacked. Studies such as Thompson (1963), Adler and Hogan (1963), Rasa (1971), Sevenster (1973), Bols (1977) Hogan and Bols (1980) and Bronstein (1981) clearly indicate that provision of the opportunity to perform aggression can be rewarding.

But what can we learn of behavioral ecology from such studies? Some interpret this to mean that animals will seek out opportunities to fight (Rasa, 1971). Priming effects found by Hogan and Bols (1980) indicated that, once in an aggressive mood, Siamese fighting fish might choose opportunities to fight rather than to feed. Hogan et al. (1970) indicated that, unlike feeding behavior, fish would not "work harder" (perform more operant tasks) to maintain a constant level of reinforcement. Even though more and more responses were required to obtain reinforcement, fish kept responding at about the same rate. Thus, as opportunities to perform aggression become more rare (fewer territory intruders, fewer neighbors, etc.), we need not expect an individual to start spoiling for a fight and show aggression in lower threshold situations. Ecological models of territorial defense can assume a more or less constant tendency to defend. But all of these studies are somewhat difficult to interpret due to the associated effects of classical conditioning. Many fish come to treat the operant stimulus as an aggressive releasing stimulus and display even before the presentation of the opponent. Hogan *et al.*'s fish may have maintained a constant level of aggressive display to the experimental apparatus itself.

2. Classical Conditioning

Hollis (1984, 1990) has conducted a creative series of experiments with gouramis (*Trichogaster trichopterus*) in which illumination of a light source signaled that an opponent was about to be presented. Not surprisingly, The fish soon began to approach and display at the light source. But, of more interest, trained fish were able to defeat untrained opponents when the conditioned stimulus preceded the fight. The conditioned stimulus had the same effect as the priming or "warming up" effects of performing aggressive acts. Conversely, conditioned stimuli that reliably indicated that an opponent would not be presented produced an inhibition of aggressive behavior.

Classical conditioning to landmarks where fights have occurred and various stimuli that predict the appearance of (or lack of) opponents must occur in the field. A new territory holder should be the least conditioned to such predictive stimuli and might show strikingly different behavior from experienced territory holders. Thorough knowledge of and ability to predict events within a territory would be of obvious advantage to a resident and could favor a sort of "social inertia" (Stamps, 1991) as a reluctance to abandon a known location. Social inertia is of obvious importance to behavioral ecologists who assume that individuals are behaving optimally.

3. Consequences of Aggression

Finally there are various sorts of approaches to the study of how experience with the consequences of aggression change an individual's aggressive behavior (Baenninger, 1984). The simplest type of experiment is to deprive young individuals of social interaction. Tooker and Miller (1989) revealed a variety of effects when young gouramis were deprived of social experience. With a total lack of social experience, the stereotyped movements were all found but the fish had difficulty combining, coordinating and correctly orienting the aggressive acts. They also lacked aggressive responsiveness, perhaps not recognizing releasing stimuli. Isolation at 60 days of age, an age when overt aggression is increasing in the social group, produced strongly aggressive fish. Their behavior lacked control by the opponent's signals. They appeared to have learned which acts were effective and how to use them, but not when to use them most effectively.

Such results suggest that animals could develop deceitful signaling. So long as there were no adverse consequences in the performance of displays or attacking an opponent, an individual could learn to use them in new situations. Unfortunately, isolation and deprivation experiments are plagued with difficulties (Fuller, 1967). We are tempted to interpret the results as meaning that some critical information that is required for normal development has been missed. But the effects of isolation can have far reaching perceptual effects. The deterioration of patterns of behavior due to their disuse can also confound the results. And even the circumstances of emergence from isolation as dishabituated individuals can cause changes in behavior.

An improvement on the isolation experiment is to rear animals under different, but not impoverished conditions. Hoelzer (1987) investigated the effects of density and provision of shelter in two species of rockfish, *Sebastes carnatus* and *S. chrysomelas*.

Both territorial and "floater" individuals are found in the field. Rearing under conditions of high density and available shelter resulted in permanently enhanced aggression and territoriality.

Stanton (1990) reared the Hawaiian sergeant, *Abudefduf abdominalis*, under two different feeding regimes, planktonic vs. benthic food. Similarly experienced individuals can be found in the field under buoys and on reefs, respectively. He found aggression to be strongly associated with food resources only when the individual was familiar with that food supply. Early experience with food could produce differences in their defense of food resources.

To stray briefly from the world of fishes, Groothuis (1989) found that black headed gulls develop slightly different forms of threat behavior in social groups of different size. Only in small social groups did variant forms of display develop. And only in small social groups were the consequences of display based mainly on social position rather than the form of the display itself.

Such studies improve on the deprivation experiment but still leave us to speculate on the actual mechanisms involved and predictions that we can make on the function of the differences produced. It is tempting to think that experience with successful vs. unsuccessful aggressive actions has shaped the behavior of the individual. However, we cannot determine just how this happens.

However, it is possible to gain direct control over the consequences of aggression in learning experiments. If displays have been adapted for truthful communication of mood and intentions, we should not be able to alter their use through conditioning. However, if animals learn to use behavior to their own maximum advantage, we should be able to control this process by altering the consequences of aggression. I will describe my work on the threespined stickleback *Gasterosteus aculeatus* that led me to feel that, for this species, modification of aggression is largely constrained (Losey and Sevenster, 1991).

We used adults of an anadromous population since juveniles and adults under winter conditions show very little aggression and have probably never displayed (Bakker, 1985; Bakker and Feuth-de Bruijn, 1988). Thus we could have completely naive fish without having to deprive them of normal development, albeit largely without aggression. Males were isolated in summer conditions and, within about one week, built and defended a nest. We placed one male on each side of a transparent partition with a sliding opaque door. When the door was raised, both males showed strong aggression across the glass and a complete repertoire of aggressive action patterns. Males selected for punishment training were given a brief electric shock that caused mild trauma each time they showed the head down display. Other males were rewarded by lowering the door each time they showed the head down and thus "removing" the opponent. We reasoned that punished fish should decrease their use of the display since it always resulted in a painful consequence. Rewarded fish should increase use of display since a logically desirous consequence of territorial aggression is to make the opponent "go away."

Punished fish showed no clear trends in the number of head down displays. There were some trends to increase the use of the display after punishment that might be argued. But, at the very least, our prediction of decreased use was not upheld. Five of the eight fish trained with a reward showed a significant increase in their use of head down and all eight fish had positive correlation coefficients. Latency to show a head down had a negative correlation coefficient for seven of the eight fish but was significant for only two. These findings were consistent with our predictions but not overwhelmingly convincing.

But are such traditional measures of learning performance the correct ones for such studies. We were interested in changing the decision rules governing when a display is shown. Study of the transitions between behavior patterns is one way to more directly examine these rules rather than looking at overall frequency of the use of a behavior. Given that the fish is performing a certain act, what is the probability that it will change its behavior to the head down display? Continuous time Markov chain models offer a powerful tool for examining changes in such probabilities (Haccou *et al.,* 1983; Haccou, 1987).

All but two of the nine punished fish and all of the rewarded fish had a higher probability of changing their current behavior to a head down display after training. Our hypothesis for punished behavior was again rejected and the predictions for the reward were upheld.

Finally, several of the reward-trained fish were exposed to opponents in a different context such as with a different partition, free-swimming in their tank, an opponent in a tube inside their territory, etc. Each time the context was changed, the performance of the fish reverted back to their pretraining behavior, head down became a rare behavior. The effects of reward training were masked.

These results provide new clues about a functional aspect of head down posturing in sticklebacks. We argued that punished fish may fail to decrease use of the punished display due to functional constraints on this behavior. Head down also appears to serve as a defensive maneuver to deflect attacks toward armored portions of the flank and tail. An individual should not learn to decrease the use of a protective maneuver merely because it results in a painful attack. Pain may well be the normal consequence of performing this behavior that is both an informative display and a tactical maneuver.

The failure of rewarded fish to show stronger and more permanent changes in the use of head down may relate more closely to truth in communication. When dealing with a well known and predictable neighbor in a familiar context, performance can change without a loss of function. Almost a short hand form of communication may suffice between "dear enemies." But when the context changes and the consequences of aggression are less certain (e.g. see Groothuis, 1989), there may be constraint against using variants of behavior. Especially for anadromous sticklebacks, who may have but one chance to reproduce, selection may have favored a largely closed developmental program for social behavior that will work the very first time it is called upon. On reaching a potential nesting site, they must fight for, and defend, this site for what is likely their single chance at reproductive success. Lorenz (1965: 46) suggested that complex adaptations could be destroyed if "... individual modification by learning were allowed to tamper with them." Huntingford and Turner (1987: 276) described the "knife-edge" balance in aggressive contests in which one attempts "...to coerce, immobilize or injure the opponent without being injured oneself."

EXPERIENCE AND INTERSPECIFIC AGGRESSIVE BEHAVIOR

Inhabitants of a coral reef are faced with a myriad of species all living in close proximity and with much overlap in their ecological niche. Aggression between species is far more common, or at least more noticeable, than in terrestrial habitats.

All forms of interspecific aggression were once thought to be the result of a maladaptive misidentification of another species as ones own. Principles of competitive exclusion were used to suggest that when interspecific aggression does occur, it must result from "incomplete evolutionary divergence" due to factors such as recent sympatry or a narrow zone of overlap between otherwise allopatric species (e.g. Murray, 1971).

Study of the proximate causes of attacks in herbivorous reef fishes were instrumental in changing our views of the adaptive value of interspecific aggression. Territorial damselfish were the showcase species for these studies. Many authors demonstrated that these damselfish focussed their attacks on competitors (e.g. Low, 1971; Myrberg and Thresher, 1974; Ebersole, 1977, 1980; Kohda, 1981; Lassuy, 1981; Robertson, 1984) They defend a territorial algal matte that is far richer than algae found elsewhere in the reef. Herbivorous predators are selected for attack according to both their location relative to the territory and their behavior. The precision with which these fishes focussed their attacks on competitors led many to doubt the misidentification hypothesis. However, doubt remained as to whether this was some special recognition system or merely a generalization of conspecific recognition cues (Kohda, 1981).

I felt it was more reasonable to expect some sort of ecological cues that would focus aggression on competitors. One way to test for such a process is to examine the responses of adults to new species. Would they learn to attack apparent competitors and ignore non-competitors? Tilapia provides an ideal model for a new species of potential competitor. This freshwater fish can live in sea water but is not found on most Hawaiian coral reefs. I was also able to train them to feed with movements that looked like a benthic herbivore, or to feed on planktonic food.

When the Hawaiian damselfish, *Stegastes fasciolatus*, was first exposed to tilapia, they were largely ignored (Losey, 1984). They look like, and were treated like, a generalized reef predator. After two weeks of experience with tilapia trained to feed at the surface of the water, the damsel fish continued to ignore their tilapia tank mates. But after two weeks of experience with tilapia that fed like an herbivore, the damselfish attacked the tilapia in the same manner as it attacked the other herbivores. The chance and severity of the attack varied with both the location of the tilapia within the territory and its behavior.

What were they learning? What was the motivationally significant event that produced the change in behavior? This has not been thoroughly investigated but must be some sort of ecological cue as to feeding type. Only reef herbivores feed in strongly clustered feed bites. Omnivores deliver only one or two bites close together. My trained tilapia had patterns of biting the were indistinguishable from the reef herbivores. Does this pattern of feeding fit some anger-producing template that eventually results in attack?

Hollis *et al.* (1984) thought it unlikely that this could result from the fitting of stimuli to an "innate template." The acquisition of attack was, instead, a gradual process more indicative of learning. However, there is no *a priori* reason why a reward system cannot be based on some template or stimulus analyzer present in naive individuals. Events that are paired with stimulation of this analyzer would follow the normal trajectory of learning performance. After all, are not all stimuli that have rewarding and punishing properties nothing but templates for pleasure, pain, anger, etc. A visual cue could favor an angry response but only trigger actual attacks after it had been established as a reliable cue that accompanied a particular type of opponent. Chance bites at algae should not result in costly and possibly dangerous attacks.

The indication of a special proximate mechanism for acquiring responses to new species is strongly supportive of the adaptive nature of the behavior. Interspecific aggression in such species must have been selected for and, at least at that time, been of positive survival value. The use of experience to shape interspecific attacks provides an efficient method for dealing with the many hundreds of species of fishes that must be discriminated on the coral reef.

EFFECTS OF EXPERIENCE ON CLEANING SYMBIOSIS

Cleaning symbiosis is a widespread feeding guild in which the cleaner feeds on mucus, tissues and ectoparasites from the skin of its hosts (see Feder, 1966; Losey, 1987). The hosts are strongly cooperative and many solicit the attentions of the cleaners with special behavior called posing. One qualitative field experiment suggested that, if cleaners are removed from a reef, fish either emigrate or suffer declines in health (Limbaugh, 1961). Despite failed attempts to repeat this experiment (Youngbluth, 1968; Losey, 1972), the *proximate* cooperation observed between cleaners and hosts has proved to be very compelling. Most authors accept the *ultimate* argument that cleaning was a mutualism and evolved because of the promotion of better health through the removal of ectoparasites (e.g. McFarland, 1985; Trivers, 1971). Cleaners and hosts were co-adapted with cooperation in the hosts having evolved for the purpose of having their parasites removed.

Study of proximate causation of host behavior suggested a very different evolutionary scenario. Reef fish quickly learned, either with classical or instrumental learning paradigms, when the reward was presentation of the model of a cleaner fish (Losey and Margules, 1974; Losey, 1977, 1979). But the fish model was unnecessary for the learning to occur. Any object, even the bare wire used to present the model, could produce the effect so long as it gently rubbed against the side of the fish. Tactile stimulation was the key to producing the hosts' responses. But the response was poorly adapted to the demands of cleaning. The presence of ectoparasites had only minimal effect (Losey, 1979). All host fish showed a strong response to tactile stimulation. No evidence could be found of a motivational system in the host fishes that served to regulate their actions associated with cleaning (review in Losey, 1987).

These findings suggested a unique scenario for the evolution of cleaning symbiosis (Losey, 1979). All sorts of small fishes occasionally pick at the side of another fish and ingest mucus, tissues and scales. Some have developed into voracious hit-and-run parasites that are specialized to feed on fish scales (e.g. Marlier and Lelup, 1954; Roberts, 1970; Major, 1974). Others were likely cleaner precursors. This is the point at which my hypothesis diverges from others. Earlier hypotheses assumed that cleaning symbiosis was a closely co-evolved system that had many independent origins (e.g. Trivers, 1971). Cleaner precursors began to specialize on ectoparasites as a food source. Host fishes that solicited the attentions of these early cleaners, and avoided preying on these early cleaners, were more successful than others due to ectoparasite removal. As a result, special behaviors to implement the mutualistic symbiosis evolved in both the cleaners and the host.

However, my work revealed no behaviors in the hosts that are not found in all vertebrates: A rewarding effect of tactile stimulation is all that is required to explain host behavior. My scenario concentrates on evolution of the cleaner as a "behavioral parasite." These early cleaners likely minimized the painful stimuli that result from occasionally feeding on the side of a host. But I envision that the important evolutionary breakthrough was the separate provision of special means of delivering tactile stimulation to their hosts. Hosts would be rewarded by this stimulation and seek out the cleaner as a hedonistic response. The probably unavoidable pain that results from some of a cleaner's feeding must be balanced against the positive effects of tactile reward.

This opening of an entirely new feeding niche could place tremendous pressure on cleaners to develop methods of delivering tactile reward. *Labroides* species of cleaners have a behavior called pelvic ride or stabilization behavior (Youngbluth, 1968; see Fig 7.4A and cover photo in McFarland, 1985) during which the pelvic fins stroke the sides of the host. *Gobiosoma* cleaners flick the host with their caudal fins. So long as there is little selection pressure against the hosts responding to these stimuli, specialized cleaners would evolve while the hosts could remain in their primitive form. Cleaning could exist as a commensalism with little or no function for the hosts. The "cooperative" behavior of the hosts may be simply the result of instrumental conditioning.

Of course, more specialized hosts might be expected in areas where ectoparasites are of critical survival cost. Recent work in Japan, in an area where ectoparasites are far more common, may have revealed the presence of more specialized hosts who alter their response to cleaners based on their ectoparasite loads (M. Chikasue, pers. comm., 1991).

What is the evolutionary origin of this positive response to tactile stimulation? It appears to occur in all vertebrates. As such, it could hardly have evolved as a result of cleaning symbiosis. Indeed, my hypothesis depends on this assumption. But I am unable to offer a universal answer. I can point to individual cases of sexual and other social behavior in which tactile stimulation is a central issue. But there are many more species in which no plausible explanation has yet been found.

CONCLUSIONS

Students of proximate causation have much to offer to the field of behavioral ecology. When demonstration of the function of a behavior is elusive, study of the proximate causes may be a valuable source of evidence. A method for learning which competitor to attack based on ecological cues is strong evidence for the adaptive nature of interspecific aggression. Estimation of the actual costs and benefits of such attacks would, however, be very difficult. The topic of truth vs. deception in communication has demanded many gallons of printer's ink but there is always room for another theoretical twist to add a different flavor to the stew (e.g. van Rijn and Vodegel, 1980; Bond, 1989a,b). Demonstration of changes in how animals use displays based on the consequences of such use could convince many that deception is possible. And, finally, knowledge of the structure and function of proximate causes for behaviors can provide valuable clues as to the evolutionary paths that could have produced such structures.

References

Adler, N. and Hogan, J. A. (1963). Classical conditioning and punishment of an instinctive response in *Betta splendens*. *Anim. Behav.*, **11**, 351–354.

Alcock, J. (1989). *Animal Behavior*, Fourth edn. Sunderland: Sinauer Assoc., Inc.

Baenninger, R. (1984). Consequences of aggressive threats by *Betta splendens*. *Aggressive Behav.*, **10**, 1–9.

Baerends, G. P. (1985). Do the dummy experiments with sticklebacks support the IRM-concept? *Behav.*, **93**, 258–277.

Bakker, Th. C. M. (1985). Two-way selection for aggression in juvenile, female and male sticklebacks (*Gasterosteus aculeatus* L.), with some notes on hormonal factors. *Behav.*, **93**, 69–81.

Bakker, Th. C. M. and Feuth-de Bruijn, E. (1988). Juvenile territoriality in stickleback *Gasterosteus aculeatus* L. *Anim. Behav.*, **36**, 1556–1558.

Barlow, G. W. (1989). Has sociobiology killed ethology or revitalized it. *Perspectives in Ethology*, **8**, 1–45.

Bindra, D. (1978). How adaptive behavior is produced: A perceptual-motivational alternative to response-reinforcement. *Beh. Brain Sci.*, **1**, 41–91.

Bond, A. B. (1989a). Toward a resolution of the paradox of aggressive displays: I. Optimal deceit in the communication of fighting ability. *Ethology*, **81**, 29–46.

Bond, A. B. (1989b). Toward a resolution of the paradox of aggressive displays: II. Behavioral efference and the communication of intentions. *Ethology*, **81**, 235–249.

Bronstein, P. M. (1981). Social reinforcement in *Betta splendens*: A reconsideration. *J. Comp. Physiol. Psychol.*, **95**, 943–950.

Dawkins, R. and Krebs, J. R. (1978). Animal signals: Information of manipulation? In: *Behavioural Ecology* (Ed. by J. R. Krebs and N. B. Davies), pp. 282–309. Oxford: Blackwell.

Drewnowski, A., Shrager, E. E. and Lipsky, C. (1989). Sugar and fat: Sensory and hedonic evaluation of liquid and solid foods. *Physiol. and Behav.*, **45**, 177–183.

Ebersole, J. P. (1977). The adaptive significance of territoriality in the reef fish *Eupomacentrus leucostictus*. *Ecol.*, **58**, 914–920.

Ebersole, J. P. (1980). Food density and territory size: an alternative model and a test on the reef fish *Eupomacentrus leucostictus*. *Am. Nat.*, **115**, 492–509.

Feder, H. M. (1966). Cleaning symbiosis in the marine environment. In: *Symbiosis, vol 1* (Ed. by S. M. Henry), pp. 327–380. New York: Academic Press.

Groothuis, T. (1989). On the ontogeny of display behaviour in black-headed gulls: II. causal links in the development of aggression, fear and display behaviour: emancipation reconsidered. *Behav.*, **110**, 161–204.

Haccou, P. (1987). Statistical Methods for Ethological Data. Ph.D. thesis. Univ. of Leiden.

Haccou, P., Dienske, H. and Meelis, E. (1983). Analysis of time-inhomogeneity in Markov chains applied to mother–infant interactions of Rhesus monkeys. *Anim. Behav.*, **31**, 927–945.

Halpin, Z. T. (1991). Introduction to the symposium: animal behavior: Past, present, and future. *Amer. Zool.*, **31**, 283–285.

Hoelzer, G. (1987). The effect of early experience on aggression in two territorial socrpaenid fishes. *Env. Biol. Fish.*, **19**, 183–194.

Hogan, J. A., Kleist, S. and Hutchings, C. S. L. (1970). Display and food as reinforcers in the Siamese fighting fish (*Betta splendens*). *J. Comp. and Physiol. Psychol.*, **70**, 351–357.

Hollis, K. L. (1984). The biological function of Pavlovian conditioning: The best defense is a good offense. *J. Exper. Psychol.*, **10**, 413–425.

Hollis, K. L. (1990). The role of Pavlovian conditioning in territorial aggression and reproduction. In: *Contemporary issues in comparative psychology* (Ed. by D. A. Dewsbury), pp. 197–219. Sunderland: Sinauer Assoc. Inc.

Hollis, K. L., Martin, K. A., Cadieux, E. L. and Colbert, M. M. (1984). The biological function of Pavlovian conditioning: Learned inhibition of aggressive behavior in territorial fish. *Learning and Motivation*, **15**, 459–478.

Huntingford, F. A. and Turner, A. K. (1987). *Animal Conflict*. New York: Chapman and Hall.

Kohda, M. (1981). Interspecific territoriality and agonistic behavior of a temperate pomacentrid fish, *Eupomacentrus altus* (Pisces: Pomacentridae). *Zeit. fur Tierpsychol.*, **56**, 376–384.

Kroodsma, D. E. and Byers, B. E. (1991). The function(s) of bird song. *Amer. Zool.*, **31**, 318–328.

Lassuy, D. R. (1980). Effects of "farming" behavior by *Eupomacentrus lividus* and *Hemiglyphidodon plagiometopon* on algal community structure. *Bull. Mar. Sci.*, **30**, 304–312.

Lorenz, K. (1965). *Evolution and Modification of Behavior*. Chicago Il: Univ. Chicago Press.

Losey, G. S. (1972). The ecological importance of cleaning symbiosis. *Copeia*, 820–833.

Losey, G. S. (1977). The validity of animal models: A test for cleaning symboisis. *Biol. Behav.*, **2**, 223–238.

Losey, G. S. (1979). Fish cleaning symbiosis: Proximate causes of host behaviour. *Anim. Behav.*, **27**, 669–685.

Losey, G. S. (1987). Cleaning symbiosis. *Symbiosis*, **4**, 229–258.

Losey, G. S. and Margules, L. (1974). Cleaning symbiosis provides a positive reinforcer for fish. *Science*, **184**, 179–180.

Losey, G. S. and Sevenster, P. (1991). Can threespine sticklebacks learn when to display? I. Punished displays. *Ethology*, **87**, 45–58.

Low, R. M. (1971). Interspecific territoriality in a pomacentrid reef fish, *Pomacentrus flavicauda* Whitely. *Ecol.*, **52**, 648–654.

Major, P. F. (1973). Scale feeding of the leatherjacket, *Scomberoides laysan* and two species of the genus *Oligoplites* (Pisces: Carangidae). *Copeia*, 151–154.

Marlier, G. and Lelup, N. (1954). A curious ecological 'niche' among fishes of Lake Tanganyika. *Nature*, **174**, 935–936.

Maynard Smith, J. (1982). *Evolution and the Theory of Games*. Cambridge: Cambridge Univ. Press.

McFarland, D. (1985). *Animal Behavior*. Menlo Park: Benjamin/Cummings Pub. Co., Inc.

Murray, B. G. (1971). The ecological consequences of interspecific territorial behavior in birds. *Ecol.*, **71**, 414–423.

Myrberg, A. A. Jr. and Thresher, R. E. (1974). Interspecific aggression and its relevance to the concept of territoriality in reef fishes. *Amer. Zool.*, **14**, 81–96.

Rasa, O. A. E. (1971). *Appetance for aggression in juvenile damsel fish*. Berlin: Paul Parey.

Rhijn, J. B. V. and Vodegel, R. (1980). Being honest about one's intentions: an evolutionary stable strategy for animal conflicts. *J. Theor. Biol.*, **85**, 623–641.

Roberts, T. R. (1970). Scale eating American characoid fishes with special reference to *Probolodus heterostomus*. *Proc. Calif. Acad. Sci., (Fourth Series)*, **38**, 383–390.

Robertson, D. R. (1984). Cohabitation of competing territorial damselfishes on a Caribbean coral reef. *Ecol.*, **65**, 1121–1135.

Schleidt, W. M. (1962). Die historische Entwicklung der Begriffe "Angeborenes auslosendes Schema" und "Angeborener Auslosmechanismus." *Zeit. für Tierpsychol.*, **19**, 697–722.

Sevenster, P. (1973). Incompatibility of response and reward. In: *Constraints on Learning*. (Ed. by R. A. Hinde and J. Stevenson-Hinde), pp. 265–283. New York: Academic Press.

Stamps, J. (1991). Why evolutionary issues are reviving interest in proximate behavioral mechanisms. *Amer. Zool.*, **31**, 338–348.

Stanton, F. (1990). The ontogeny of social behavior in a Hawaiian damselfish, *Abudefduf abdominalis*. Doctoral dissertation. Univ. Hawaii.

Tinbergen, N. (1950). *The Study of Instinct*. New York: Oxford Univ. Press.

Tooker, C. P. and Miller, R. J. (1980). The ontogeny of agonistic behaviour in the blue gourami, *Trichogaster trichopterus* (Pisces: Anabantoidei). *Anim. Behav.*, **28**, 973–988.

Trivers, R. L. (1971). The evolution of reciprocal altruism. *Q. Rev. Biol.*, **46**, 35–57.
Youngbluth, M. J. (1968). Aspects of the ecology and ethology of the cleaning fish, *Labroides phthirophagus* Randall. *Zeit. fur Tierpsychol.*, **25**, 915–932.

BEHAVIOURAL IMPLICATIONS OF INTRASPECIFIC LIFE HISTORY VARIATION

JEFFREY A. HUTCHINGS

Institute of Cell, Animal and Population Biology, University of Edinburgh, West Mains Road, Edinburgh EH9 3JT, Scotland, UK

INTRODUCTION

In relative terms, life history theory predicts the age at which an organism should reproduce and the effort that it should expend on reproduction at that age. Although behaviour is a constituent part of reproductive effort, life history theory provides no predictions about specific changes in behaviour that can be expected from specific changes in life history. How might evolutionary changes in life history bring about evolutionary changes in behaviour? Is the implicit assumption of life history theory that sufficient behavioural variability exists to effect changes in reproductive effort always a valid one? My objectives are to use empirical data on brook trout, *Salvelinus fontinalis*, and Atlantic salmon, *Salmo salar*, i) to test two predictions of life history theory, ii) to predict how evolutionary changes in life history can effect evolutionary changes in behaviour, and iii) to show how empirical tests of a theory in behavioural ecology can be compromised by excluding information on life history.

The relationship between life history and behaviour has been examined in fish although the intent of much of this work has been one of assessing how behaviour early in life influences some aspect of life history later in life. For example, Metcalfe *et al.* (1989) demonstrated that larger and behaviourally dominant juvenile Atlantic salmon generally migrate to sea, and possibly mature, earlier than smaller, subordinate individuals. Rather than ask how behaviour influences life history, I consider the changes in behaviour that might be expected to result from specific changes in life history. This approach shifts the emphasis from the behavioural aspects associated with foraging, territorial defence (as it relates to food acquisition), and predator avoidance to those associated with reproduction, e.g. selection of nest site, mate choice, and mate competition. Gross (1991) recently used such an approach to assess the conditional behavioural tactics employed by small, early-maturing ("jack") and large, late-maturing ("hooknose") coho salmon, *Oncorhynchus kisutch*, during breeding.

I consider the behavioural implications of intraspecific life history variation at the level of the population and at the level of the individual. At the population level, I use empirical data on unexploited brook trout populations to show how selection of nest site, selection of mates, and levels of intrasexual competition can be influenced by evolutionary changes in life history. Within populations, I examine two ways in which increased growth rate influences life history and behaviour. For female brook trout, increased growth rate favours earlier reproduction and increased reproductive effort (Hutchings, in press a). For male Atlantic salmon, increased growth rate also favours early maturation—as mature male parr (Myers et al., 1986). It is commonly thought that alternative maturation strategies in male Atlantic salmon coexist in an evolutionarily stable state through frequency dependent selection (Myers, 1986). By incorporating a measure of behavioural (varying numbers of male parr and anadromous males per anadromous female) and life historical reality (varying age at maturity among parr and among anadromous males) into fitness calculations for the two strategies, I show that the evolutionary stability of alternative male maturation strategies in Atlantic salmon, and perhaps other fish as well, is much more complicated than is currently believed.

LIFE HISTORY THEORY

Life histories reflect the ways in which individuals vary their age-specific expenditures of reproductive effort in response to physiologically and environmentally-induced changes in age-specific mortality. Strictly speaking, life history traits include age-specific rates of mortality and fecundity although this definition is usually extended to include characteristics that directly influence these parameters, e.g. age and size at first reproduction, egg size and parity. Age-specific fecundity and survival differ from these other traits in one important respect. They are direct components of fitness in the sense that an increase in either of these values increases fitness whereas maximum values of the other traits can be associated with intermediate levels of fitness (Charlesworth, 1980).

Life history theory assumes that natural selection favours those genotypes whose age-specific schedules of survival and fecundity generate the highest per capita rate of increase relative to other genotypes in the population. That is, selection tends to maximize r (Fisher, 1930; Charlesworth, 1980)—the intrinsic rate of natural increase—as given by the Euler-Lotka equation

$$1 = \sum l_x m_x e^{-rx} \tag{1}$$

where l_x represents survival from birth to the beginning of the breeding season at age x and m_x is the number of female zygotes produced by an individual at age x. The left hand side of the equation equals one because each female is assumed to reproduce itself at equilibrium.

Life history theory seeks to predict the age-specific reproductive pattern that will evolve in environments that have defined effects on age-specific schedules of mortality and fecundity. Consider the life-cycle of an annual breeder that starts breeding at age=1, where m offspring are produced at each breeding attempt, s_j is the survivorship from birth to breeding at age 1, and s_a represents adult survival between breeding events. Following Schaffer (1974) and Charlesworth (1980), the Euler-Lotka equation can be simplified to

$$e^r = s_j m + s_a \tag{2}$$

Re-arranging this equation yields

$$s_a = e^r - s_j m \qquad (3)$$

which expresses the central tenet of life history theory that future adult survival is negatively associated with present fecundity (Williams, 1966).

The combinations of adult survival and fecundity that an organism can potentially achieve is assumed to be limited and to be constrained to a set of evolutionary options (Levins, 1968; Partridge and Sibly, 1991). The optimal life history is that combination of adult survival and fecundity that maximizes e^r. Graphically, these optimal rates correspond to the point in the options set through which the fitness contour, i.e. Equation (3), passes and at which fitness, e^r, the y-intercept, is maximized. Given an options set whose boundary is concave with respect to the origin (Figure 1), as juvenile survival increases, holding adult survival constant, the slope of the fitness contour becomes increasingly negative, and selection favours increasing fecundity, or reproductive effort, with a concomitant reduction in adult survival. Similarly, decreased s_j favours reduced effort.

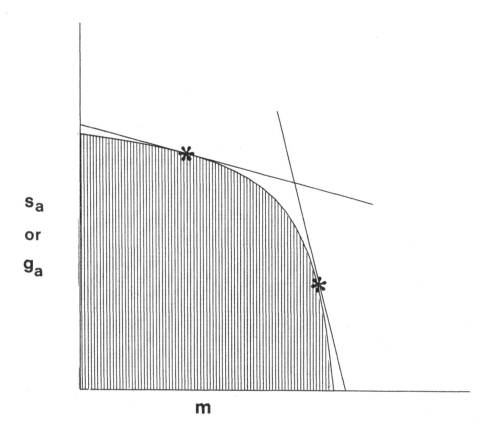

Figure 1   Options set for present fecundity, m, with adult survival, s_a, and adult growth rate, g_a. Straight line fitness contours correspond to Equation (3) for the survival fitness function and to Equation (5) for the growth fitness function. The slope of the contour equals juvenile survival, s_j, when the ordinate equals s_a and juvenile growth rate, g_j, when the ordinate equals g_a.

Relatively high adult mortality also selects for genes that increase reproduction earlier rather than later in life (Charlesworth, 1980). Thus, life history theory predicts that high juvenile survival, relative to adult survival, favours increased reproductive effort and decreased age at first reproduction (assuming that fecundity does not change with age).

Life history theory makes a second main prediction—one related to growth rate. In addition to the trade-off between current fecundity and future survival, Schaffer (1974) considered a trade-off between present fecundity and future growth. The importance of this trade-off to the life histories of organisms that continue to grow following maturation (e.g. many species of fish, reptiles, amphibians, and plants) lies in the positive associations (generated by correlations with body size) that generally exist between growth rate and both fecundity and survival (cf. Law, 1979; Bell, 1980; Roff, 1984; Stearns and Koella, 1986). Assuming equal constants of proportionality between growth rate and survival, Equation (2) becomes

$$e^{r} = g_{j}m + g_{a} \qquad (4)$$

or
$$g_{a} = e^{r} - g_{j}m \qquad (5)$$

where g_{a} and g_{j} represent adult and juvenile growth, respectively. Assume that the options set for combinations of present fecundity and post-reproductive or adult growth is that given in Figure 1. Rapid juvenile growth, relative to adult growth, favours increased effort, m, whereas low values of g_{j} select for decreased effort. Increased juvenile growth rate also results in increased fecundity in early life and this favours earlier age at first reproduction (Charlesworth, 1980). Thus, the second main prediction of life history theory is that, relative to adult growth rate, increased juvenile growth rate favours increased effort at an earlier age at first reproduction (assuming constant age-specific rates of survival) (cf. Hutchings, in press a).

INTRASPECIFIC VARIATION AMONG POPULATIONS

Life History Differences Among Populations of Brook Trout

Empirical life history data were collected from several brook trout populations on Cape Race, Newfoundland, Canada, in 1987 and 1988 (see Hutchings, 1990 for sampling protocol and river descriptions). Cape Race is located on the southeastern tip of the Avalon Peninsula and is bounded by 53°16'W, 46°45'N, 53°04'E, and 46°38'S. Brook trout are the only salmonids and, with few exceptions, are the only fish in these rivers. This removes the potentially confounding influences of interspecific competition and predation (birds do not prey on these populations) on life history. The 10–30 metre cliffs at the river mouths prevent migration between rivers, precluding gene flow between populations. The populations are electrophortically distinguishable (Ferguson et al., 1991) and have probably been isolated from one another since the Wisconsin glaciation, 10–12 000 years ago (cf. Rogerson, 1981).

In terms of life history, Cripple Cove (CC) and Freshwater (FW) are the most divergent populations on Cape Race. Relative to CC females, FW females matured, on average, at a 39% smaller size (99.8±17.8 mm fork length) but allocated 55% more body tissue to gonads. The ratio of gonad weight to total body weight, expressed as the gonadosomatic index (GSI), was 0.17 ± 0.03 and 0.11 ± 0.02 for FW and CC females, respectively. In part, population differences in GSI were due to differences in size-specific fecundity—FW females produced more eggs per unit body mass than similarly sized CC females. However, most of the differences in GSI were due to differences in egg size. FW females produced 35% larger eggs (volume=46.5 ± 12.1 mm^3) than their CC counterparts.

Fitness Calculations

If brook trout life histories on Cape Race are adaptive, the observed patterns of age-specific survival and fecundity should generate the highest fitness relative to other potential age-specific combinations of survival and fecundity. To calculate fitness, r, it is necessary to have reliable, empirically-based estimates of age-specific fecundity and survival as well as estimates of the degree to which these parameters are influenced by other traits, e.g. egg size, body size, and growth rate (see Hutchings, in press a, for details of these analyses).

Survival was calculated separately for the summer (May–September) and winter (October–April) months (details of the survival data are given in Hutchings, 1990 and Hutchings, unpublished data). Summer survival rates were determined from a static life table and were assumed to be independent of body size. (Violation of this assumption will have little effect on the analysis given that 70 to 80% of the annual mortality at the population level occurs during winter.) Mean overwinter survival ranged from 0.13 for FW females to 0.40 for CC females.

Overwinter survival generally increased with body size. This association probably has a physiological basis. Brook trout rely upon lipid reserves to survive the winter (Cunjak, 1988). Small individuals are disadvantaged primarily because, relative to large individuals, they utilize their proportionately smaller fat stores (Brett et al., 1969; Elliott, 1976) at a faster rate (Schmidt-Nelson, 1984; Dunbrack and Ramsay, unpubl.).

The dependence on lipid reserves during winter is the basis for the primary cost of reproduction in brook trout on Cape Race. This survival cost, defined as the proportional reduction in overwinter survival of mature females relative to that of immature females of the same size and age, increases with effort (approximated as GSI) within and among populations and is dependent on body size—the smallest individuals experience disproportionately high costs relative to their effort (Hutchings, unpublished data).

To determine body sizes at different ages, I used empirical data for immature individuals (Hutchings, in press a). I assumed that maturation reduced growth rate during the year preceding reproduction by an amount equal to expected GSI, i.e. the proportion of surplus energy allocated to gonads was directly related to the proportional loss of surplus energy devoted to somatic growth. Survival in early life was directly related to the size of egg from which an individual was produced (Hutchings, 1991).

Age-specific fecundity rates were calculated from the empirical data on the Cape Race populations. Fecundity depended upon egg size and GSI, all three of which were positively associated with body size.

Variation in Age at Maturity

To determine whether the inter-population differences in age at maturity were adaptive, I calculated the fitness (r) associated with the observed age-specific schedules of fecundity and survival and compared this with the fitness of potentially alternative life histories within three populations (FW, CC, and Watern Cove [WC]). The fitness calculations indicated that the optimal age at first reproduction (i.e. the age of initial maturity that maximized fitness) for females in FW, WC, and CC was 3, 3, and 4 years, respectively. These accord well with observed modal ages at maturity which were 3, 3, and 5 years for FW, WC, and CC, respectively (modal ages were used because ages at first reproduction, revealed as spawning checks in some species, cannot be discerned reliably from calcified structures in brook trout). I calculated predicted mean ages at maturity by assuming that individuals matured initially at the calculated optimal age and survived according to observed rates of age-specific survival. The differences between expected and observed ages at maturity were small, being 6% for WC, 7% for FW, and 8% for CC. The similarity between observed and predicted data support the hypothesis that population variation in age at maturity among Cape Race brook trout is adaptive (Hutchings, in press a).

Variation in Reproductive Effort (GSI)

To assess the degree to which observed levels of effort were optimal, I calculated fitness as a function of age at first reproduction for CC and FW populations at two levels of effort—0.11 and 0.17, the observed mean GSIs for CC and FW, respectively. (The fitness calculations controlled for the changes in post-reproductive survival effected by changes in GSI.)

The results support the hypothesis that observed levels of effort are adaptive. The fitness functions that incorporated the observed GSI values yielded the greater fitness for all potential ages at first reproduction (2–5 years). The fitness of FW females was greater (14–22%) at high effort whereas the fitness of CC females was greater (9–25%) at low effort.

Variation in Egg Size

The adaptive significance of inter-population variation in egg size can be assessed by considering the differences in food abundance among rivers. The biomass of aquatic stream invertebrates—the main source of food for brook trout (Power, 1980)—differed almost three-fold between CC and FW. The mean biomass of invertebrates per colonization bag in CC was 1.10 ± 0.59 g m^{-2} as compared with 0.39 ± 0.15 g m^{-2} in FW.

Smith and Fretwell (1974) and Kaplan and Cooper (1984) predicted that natural selection favours increased offspring size with reductions in food abundance. This prediction is based on the observation that 1) larger offspring are produced from larger eggs, and 2) competitive success for scarce resources is directly related to body size. Hutchings (1991) evaluated the fitness consequences for female brook trout of producing eggs of different sizes in environments having different amounts of food. Maternal fitness was optimized when females produced small numbers of large eggs under low food conditions and large numbers of small eggs under high food conditions. Thus, the population differences in egg size among the Cape Race populations can be explained as adaptive responses to local differences in food abundance.

Tests of Life History Theory

To test the predictions of life history theory, it was necessary to calculate juvenile and adult values of survival and growth rate (details in Hutchings, in press a). For the survival calculations, I assumed that the juvenile stage extended from birth to the optimal age at first reproduction and that the adult stage extended from the optimal age at first reproduction to death. As I have estimates of the survival costs of reproduction for these populations (Hutchings, unpublished data), I used the survival probabilities of non-reproductive adults in these calculations. Because of population differences in ages at first reproduction and death, mean annual juvenile and adult survival probabilities were used in the analysis. Juvenile growth rate was estimated as the length of individuals at the end of their first year of feeding. Adult growth rate was estimated as the difference in the length of individuals between the end of their first year of feeding and the end of their third year of feeding. This approximation of adult growth rate ensured that all fish were larger than their physiologically minimum size of reproduction (Hutchings, 1990) and minimized the probability of including previously mature adults in the sample.

The ratios of juvenile to adult survival and of juvenile to adult growth differed dramatically among populations (Table 1). The empirical data support the predictions of life history theory. Relative to adults, both high juvenile survival and high juvenile growth rate were associated with early reproduction, high effort, and a high cost of reproduction.

Reznick et al. (1990) found that high adult mortality in guppies, Poecilia reticulata, favoured early reproduction and high effort. Their experiments were supported by a transplant experiment in the field and by laboratory work which indicated that the observed life histories had a genetic basis. Schaffer and Elson (1975) reported that high adult growth rate favoured delayed reproduction in Atlantic salmon although their data base was not a reliable one (Myers and Hutchings, 1987a).

Table 1   Ratios of juvenile:adult survival (s_j/s_a) and juvenile:adult growth rate (g_j/g_a), mean age at maturity (yr), gonadosomatic index (GSI), and mean survival cost of reproduction. See text for parameter definitions

Population	Trait				
	s_j/s_a	g_j/g_a	Age	GSI	Cost
Freshwater	6.25	1.85	3.15	0.17	0.58
Watern Cove	0.39	1.35	3.12	0.12	0.38
Cripple Cove	0.31	1.15	4.09	0.11	0.31

Behavioural Implications of Adaptive Variation in Age at Maturity

Selection of mates

Evolutionary changes in age at maturity will effect concomitant changes in body size at maturity. Body size is an important determinant of breeding success in salmonids (e.g. van den Berghe and Gross, 1989), making it a likely candidate for sexual selection. For example, there is evidence that male sockeye salmon and kokanee, *O. nerka*, (Foote, 1988), coho salmon and threespine sticklebacks, *Gasterosteus aculeatus*, (Sargent *et al.*, 1986) use body size as a proximate cue to assess the reproductive quality of potential mates. Selection favouring reduced age at maturity results in reduced mean size at maturity and reduced variation in size at maturity. Reduced variation in body size among spawning individuals should *ceteris paribus* reduce the scope for mate choice on the basis of body size.

One consequence of life history variation at the population level may be a shift from one potentially sexually selected trait to another (Partridge and Endler, 1987). Male brook trout compete with one another for access to a single female who selects the nest site and defends it against other females (see Power, 1980). Male trout take on an orange-red coloration during the autumn (Ricker, 1932). Although there are no empirical data on interpopulation variation in body coloration in brook trout (Power, 1980 does indicate that colour intensity may increase with body size), visual inspection of trout from Cape Race indicates that such variation exists (pers. obs.). One would predict that males in those populations for which early age at maturity is favoured should express greater intensities or amounts of the orange-red colour than populations for which delayed reproduction is favoured. This is because early-maturing males are less variable in body size and because their reproductive strategy is one that favours semelparity. Similarly, female preference for males in early-maturing populations would be predicted to be more dependent on body colour than in late-maturing populations. The existence of adaptive variation in male body coloration and genetic variation in female choice among guppy populations (Houde, 1988; Houde and Endler, 1990) suggests that life history evolution can effect similar variation among salmonid populations.

Selection of nest sites

Decreased size at maturity also has important consequences for selection of nest sites. Brook trout bury their eggs in the stream substrate but the size of substrate particle that females are able to displace is directly related to their body size (cf. van den Berghe and Gross, 1984). Thus, small females can only bury their eggs in substrate comprised of particles of small size. Such substrate is at a premium in short, northern boreal streams in which water velocities are generally high. Thus, reduced availability of suitable spawning sites may effect increased densities of spawning fish. This is indeed the case in Freshwater and Watern Cove populations for which early age at maturity is favoured. Most of the spawning in these rivers occurs in relatively calm channels separate from the main stem of the rivers. Spawning densities can reach 7.8 and 16.0 fish m^{-2} in WC and FW populations, respectively, and result in a high incidence of nest site superimposition (pers. obs.).

From a behavioural perspective, high-density spawning environments were forced upon these fish by life history pressures through selection for early age at maturity. Such a spawning environment, relative to that typically encountered by members of this species, must result in substantially different decision-making processes and behavioural tactics with regard to selection of nest sites, selection of mates, and levels of aggression during mate competition.

INTRASPECIFIC VARIATION IN LIFE HISTORY WITHIN POPULATIONS

Adaptive Phenotypic Plasticity and Individual Optimization

The capacity of a genotype to alter its phenotype in response to environmental changes is termed phenotypic plasticity (Bradshaw, 1965). Plasticity can reflect either an environmentally-induced constraint on an individual's physiological state or it can represent an adaptive response to environmental change (e.g. varying population density, resource supply, temperature). Adaptive phenotypic plasticity can evolve as a norm of reaction, i.e. the systematic change with which a genotype alters its phenotypic expression across an environmental gradient (Schmaulhausen, 1949; Stearns and Koella, 1986). Reaction norms have been shown to be heritable and to respond to selection (reviewed by Thompson, 1991).

One consequence of adaptive phenotypic plasticity is individual optimization of life history. For example, there is evidence that avian clutch size depends largely on individual ability to acquire food such that the clutches represent optimal responses by individuals to their local environment (reviewed by Lessells, 1991).

Adaptive phenotypic plasticity, or individual optimization, can be expected to evolve when (i) the spatial distribution of individuals across habitats is random with respect to genotype and when (ii) habitat significantly influences fitness. These conditions probably exist for many stream-dwelling salmonid fish. Following yolk sac resorption, juvenile brook trout emerge from the stream substrate in spring and are carried downstream to small, relatively discrete sections of slow-moving water where they spend most, occasionally all, of the summer months (Hutchings, 1990). The environment inhabited by these juveniles is probably random with respect to genotype because emerging individuals are unable to swim against all but the weakest of currents. And, as growth rate can vary significantly among individuals within and among juvenile habitats (Hutchings, unpubl. data), environmental variation can influence fitness through size-dependent effects on life history traits (Hutchings and Morris, 1985).

If body size influences the relationship between cost and effort, individuals may allocate their own optimal effort subject to their ability to survive reproduction and to their prospects for increasing fecundity in subsequent spawnings. To explore this possibility, I used empirical data from the Cape Race brook trout populations to construct age-specific reaction norms between reproductive effort at age of first reproduction and female body size. My approach was to calculate the fitness associated with different growth rates, different levels of effort, and different ages at first reproduction for each of the populations (details in Hutchings, in press b).

Reaction norms for body size and reproductive effort varied with age at first reproduction. The general pattern that emerged was that selection favoured maximal effort independent of body size when age at first reproduction was low (fastest growing females mature earliest) but favoured an increase in effort with size from minimal to sub-maximal effort when age at first reproduction was high (slowest growing females matured latest in life; Figure 2). Thus, when growth rates vary among individuals, selection should modify individual effort on a size-dependent basis according to the shape of the effort:size reaction norm.

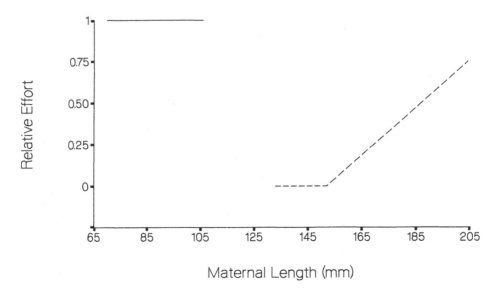

Maternal Length (mm)

Figure 2 Reaction norms for reproductive effort and body size for female brook trout in Cripple Cove River. Solid and dashed reaction norms refer to females maturing initially at ages 2 and 5, respectively.

Is there any evidence that individual fish optimize their effort in this manner? The reaction norms predict that females reproducing for the first time early in life should maximize effort. Thus, the GSI values for such fish should fall above the regression line relating GSI to body length. Although the data are few, they do support the prediction that the effort of FW females reproducing for the first time is greater than that which would have been predicted by body size alone (Figure 3).

Behavioural Implications of Intrapopulation Variation in Effort

Does a fish for which selection favours maximal effort behave differently from a fish for which selection favours a sub-maximal expenditure of effort? At the species level, semelparous salmonids defend their nest sites for more than a week following reproduction (van den Berghe and Gross, 1986) whereas iteroparous brook trout typically abandon their nests within hours after spawning (Power, 1980). Body coloration and aggression are much more intense among semelparous species of killifish than they are among iteroparous species (van Ramshorst and van den Nieuwenhuisen, 1978).

Magnhagen (1990) compared the nest-building behaviour of an iteroparous and a semelparous species of goby in the presence and absence of a predator. She found that the iteroparous goby significantly reduced the numbers of nests built in the presence of a predator whereas the nest-building behaviour of the semelparous goby was unaffected by the predator. Magnhagen (1990) also reported age-related, within-population variation in reproductive behaviour. For the iteroparous goby, older males built significantly more nests in the presence of the predator than younger

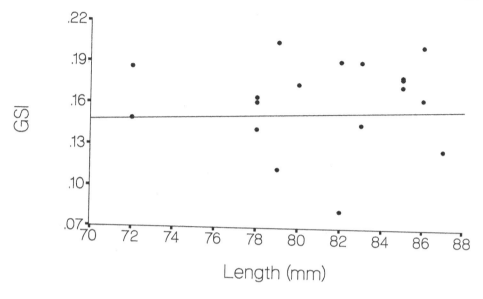

Figure 3 Gonadosomatic index (GSI=gonad weight/total body weight) as a function of body size for virgin female brook trout (age = 2yr) in Freshwater River (FW). Regression between GSI and length for all females in FW is indicated by the solid line.

males. This suggests that individuals are able to modify their reproductive behaviour in relation to their future expectation of producing offspring, i.e. their reproductive value (Fisher, 1930). Intuitively, it seems reasonable to expect adaptive plasticity in behaviour to co-evolve with adaptive plasticity in effort.

Alternative Mating Strategies

Phenotypically plastic spawning behaviour exists for many male salmonids. For one species, the iteroparous Atlantic salmon, males commonly mature as one of two life history forms (Jones, 1959). Parr mature in fresh water and are usually 2–4 years younger and 50–60 cm smaller than anadromous males which mature following a feeding migration to sea. Prior to spawning, a dominant anadromous male defends access to an anadromous female while mature male parr establish a linear dominance hierarchy immediately downstream of the courting pair with the largest parr usually nearest the female (Jones, 1959; Myers and Hutchings, 1987b). Mature male parr dart in close to the anadromous pair and shed sperm at the time of egg extrusion (Jones, 1959). Several males fertilize the eggs of a single female and males spawn with more than one female. The sex ratio of anadromous females to anadromous males usually exceeds unity. The ratio of mature male parr to anadromous males exceeds 20:1 at the spawning site in some populations (Hutchings, 1986). The incidence of parr maturation among male Atlantic salmon ranges from 1 to 100% in 28 Canadian populations (Myers et al., 1986).

Early maturity in male Atlantic salmon appears to be under both genetic and environmental control. Additive genetic variation (i.e. narrow-sense heritability;

Falconer, 1989) has been detected for parr maturation in hatchery environments (Naevdal *et al.*, 1976). Thus, parr maturation appears to be a polygenic trait yet its expression is discrete—a male either matures as a parr or it does not. Myers and Hutchings (1986) modelled parr maturation as a threshold character having an underlying normal distribution for liability (Falconer, 1989). This implies that there is a genetically-determined threshold, perhaps in the levels of specific proteins, hormones or lipids (cf. Rowe *et al.*, 1991), above which males mature as parr. For males whose genotype includes such a threshold, the expression of early maturity appears to depend on growth rate (Thorpe, 1986). Within and among Canadian populations, the covariation between growth rate and parr maturity produces a relationship characterized by a size or growth rate threshold below which males do not mature as parr (Myers *et al.*, 1986). The logistic relationship between growth rate and parr maturity provides support for the genetic threshold hypothesis. The observation that the fastest-growing males mature earliest in life supports the individual optimization hypothesis and one of the predictions of life history theory.

Alternative male maturation strategies are thought to exist in evolutionarily stable proportions through frequency-dependent selection (Gross, 1985; Myers, 1986). The evolutionarily stable proportion of males in a population that mature as parr is that proportion which yields equal fitness for parr and anadromous males (Maynard Smith, 1982). Two implicit assumptions of this model for Atlantic salmon are that the fertilization success of parr is independent of the number of anadromous males present and that age at first reproduction among male parr does not vary within populations.

Using age-specific survival data for a Newfoundland population (Hutchings and Myers, 1986) and using egg fertilization rates calculated from the laboratory (Hutchings and Myers, 1988), I calculated the fitness, r, associated with adopting parr and anadromous male strategies under different densities of parr and anadromous males (details in Hutchings, unpublished data). I assumed here that total parr fertilization success would be reduced by a proportion equal to $1/n$, where n was the number of anadromous males per female. Equilibrium parr densities were calculated to be the parr densities at which the fitness of individuals from the same cohort (i.e. born at the same time) adopting parr and anadromous male strategies were equal. That is, the "decision" either to reproduce during the autumn or to migrate to sea the following spring was made by all males at the same age such that parr matured initially at age x and anadromous males at $x+2$.

Parr and anadromous male behaviour, regarding mate choice and mate competition, depends on their respective densities. Equilibrium mature parr densities (an approximation of the proportion of males in a population maturing as parr) decline as the density of anadromous males increases (Figure 4). This is because the presence of a second, third, or fourth anadromous male probably displaces the hierarchical group of male parr further downstream, further away from the female, and reduces their chances of fertilizing eggs. From a behavioural perspective, parr mate choice decisions will not be clear-cut, such as "always choose the female with the fewest number of parr" or "always choose the female with the fewest number of anadromous males", because parr fitness is conditional upon the numbers of both parr and anadromous males present during spawning.

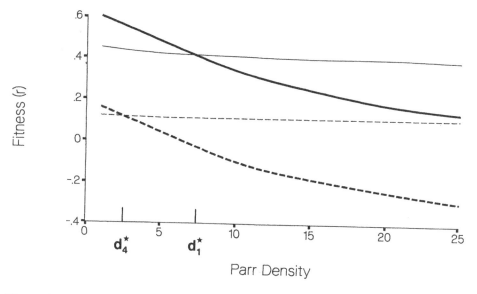

Figure 4 Fitness functions for mature male parr (bold lines) and anadromous male Atlantic salmon (assumed to be of the same cohort) at spawning densities of 1 (—) and 4 (---) anadromous males per spawning female. The equilibrium parr densities at which the fitness of parr and anadromous males is equal are indicated by d*; d_1*=7 and d_4*=2.5 parr per anadromous female.

From a life history perspective, variation in age at first reproduction among parr and anadromous males imposes additional complexity upon equilibrium proportions of parr and anadromous males within populations (Figure 5). The earlier reproduction occurs in life, the higher the fitness for both parr and anadromous males but, more importantly, equilibrium parr density increases as age at maturity decreases (ages refer here to age at first reproduction for parr and to age at seaward migration by anadromous males; males were assumed to spend one winter at sea). From an individual parr's perspective, his choice of female and his choice of the level of mate competition he is willing to engage will depend on his age. Evolutionarily, a male parr reproducing initially at a young age can tolerate a much larger group than parr mating initially at an older age because of the decline in equilibrium parr density with increased age at first reproduction.

There are no empirical data to test the predictions that parr reproductive success is influenced by the spawning density of anadromous males or that mate choice by parr is conditional in an age-specific manner. The potential for such individual optimization of behaviour and the implications it has for the stable coexistence of alternative mating strategies warrants experimental study.

SUMMARY

The behavioural implications of intraspecific life history variation are examined within and among populations. At the population level, brook trout life history variation appears to be adaptive and it supports the predictions of life history

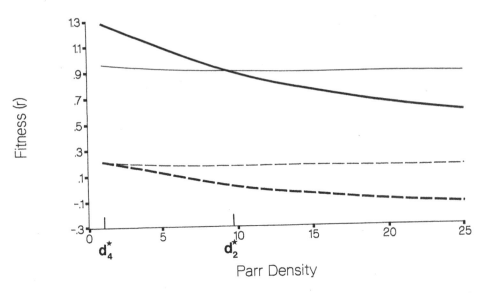

Figure 5 Fitness functions for mature male parr (bold lines) and anadromous male Atlantic salmon at ages of first reproduction (parr) and seaward migration (anadromous males) of 2 (—) and 4 years (----). Equilibrium parr densities per anadromous female equal d^*; $d_2^*=9.5$ and $d_4^*=1$.

theory that high reproductive effort and early reproduction are favoured by high juvenile:adult survival and by high juvenile:adult growth rate. Selection for reduced age at maturity can influence the selection of mates by changing the relative importance of traits under sexual selection and by affecting the selection criteria for choosing nest sites.

Within populations, variation in growth rate should result in individual optimization in life history and in the behavioural tactics associated with reproduction. There is some evidence that individual behaviour and reproductive effort is related to reproductive value. By incorporating realistic behavioural and life historical assumptions in fitness calculations for alternative maturation phenotypes in male Atlantic salmon, the evolutionary stability of alternative maturation phenotypes and the behavioural tactics adopted during spawning are shown to be considerably dynamic and complex phenomena.

ACKNOWLEDGMENTS

I gratefully acknowledge the financial support provided by a Postdoctoral Fellowship and a Postgraduate Scholarship from the Natural Sciences and Engineering Research Council (NSERC) of Canada, a NSERC Operating Grant to Douglas W. Morris, a Sigma Xi Grant-In-Aid of Research, an Atlantic Salmon Federation Olin Fellowship, and a Department of Supply and Services contract (Unsolicited Proposal UN-N-222). I have benefited from discussion with Joe

Brown, Doug Morris, Ram Myers and Linda Partridge and I am grateful to Neil Metcalfe, Linda Partridge and Gunilla Rosenqvist for comments on an earlier draft.

References

Bell, G. (1980). The costs of reproduction and their consequences. *Am. Nat.*, **116**, 45–76.

Bradshaw, A. D. (1965). Evolutionary significance of phenotypic plasticity in plants. *Adv. Genet.*, **13**, 115–155.

Brett, J. R., Shelbourn, J. E. and Shoop, C. T. (1969). Growth rate and body composition of fingerling sockeye salmon, *Oncorhynchus nerka*, in relation to temperature and body size. *J. Fish. Res. Board Can.*, **26**, 2363–2394.

Charlesworth, B. (1980). *Evolution in Age-Structured Populations*. Cambridge: Cambridge Univ. Press.

Cunjak, R. A. (1988). Physiological consequences of overwintering in streams: the cost of acclimitization? *Can. J. Fish. Aquat. Sci.*, **45**, 443–452.

Elliott, J. M. (1976). The energetics of feeding, metabolism and growth of brown trout (*Salmo trutta* L.) in relation to body weight, water temperature and ration size. *J. Anim. Ecol.*, **45**, 923–948.

Falconer, D. S. (1989). *Introduction to Quantitative Genetics*. London: Longman.

Ferguson, M. M., Danzmann, R. G. and Hutchings, J. A. (1991). Incongruent estimates of population differentiation among brook charr, *Salvelinus fontinalis*, from Cape Race, Newfoundland, Canada, based upon allozyme and mitochondrial DNA variation. *J. Fish Biol.*, **39** (Supplement A), 79–85.

Fisher, R. A. (1930). *The Genetical Theory of Natural Selection*. New York: Dover.

Foote, C. J. (1988). Male mate choice dependent on male size in salmon. *Behaviour*, **106**, 63–80.

Gross, M. R. (1985). Disruptive selection for alternative life histories in salmon. *Nature*, **313**, 47–48.

Gross, M. R. (1991). Salmon breeding behavior and life history evolution in changing environments. *Ecology*, **72**, 1180–1186.

Houde, A. E. (1988). Genetic difference in female choice between two guppy populations. *Anim. Behav.*, **36**, 510–516.

Houde, A. E. and Endler, J. A. (1990). Correlated evolution of female mating preferences and male color patterns in the guppy *Poecilia reticulata*. *Science*, **248**, 1405–1408.

Hutchings, J. A. (1986). Lakeward migrations by juvenile Atlantic salmon, *Salmo salar*. *Can. J. Fish. Aquat. Sci.*, **43**, 732–741.

Hutchings, J. A. (1990). The evolutionary significance of life history divergence among brook trout, *Salvelinus fontinalis*, populations. Ph.D. thesis. Memorial University of Newfoundland, St. John's, NF, Canada.

Hutchings, J. A. (1991). Fitness consequences of variation in egg size and food abundance in brook trout, *Salvelinus fontinalis*. *Evolution*, **45**, 1162–1168.

Hutchings, J. A. (In press a). Adaptive life histories effected by age-specific survival and growth rate. *Ecology*.

Hutchings, J. A. (In press b). Reaction norms for reproductive effort and body size in brook trout, *Salvelinus fontinalis*, and their influence on evolutionary changes in life history effected by size-selective harvesting. In: *The Evolution of Exploited Populations* (Ed. by K. Stokes, R. Law and J. McGlade). Berlin: Springer-Verlag.

Hutchings, J. A. and Morris, D. W. (1985). The influence of phylogeny, size and behaviour on patterns of covariation in salmonid life histories. *Oikos*, **45**, 118–124.

Hutchings, J. A. and Myers, R. A. (1986). The economics of artificial selection for reducing the proportion of mature male parr in natural populations of Atlantic salmon, *Salmo salar*. Int. Counc. Explor. Sea, C.M.1986/Mini-Symposium/No.3.

Hutchings, J. A. and Myers, R. A. (1988). Mating success of alternative maturation phenotypes in male Atlantic salmon, *Salmo salar*. *Oecologia*, **75**, 169–174.

Jones, J. W. 1959. *The Salmon*. London: Collins.

Kaplan, R. H. and Cooper, W. S. (1984). The evolution of developmental plasticity in reproductive characteristics: An application of the "adaptive coin flipping" principle. *Am. Nat.*, **123**, 393–410.

Law, R. (1979). The cost of reproduction in annual meadow grass. *Am. Nat.*, **113**, 3–16.

Lessels, C. M. (1991). The evolution of life histories. In: *Behavioural Ecology*, 3rd edn (Ed. by J. R. Krebs and N. B. Davies), pp. 32–68. Oxford: Blackwell.

Levins, R. (1968). *Evolution in Changing Environments*. Princeton: Princeton Univ. Press.

Magnhagen, C. (1990). Reproduction under predation risk in the sand goby, *Pomatoschistus minutus*, and the black goby, *Gobius niger*: The effect of age and longevity. *Behav. Ecol. Sociobiol.*, **26**, 331–335.

Maynard Smith, J. (1982). *Evolution and the Theory of Games*. Cambridge: Cambridge Univ. Press.

Metcalfe, N. B., Huntingford, F. A., Graham, W. D. and Thorpe, J. E. (1989). Early social status and the development of life-history strategies in Atlantic salmon. *Proc. R. Soc. Lond. B*, **236**, 7–19.

Myers, R. A. (1986). Game theory and the evolution of Atlantic salmon (*Salmo salar*) age at maturation. In: *Salmonid Age at Maturity* (Ed. by D. J. Meerburg), pp. 53–61. *Can. Spec. Publ. Fish. Aquat. Sci.*, **89**.

Myers, R. A. and Hutchings, J. A. (1986). Selection against parr maturation in Atlantic salmon. *Aquaculture*, **53**, 313–320.

Myers, R. A. and Hutchings, J. A. (1987a). A spurious correlation in an interpopulation comparison of Atlantic salmon life histories. *Ecology*, **68**, 1839–1843.

Myers, R. A. and Hutchings, J. A. (1987b). Mating of anadromous Atlantic salmon, *Salmo salar* L., with mature male parr. *J. Fish. Biol.*, **31**, 143–146.

Myers, R. A., Hutchings, J. A. and Gibson, R. J. (1986). Variation in male parr maturation within and among populations of Atlantic salmon, *Salmo salar*. *Can. J. Fish. Aquat. Sci.*, **43**, 1242–1248.

Naevdal, G., Holm, M., Moller, D. and Osthus, O. D. (1976). Variation in growth rate and age at sexual maturity in Atlantic salmon. *Int. Counc. Explor. Sea*, C.M.1976/E:40.

Partridge, L. and Endler, J. A. (1987). Life history constraints on sexual selection. In: *Sexual Selection: Testing the Alternatives* (Ed. by J. W. Bradbury and M. B. Andersson), pp. 265–277. New York: John Wiley and Sons.

Partridge, L. and Sibly, R. (1991). Constraints in the evolution of life histories. *Phil. Trans. R. Soc. Lond. B*, **332**, 3–13.

Power, G. (1980). The brook charr, *Salvelinus fontinalis*. In: *Charrs: Salmonid Fishes of the Genus Salvelinus* (Ed. by E. Balon), pp. 141–203. The Hague: Dr. W. Junk.

Reznick, D. A., Bryga, H. and Endler, J. A. (1990). Experimentally-induced life history evolution in a natural population. *Nature*, **346**, 357–359.

Ricker, W. E. (1932). Studies of speckled trout (*Salvelinus fontinalis*) in Ontario. *Univ. Toronto Stud., Ont. Fish. Res. Lab.*, **44**, 67–110.

Roff, D. A. (1984). The evolution of life history parameters. *Can. J. Fish. Aquat. Sci.*, **41**, 989–1000.

Rogerson, R. J. (1981). The tectonic evolution and surface morphology of Newfoundland. In: *The Natural Environment of Newfoundland, Past and Present* (Ed. by A. G. Macpherson and J. B. Macpherson), pp. 24–55. St. John's: Memorial Univ. Print. Serv.

Rowe, D. K., Thorpe, J. E. and Shanks, A. M. (1991). Role of fat stores in the maturation of male Atlantic salmon (*Salmo salar*) parr. *Can. J. Fish. Aquat. Sci.*, **48**, 405–413.

Sargent, R. C., Gross, M. R. and van den Berghe, E. P. (1986). Male mate choice in fishes. *Anim. Behav.*, **34**, 545–550.

Schaffer, W. M. (1974). Selection for optimal life histories: the effects of age structure. *Ecology*, **55**, 291–303.

Schaffer, W. M. and Elson, P. F. (1975). The adaptive significance of variations in life history among local populations of Atlantic salmon in North America. *Ecology*, **56**, 577–590.

Schmalhausen, I. I. (1949). *Factors of Evolution*. Philadelphia: Blakiston.

Schmidt-Nielson, K. (1984). *Scaling: Why is Animal Size So Important?* Cambridge: Cambridge Univ. Press.

Smith, C. C. and Fretwell, S. D. (1974). The optimal balance between size and number of offspring. *Am. Nat.*, **108**, 499–506.

Stearns, S. C. and Koella, J. C. (1986). The evolution of phenotypic plasticity in life-history traits: predictions of reaction norms for age and size at maturity. *Evolution*, **40**, 893–913.

Thompson, J. D. (1991). Phenotypic plasticity as a component of evolutionary change. *Trends Ecol. Evol.*, **6**, 246–249.

Thorpe, J. E. (1986). Age at first maturity in Atlantic salmon, *Salmo salar*: freshwater period influences and conflicts with smolting. In: *Salmonid Age at Maturity* (Ed. by D. J. Meerburg), pp. 7–14. *Can. Spec. Publ. Fish. Aquat. Sci.*, **89**.

van den Berghe, E. P. and Gross, M. R. (1984). Female size and nest depth in coho salmon (*Oncorhynchus kisutch*). *Can. J. Fish. Aquat. Sci.*, **41**, 204–206.

van den Berghe, E. P. and Gross, M. R. (1986). Length of breeding life in coho salmon (*Oncorhynchus kisutch*). *Can. J. Zool.*, **64**, 1482–1486.

van den Berghe, E. P. and Gross, M. R. (1989). Natural selection resulting from female breeding competition in a Pacific salmon (coho: *Oncorhynchus kisutch*). *Evolution*, **43**, 125–140.

van Ramshorst, J. D. and van den Nieuwenhuisen, A. (1978). *Aquarium Encyclopedia of Tropical Freshwater Fish*. Lausanne: HP Books/Elsevier.

Williams, G. C. (1966). Natural selection, the costs of reproduction, and a refinement of Lack's principle. *Am. Nat.*, **100**, 687–690.

BEHAVIOURAL CAUSES AND CONSEQUENCES OF LIFE HISTORY VARIATION IN FISH

NEIL B. METCALFE

*Fish Behaviour and Ecology Group, Department of Zoology,
University of Glasgow, G12 8QQ, UK.*

WHAT IS LIFE HISTORY VARIATION?

Intraspecific variation in life histories can take many forms, for example variation in the age at which an individual metamorphoses, migrates or first reproduces, or in the strategies or tactics it adopts whilst reproducing. Such variability can be under environmental and/or genetic control. Evidence from a range of organisms suggests that it is rare for there to be no environmental input, presumably because a life-history trajectory that takes account of the local environmental conditions experienced by the organism allows greater flexibility than one under fixed allelic control, and will therefore have a selective advantage (Levins, 1968; West-Eberhard, 1989).

Environmentally-modulated traits are termed phenotypically plastic, i.e. a given genotype can express more than one phenotype across a range of environmental conditions. The variation in phenotypic expression can be either continuous or take the form of two or more discrete polymorphisms (e.g. alternative reproductive strategies). This paper will focus primarily on the latter of which an abundance of examples exist in fish (see Bruton, 1990). It will also be biased towards those polymorphisms where an adaptive explanation for the plasticity is apparent. My aim is to give an overview of the types of plasticity that occur and the variety of ways in which their expression is controlled. To illustrate general principles I will consider discrete polymorphisms in organisms other than fish, since the details of the processes involved in fish are often poorly understood. I will first explore the general scope of life history variation and how this relates to behaviour. I will then consider two systems (the energetics of breeding in Arctic charr *Salvelinus alpina* and the early life history of Atlantic salmon *Salmo salar*) in detail to show how behaviour, physiology and variable life history patterns interact with one another.

CONTROL MECHANISMS

The environment exerts its influence on a phenotypically plastic life history trait through a control mechanism. Mechanisms are designed to respond to one of two types of stimulus: either a specific cue or a threshold level. A mechanism based around a response to a specific cue is triggered once that cue is detected. The cue usually takes the form of a particular chemical in the environment, which when detected by the organism causes a developmental 'switch' to be thrown. One of the clearest examples comes from marine nudibranchs. As planktonic larvae, they will only commence metamorphosis into the sessile juvenile phase of the life cycle when brought in contact with chemical cues unwittingly given off by corals (upon which the juvenile nudibranchs subsequently feed). This mechanism is clearly adaptive. It ensures that the switch from a mobile to a relatively immobile bodyform occurs only in the vicinity of a food supply. But it also results in a variable and unpredictable duration of the planktonic phase. It has recently been shown that the duration of this phase has no repercussions on subsequent life history (e.g. duration of the remaining phases of the life cycle or reproductive output), suggesting that the larvae enter a developmental hiatus, during which they do not age, while awaiting the coral cue (Miller and Hadfield, 1990).

Relatively few examples of cue-based control mechanisms have been discovered. Far more common is the threshold system, where some feature of either the environment or the organism itself has to exceed a threshold level for the mechanism to be triggered. Locusts *Locusta migratoria* provide a dramatic example. These exist in two quite distinct morphs, the solitary and the gregarious forms. At low population densities, individuals develop into the solitary form which is green, has small wings and lays many small eggs. If densities increase above a threshold level, an alternative developmental pathway is triggered, so producing the gregarious form. This is brown, has larger wings (allowing it to disperse) and lays fewer but larger eggs (which give rise to larger young that are therefore better able to compete). The density effect is thought to be mediated by a pheromone (Nijhout and Wheeler, 1982).

A well-documented example of a threshold control mechanism in fish concerns sex reversal in cleaner wrasse. All females have the capability of becoming males, but whether or not they do so depends on the local sex ratio and their position in the female dominance hierarchy—females are inhibited from changing sex until a male dies, and then it is only the most dominant female that will undergo the change (Robertson, 1972; see also Warner, 1988).

TYPES OF LIFE HISTORY VARIATION

Examples of life history variation can be categorised into cases where individuals adopt different (i.e. alternative) strategies, and those where they adopt the same strategy but show variation in the time taken to complete developmental stages (such as time to metamorphosis or to sexual maturation). In developmental terms, these two categories correspond to the 'developmental conversion' and 'phenotypic modulation' groupings of Smith-Gill (1983). Alternative strategies and polymorphisms were once thought of as developmental mistakes or freaks—even R. A. Fisher once said in this context that "it is not surprising that such elaborate machinery should sometimes go wrong" (Wigglesworth, 1961). In the last 15 years this view has been completely revised; it is now accepted that the observed plasticity often has an adaptive basis (see Arak, 1984; Potts and Wootton, 1984; Caro and Bateson, 1986; West-Eberhard, 1989; Bruton, 1989, 1990).

Alternative Strategies

Fish provide some of the best illustrations of alternative reproductive strategies, almost always in terms of variation amongst males. A common scenario is for some males to defend overtly either females or nesting sites (to which females are attracted), while others attempt to obtain fertilizations by stealth, sneaking in from a nearby waiting position at the moment of egg release and releasing sperm at the same time as the 'territorial' male. These 'sneaker' males may be more cryptically coloured than the territorials, and may actually mimic females in appearance and behaviour. Variations on this basic theme have been described in species as diverse as bluegill sunfish *Lepomis macrochirus* (Dominey, 1980; Gross and Charnov, 1980), salmon *Salmo* spp. and *Oncorhynchus* spp. (Jones and King, 1952; Gross, 1985; Noltie, 1990) and cichlids *Pseudocrenilabrus philander* (Chan and Ribbink, 1990). However, it is important to realise that the processes that arrive at this same end-point may be very different. In the case of *Oncorhynchus* species of salmon, which only breed once and then die, each male can only adopt one strategy during its life. This is especially evident in coho salmon *Oncorhynchus kisutch*. Males mature at either two or three years of age. Two-year-olds are only successful as sneakers (since they are too small to be successful in defending females) while 3-year-olds are too large to escape detection when sneaking and so must compete for females directly (Gross, 1985, 1991). In contrast, the male cichlid *P. philander* can spawn several times, and will switch from being a sneaker to holding a territory as it grows larger and thus rises in the dominance hierarchy; these males can therefore exhibit the different strategies sequentially. In fact, they have the ability to be completely flexible and switch rapidly from one strategy to the other in response to changes in the local density of more dominant fish (Chan and Ribbink, 1990). A more extreme form of a sequential pattern of alternative strategies occurs in those fish that change sex, either from male to female (protandry) or vice versa (protogyny); there are now many documented cases of fish changing sex in response to local population demography and to the individual's position in a dominance hierarchy (see Warner, 1988 and references therein).

Differences in Developmental Rate

There is somewhat of a fine line to be drawn between the normal 'random' variation in timing between individuals that one would expect for any developmental process and the variation that deserves to be singled out as an example of phenotypic plasticity. However, in many cases the variation is so extreme and discontinuous (e.g. as in age at first reproduction for seasonal breeders) that we can genuinely describe the population as adopting different life history patterns. A particularly striking example is the freshwater snail *Physella virgata virgata*, which in predator-free environments normally starts to reproduce at an age of 45–50 days and at a size of 5 mm (shell length). Reproduction leads to the virtual cessation of growth, so that the snails remain this size until they die some 40 days later. However, when in the presence of predatory crayfish *Orconectes virilis* (which only eat small snails) they delay maturation until they are about 60 days old, by which time they are of a size that suffers from minimal predation (Crowl and Covich, 1990). It appears that the deferring of maturation is triggered by a specific chemical cue released when crayfish attack snails, since the same delay occurs in snails merely exposed to water coming from tanks that contain feeding crayfish.

BEHAVIOURAL CAUSE OR CONSEQUENCE?

The title of this article is perhaps misleading, in that it implies that we can separate causes from consequences of life-history variation. In fact, it is often very difficult to tell which is which. Where reproductive strategies are involved it is usually the case that the behaviour (e.g. whether to sneak or to defend) is a *consequence* of the fish having followed a particular developmental pathway: in the earlier example of coho salmon, the two-year-old males have little option but to behave as sneakers once they have adopted the strategy of maturing at a small size.

It is more difficult to demonstrate that a behaviour was itself a direct *cause* of a subsequent life-history strategy. One clear-cut situation arises in larval *Ambystoma tigrinum* salamanders. The larvae normally eat small aquatic invertebrates, but occasionally some individuals turn cannibalistic, eating sibs that are slightly smaller than themselves. Cannibalism leads to a growth spurt, and reduces the time taken to reach the size threshold for metamorphosis by almost one third (Lannoo *et al.,* 1989); thus the duration of the most vulnerable stage of the life-cycle can apparently be compressed merely by a change to a more gruesome diet. There are also documented cases of diet-induced physiological changes in fishes. *Cichlasoma* cichlids can be categorised into two forms with respect to their method of feeding, either biting or sucking. Adaptations to the two separate feeding methods have led to their being recognisable on the basis of mouthpart morphology. However, it has been shown in *C. managuense* that this is a phenotypically plastic trait, since mouthparts specialised for either biting or sucking can be induced by providing a diet appropriate for one or the other technique. It is also a flexible response, since a diet switch will produce an analogous change in the structure of the mouthparts over the course of several months (Meyer, 1987).

In fishes, a major driving force behind life-history variation is variability in growth rates. It has long been recognised that age at sexual maturation is closely connected to growth rates (Alm, 1959), faster growing individuals tending to mature at an earlier age (see also Policansky, 1983). This is dealt with in more detail by Hutchings (this volume), and so will not be discussed further here. However, it is worth noting that growth rate or size may not be the only determining factor: successful reproduction is also dependent on adequate body reserves, and there is some evidence from Atlantic salmon that the decision on whether or not to proceed with maturation in a given year is influenced by lipid levels 8–10 months prior to actual reproduction (Thorpe *et al.,* 1990; Rowe *et al.,* 1991).

Lipid reserves may influence not only the age at first reproduction, but also the interval between reproductive attempts. Most habitats are productive enough to allow seasonally-breeding fish to recoup the energy and resources expended in one breeding season by the onset of the next. However, in the arctic there is such a short period of the year when both temperatures and food availability are sufficiently high to allow fish to feed and grow that they may not be able to replenish reserves lost through breeding in the course of a single year, and so cannot breed annually. This has been demonstrated in a study of the energetics of reproduction in arctic charr *Salvelinus alpinus* in northern Canada (Dutil, 1986). These fish obtain virtually all of their food during a two month spell in sea water during the summer. They then return to barren fresh water lakes for the rest of the year, during which time they lose 30% of their energy reserves. However, when preparing to reproduce they migrate into a small spawning lake rather than migrating to feeding areas at sea. The combined effects of breeding and a missed year of feeding results in their losing, on average, an additional 53% of their reserves before next being able to feed at sea the following summer. This results in a *minimum* of a two-year reproductive cycle; moreover, the ability to recoup lost reserves within a single two-month feeding season decreases as the fish get larger (Figure 1), so that older fish may only be able to reproduce once every 3–4 years. A similar (but less extreme) situation has recently been reported in Atlantic salmon, with larger repeat-spawning fish being more likely

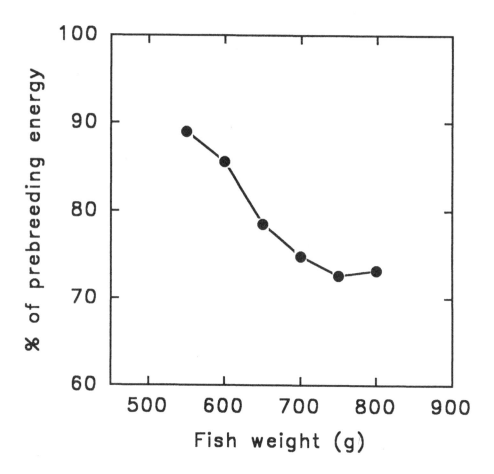

Figure 1   The cost of breeding in arctic charr, in relation to fish size. Cost is measured here as the percentage of energy they had prior to breeding that can be regained after one summer of feeding. On average, small breeders can reach almost 90% of their pre-breeding level of energy reserves within one year, whereas larger fish only attain 70%, and so will require longer before being able to breed again. Data taken from Dutil (1986).

to skip a year between breeding attempts, presumably because the quantity of lost resources that must be restored is greater in absolute terms (Jonsson *et al.*, 1991).

MIGRATION BETWEEN HABITATS

The charr example given above raises the question of why some fish make extensive migrations between fresh and salt water (i.e. are diadromous), given the problem that it involves a complex physiological switch between different modes of osmoregulation and a switch between vastly different predator-induced mortality regimes. A clue is given by the fact that the proportion of fish species that are anadromous (i.e. are born in fresh water but subsequently migrate seawards at some point in the life-cycle in order to feed) increases with latitude (Figure 2). This corresponds with a decline in the relative productivity of freshwater ecosystems compared to marine ones, and it has been shown that fish generally migrate towards the more productive medium, so that they migrate from marine to fresh waters at low latitudes but switch to the opposite direction nearer the poles (Gross *et al.*, 1988).

There are intraspecific as well as interspecific variations in migratory tendency. The brown trout *Salmo trutta* tends to be a freshwater resident in productive ecosystems, but an anadromous migrant when the freshwater environment is less suitable for growth; in intermediate habitats both forms can coexist, with females (which have a greater requirement for body reserves when reproducing) tending to migrate while males are more likely to remain as residents (Jonsson, 1985). In Atlantic salmon the majority of individuals in most populations are migratory (although see Thorpe (1986) for sex biases in the composition of migrants), but there is still considerable variability in the timing of this migration, and it is the causes and

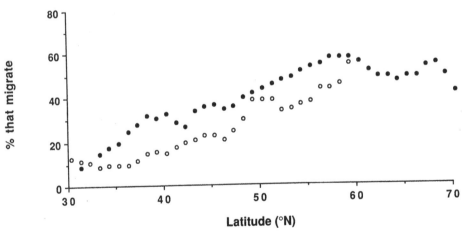

Figure 2   The percentage of North American fish species born in fresh water that migrate to sea at some point in their life cycle, in relation to latitude. The data are separated into Pacific (closed circles) and Atlantic (open circles) river systems; on both coasts the incidence of anadromy increases with latitude. Data taken from McDowall (1988).

consequences of this life-history variation that I will concentrate on in the rest of this article.

AGE AT SEAWARD MIGRATION IN ATLANTIC SALMON

Atlantic salmon eggs are laid in fast-running freshwater streams and rivers, and the young (called parr) that emerge in the spring remain in this habitat until they undertake the spring migration to sea (at which point they are called smolts). They remain at sea for between 1–3 years, before returning to their natal river to spawn. The duration of the parr phase of the life-cycle is very variable: the majority of smolts migrating down the rivers of northern Spain are one year old, whereas those in northern Canada may be aged up to 8. This between-population variability has been shown to be related to growth rates (faster-growing fish migrating at a younger age). An index of growth opportunity based on mean monthly temperatures and hours of daylight (since salmon parr do not feed in darkness, Higgins and Talbot, 1985) has been shown to explain over 82% of the variation in mean ages of smolts from rivers throughout the geographic range of the species (Figure 3).

Large-scale environmental influences therefore account for most of the between-population differences in smolt ages, but do not explain why there is variability within single rivers—most rivers produce two or more different age-classes of smolt. Even a single family of parr reared in the same hatchery or laboratory tank and given excess food will still usually produce both 1- and 2-year-old smolts. A series of laboratory experiments over the last 15 years (summarised in Thorpe, 1989; Thorpe et al., 1992) has demonstrated that fish reach a life-history 'decision' in late July or early August of their first year (i.e. three months after first feeding) as to whether or not they will migrate some 10 months later.

What Consequences Does This Decision Have?

The decision on when to migrate has important consequences for the subsequent behaviour of the fish. Salmon that will migrate the following spring exhibit an appetite and growth spurt in their first autumn and continue feeding throughout the winter (Figure 4). However, those fish that have elected to defer migration until they are two years old rapidly lose their appetite after having made the decision, until they are consuming no more than a maintenance ration (Figure 4). They remain in this state of 'anorexia' throughout the winter, only increasing their appetite if their lipid reserves drop to low levels (Metcalfe and Thorpe, 1992). These marked differences in feeding and growth rates between fish adopting the two strategies result in the development of a bimodal size distribution in the first autumn, with future one-year-old smolts (S1 fish) forming the upper modal group and 2-year-old smolts (S2's) the lower modal group (Figure 5). A possible reason for the divergence in feeding rates is that S2 fish are minimising the risk of predation by not feeding (and instead remaining hidden in the stream bed) whereas S1 fish must feed in order to reach a large size by the time of the smolt migration. This dramatic illustration of a divergence in life-history patterns, first described (and understandably more evident) in laboratory populations (Thorpe, 1977), has also been documented in the wild (e.g. Heggenes and Metcalfe, 1991).

212

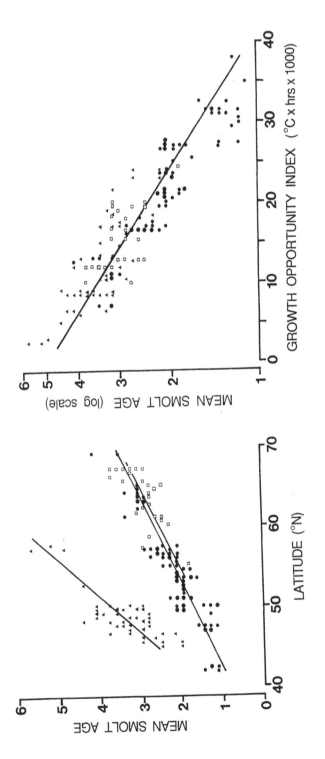

Figure 3 Mean age of Atlantic salmon smolts in different rivers in relation to (left) latitude and (right) an index of growth opportunity. The data are divided into three geographical regions: North America (closed triangles), East Europe (open squares) and West Europe (closed circles). Average smolt age is related to latitude within regions but not overall; however, differences between regions disappear when growth opportunity is taken into account (overall regression: $r^2=0.823$). Larger symbols represent 2–5 data points. Adapted from Metcalfe and Thorpe (1990).

PERCENTAGE OF PELLETS THAT WERE EATEN

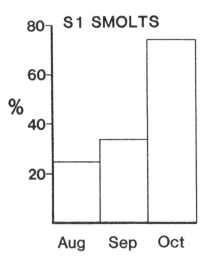

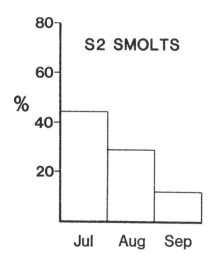

Figure 4   Differences in appetite of Atlantic salmon that will migrate to sea aged 1 (S1 smolts) and 2 (S2 smolts). Data presented are the percentage of pellets that isolated fish consumed in standardised trials. Adapted from Metcalfe *et al.* (1986, 1988).

What Causes a Salmon to Adopt a Particular Strategy?

The divergence in the population is not simply a sex- or maturation-related phenomenon, since both modal groups contain both sexes and the split occurs even when none of the fish are yet ready to mature. Nor is it genetically fixed, since the proportion of fish entering the two modes can be radically altered by manipulating growth conditions in the months leading up to the decision (Thorpe *et al.*, 1989). It appears that a fish must exceed a threshold growth rate or size at the time of the decision (July–August) if it is to carry on the trajectory towards migrating the following spring; if it falls below this threshold it 'switches off' and enters a dormant, anorexic phase for the rest of the year. It will then resume growth the following spring and re-assess its performance when the decision time comes again (Figure 6). This model can incorporate genetic differences between fish, since the reference threshold that must be exceeded will differ slightly between individuals. However, any environmental factor that affects early growth rates may influence the ability of an individual fish to exceed its own threshold and so migrate at an early age. Social as well as physical aspects of the environment should be important, leading to the prediction that a fish's social status and competitive ability should influence the age at which it migrates, since they will have an impact on its ability to acquire food. These predictions have been upheld in experiments that demonstrated that the more dominant and competitive fish were indeed more likely to migrate aged one rather than two years old (Figure 7; Metcalfe *et al.*, 1989, 1990; Metcalfe, 1991).

   What makes a salmon dominant? Salmon fry start to compete for space and food within days of first feeding. Their ability to win aggressive encounters at this stage

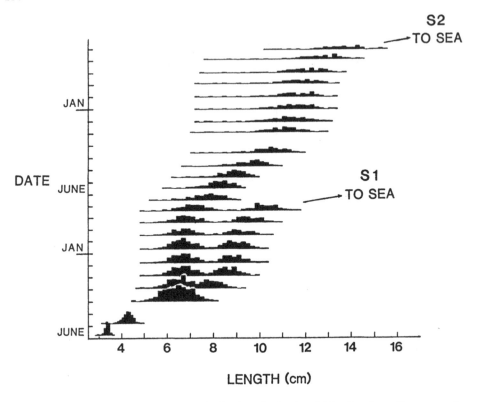

Figure 5 Changes in the size-frequency distribution of a family of Atlantic salmon from soon after first feeding. The upper mode of the bimodal size distribution contains S1 fish, the lower mode S2 fish. Adapted from Thorpe (1977).

is not dependent on their size relative to their opponent (Metcalfe *et al.*, in press)— in fact, size and status are poorly correlated for some months, and it is more a case of dominants becoming the largest fish rather than the largest fish becoming dominants (Huntingford *et al.*, 1990). The latest evidence suggests that a fry's relative metabolic rate may have a crucial influence on its future growth, possibly through an effect on social status (Metcalfe *et al.*, in press). Whatever the primary motivating force, it is clear that very small individual differences in behaviour and physiology in the first few weeks of life can have major repercussions for subsequent life-history patterns, with a small difference in competitive ability potentially being enough to tip the balance as to the number of years a salmon takes to complete the freshwater stage of the life cycle.

ACKNOWLEDGMENTS

I would like to thank Jeff Hutchings and Jim Grant for their helpful comments on the manuscript. It was written while in receipt of a NERC Advanced Fellowship.

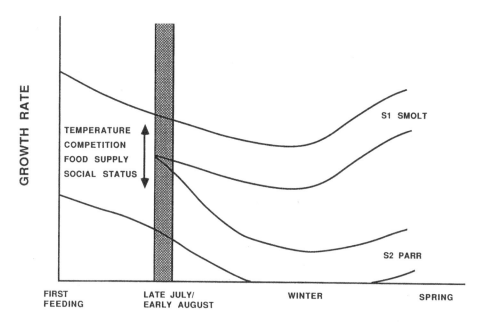

Figure 6   Model of the relationships between growth rate and age of smolting in juvenile Atlantic salmon. Fish must exceed a threshold growth rate during a critical period in late July/early August if they are to become S1 smolts; otherwise they become anorexic and their growth rate drops. Various factors will affect growth rates during the period leading up to this life-history decision, and so influence the proportion of fish entering the S1 pathway.

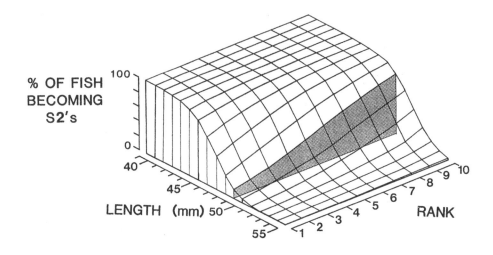

Figure 7   Logistic regression curves showing the relationship between fish length in mid-July, dominance rank (1 = most dominant) and likelihood of becoming an S2 (rather than S1) smolt. Size and rank have independent effects; the shaded section shows the influence of rank for fish of a given size. Adapted from Metcalfe *et al.* (1989).

References

Alm, G. (1959). Connection between maturity, size and age in fishes. *Rep. Inst. Freshw. Res., Drottningholm,* **40**, 5–145.

Arak, A. (1984). Sneaky breeders. In: *Producers and Scroungers, Strategies of Exploitation and Parasitism* (Ed. by C. J. Barnard), pp. 154–194. London: Croom Helm.

Bruton, M. N. (1989). *Alternative Life-history Styles of Animals* (Ed.). Dordrecht, Holland: Kluwer Academic Publishers.

Bruton, M. N. (1990). *Alternative Life-history Styles of Fishes* (Ed.). *Env. Biol. Fishes,* **28**.

Caro, T. M. and Bateson, P. P. G. (1986). Organization and ontogeny of alternative tactics. *Anim. Behav.,* **34**, 1483–1499.

Chan, T-Y. and Ribbink, A. J. (1990). Alternative reproductive behaviour in fishes, with particular reference to *Lepomis macrochira* and *Pseudocrenilabrus philander. Env. Biol. Fishes,* **28**, 249–256.

Crowl, T. A. and Covich, A. P. (1990). Predator-induced life-history shifts in a freshwater snail. *Science,* **247**, 949–951.

Dominey, W. J. (1980). Female mimicry in male bluegill sunfish—a genetic polymorphism? *Nature, Lond.,* **284**, 546–548.

Dutil, J-D. (1986). Energetic constraints and spawning interval in the anadromous arctic charr (*Salvelinus alpinus*). *Copeia,* 945–955.

Gross, M. R. (1985). Disruptive selection for alternative life histories in salmon. . *Nature,* **313**, 47–48.

Gross, M. R. (1991). Salmon breeding behavior and life history evolution in changing environments. *Ecology,* **72**, 1180–1186.

Gross, M. R. and Charnov, E. L. (1980). Alternative male life history strategies in bluegill sunfish. *Proc. Natl. Acad. Sci. U.S.A.,* **77**, 6937–6940.

Gross, M. R., Coleman, R. M. and McDowall, R. M. (1988). Aquatic productivity and the evolution of diadromous fish migration. *Science,* **239**, 1291–1293.

Heggenes, J. and Metcalfe, N. B. (1991). Bimodal size distributions in wild juvenile Atlantic salmon populations and their relationship with age at smolt migration. *J. Fish Biol.,* **39**, 905–907.

Higgins, P. J. and Talbot, C. (1985). Growth and feeding in juvenile Atlantic salmon. In: *Nutrition and Feeding in Fish* (Ed. by C. B. Cowey, A. M. Mackie and J. G. Bell), pp. 243–263. London: Academic Press.

Huntingford, F. A., Metcalfe, N. B., Thorpe, J. E., Graham, W. D. and Adams, C. E. (1990). Social dominance and body size in Atlantic salmon parr (*Salmo salar* L.). *J. Fish Biol.,* **36**, 877–881.

Jones, J. W. and King, G. M. (1952). Spawning of the male salmon parr. *Nature,* **169**, 882.

Jonsson, B. 1985. Life history patterns of freshwater resident and sea-run migrant brown trout in Norway. *Trans. Am. Fish. Soc.,* **114**, 182–194.

Jonsson, N., Hansen, L. P. and Jonsson, B. (1991). Variation in age, size and repeat spawning of adult Atlantic salmon in relation to river discharge. *J. Anim. Ecol.,* **60**, 937–947.

Lannoo, M. J., Lowcock, L. and Bogart, J. P. (1989). Sibling cannibalism in noncannibal morph *Ambystoma tigrinum* larvae and its correlation with high growth rates and early metamorphosis. *Can. J. Zool.,* **67**, 1911–1914.

Levins, R. (1968). *Evolution in Changing Environments.* Princeton: Princeton University Press.

McDowall, R. M.(1988). *Diadromy in Fishes.* London: Croom Helm.

Metcalfe, N. B. (1991). Competitive ability influences seaward migration age in Atlantic salmon. *Can. J. Zool.,* **69**, 815–817.

Metcalfe, N. B. and Thorpe, J. E. (1990). Determinants of geographical variation in the age of seaward-migrating salmon, *Salmo salar. J. Anim. Ecol.,* **59**, 135–145.

Metcalfe, N. B. and Thorpe, J. E. (1992). Anorexia and defended energy levels in over-wintering juvenile salmon. *J. Anim. Ecol.,* **61**, 175–181.

Metcalfe, N. B., Huntingford, F. A. and Thorpe, J. E. (1986). Seasonal changes in feeding motivation of juvenile Atlantic salmon (*Salmo salar*). *Can. J. Zool.*, **64**, 2439–2446.

Metcalfe, N. B., Huntingford, F. A. and Thorpe, J. E. (1988). Feeding intensity, growth rates, and the establishment of life-history patterns in juvenile Atlantic salmon *Salmo salar*. *J. Anim. Ecol.*, **57**, 463–474.

Metcalfe, N. B., Huntingford, F. A., Graham, W. D. and Thorpe, J. E. (1989). Early social status and the development of life-history strategies in Atlantic salmon. *Proc. R. Soc. Lond.*, **B236**, 7–19.

Metcalfe, N. B., Huntingford, F. A., Thorpe, J. E. and Adams, C. E. (1990). The effects of social status on life history variation in juvenile salmon. *Can. J. Zool.*, **68**, 2630–2636.

Metcalfe, N. B., Wright, P. J. and Thorpe, J. E. (In press). Relationships between social status, otolith size at first feeding and subsequent growth in Atlantic salmon (*Salmo salar* L.). *J. Anim. Ecol.*, **61**, 585–589.

Meyer, A. (1987). Phenotypic plasticity and heterochrony in *Ciclasoma managuense* (Pisces, Cichlidae) and their implications for speciation in cichlid fishes.. *Evolution*, **41**, 1357–1369.

Miller, S. E. and Hadfield, M. G. (1990). Developmental arrest during larval life and life-span extension in a marine mollusc. *Science*, **248**, 356–357.

Nijhout, H. F. and Wheeler, D. E. (1982). Juvenile hormone and the physiological basis of insect polymorphisms. *Q. Rev. Biol.*, **57**, 109–133.

Noltie, D. B. (1990). Intrapopulation variation in the breeding of male Pink salmon (*Oncorhynchus gorbuscha*) from a Lake Superior tributary. *Can. J. Fish. Aquat. Sci.*, **47**, 174–179.

Policansky, D. (1983). Size, age and demography of metamorphosis and sexual maturation in fishes. *Amer. Zool.*, **23**, 57–63.

Potts, G. W. and Wootton, R. J. (Eds) (1984). *Fish Reproduction, Strategies and Tactics*. London: Academic Press.

Robertson, D. R. (1972). Social control of sex reversal in a coral reef fish. *Science*, **177**, 1007–1009.

Rowe, D. K., Thorpe, J. E. and Shanks, A. M. (1991). Role of fat stores in the maturation of male Atlantic salmon (*Salmo salar*) parr. *Can. J. Fish. Aquat. Sci.*, **48**, 405–413.

Smith-Gill, S. J. (1983). Developmental plasticity: developmental conversion versus phenotypic modulation. *Amer. Zool.*, **23**, 47–55.

Thorpe, J. E. (1977). Bimodal distribution of length of juvenile Atlantic salmon (*Salmo salar* L.) under artifical rearing conditions. *J. Fish Biol.*, **11**, 175–184.

Thorpe, J. E. (1986). Age at first maturity in Atlantic salmon, *Salmo salar*: freshwater period influences and conflicts with smolting. *Can. Spec. Publ. Fish. Aquat. Sci.*, **89**, 7–14.

Thorpe, J. E. (1989). Developmental variation in salmonid populations. *J. Fish Biol.*, **35** (Suppl. A), 295–303.

Thorpe, J. E., Adams, C. E., Miles, M. S. and Keay, D. S. (1989). Some photoperiod and temperature influences on growth opportunity in juvenile Atlantic salmon, *Salmo salar*. *Aquaculture*, **82**, 119–126.

Thorpe, J. E., Talbot, C., Miles, M. S. and Keay, D. S. (1990). Control of maturation in cultured Atlantic salmon *Salmo salar* in pumped seawater tanks, by restricting food intake. *Aquaculture*, **86**, 315–326.

Warner, R. R. (1988). Sex change and the size-advantage model. *TREE*, **3**, 133–136.

West-Eberhard, M. J. (1989). Phenotypic plasticity and the origins of diversity. *Ann. Rev. Ecol. Syst.*, **20**, 249–278.

Wigglesworth, V. B. (1961). Insect polymorphism—a tentative synthesis. *R. Entomol. Soc. Lond. Symp.*, **1**, 103–113.

SEX ROLE REVERSAL IN A PIPEFISH

GUNILLA ROSENQVIST

Department of Zoology, Uppsala University, Box 561, S-751 22 Uppsala, Sweden

INTRODUCTION

In most animals, the rate of female reproduction limits male reproductive rate and males compete to mate with females. This sexual competition promotes traits which enhance male mating success, either by competitiveness between males or by attracting females. The resulting sexual selection is thought to explain why males are generally larger, more colourful, more vocal and more conspicuous, as well as more aggressive than females. In contrast, role reversed species are those in which females are the predominant competitors for access to mates. One consequence of role reversal is that females, rather than males, are likely to be more modified by sexual selection (Vincent *et al.*, 1992). The pipefish, *Nerophis ophidion* (Figure 1), is an example of a role reversed species. Here we find competing females with elaborate secondary sexual characters and choosy males. This article examines the causes and consequences of sex role reversal in the Syngnathid fish, *Nerophis ophidion*.

Much of our understanding of sex differences in morphology and behaviour arises from studies of birds and mammals, but these may not always be relevant to fishes. For example, mammals are at one end of a spectrum with predominant female parental care, and the teleost fishes are at the other end with predominantly male care. Parental care occurs in 29% and males are the care giver in 58% of these (Breder and Rosen, 1966; Blumer, 1982; Gross and Shine, 1981; Thresher, 1984; Gross and Sargent, 1985). This makes fishes ideal subjects for studying the behavioural dynamics of parental care within the context of life-history evolution. First they exhibit a great phylogenetic diversity of parental care. Second, many species can easily be studied in the field. Third, many species adapt readily to the laboratory where one can conveniently control or manipulate different variables of interest.

The biology of the pipefish *Nerophis ophidion* offers insights into the evolution of parental care and sex roles, and sex role reversal. Pipefishes have exclusive male care, with males that undergo a long pregnancy and females that provide no care of the young at all. This extreme case of parental investment by males leads to the prediction that males should limit female reproductive rate and that females should be the predominant competitors for mates (Trivers, 1972).

SEX ROLES

Here I will summarize some of the theory concerning the evolution of female and male sexual differences in behaviour and appearance, define the term sex role

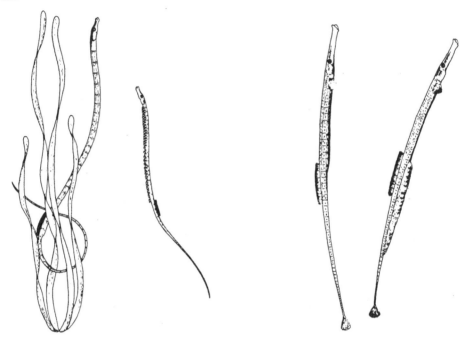

Figure 1 Female and brooding male of *Nerophis ophidion* (left) and *Syngnathus typhle* (right). Drawing by Annika Robertson.

reversal, and show that sex role reversal can be a useful term to describe patterns found between females and males in nature.

Most of us probably have some kind of picture which we would call an "ordinary role" for females and males. However, no unambiguous definition of a sex role exists, only descriptions of character assemblages claimed to be peculiar to the biology of males and females, respectively, or hypotheses about why some traits should occur together in one or the other sex. The term "sex role reversal" is thus ambiguous and often provokes considerable disagreement about its correctness or usefulness. The term has occasionally even been applied to situations in which males are performing behaviours that are culturally or politically considered to be a female prerogative and vice versa.

Williams (1975) asked "why are males masculine and females feminine and occasionally vice-versa?" His definition of masculinity was "taking less interest in the young, being more active in courtship, being less discriminating in mate choice, being more inclined towards promiscuity and polygamy", and as males "being more contentious among themselves." But why, if each individual is maximizing its own genetic survival, should a female be less anxious to have her eggs fertilized than a male is to fertilize them and why should the young be of greater interest to one sex than to the other sex (Williams, 1975)? The answer requires a closer look at the differences between females and males and the theory that relates these differences to their mating and parental behaviour.

In sexually reproducing plants and animals, the female produces large, immobile and energy rich gametes, called eggs, whereas male gametes, or sperm, are numerous, mostly mobile and consist of little more than a piece of self-propelled DNA. This fundamental sexual difference in gamete size is often, but misleadingly, used to explain all other differences between the sexes in their reproductive behaviour, dimorphism and sexual roles. A more accurate statement of the theory is that females, by virtue of producing larger gametes, initially make more of an investment in resources in the zygote.

The differential investment in offspring between the sexes probably originated as gametes diverged in size and function, (i.e., with the origin of anisogamy Bateman, 1948; Parker et al., 1972). In addition to differential investment in the zygotes, there is a difference in which sex can produce gametes most quickly (Baylis, 1978, 1981). The production of fewer larger eggs by females means, all else being equal, that females may be more limited in their reproduction by their rate of gamete production than males would be. Given that a female's reproduction is limited by her own capacity to produce gametes, a male's reproductive success will be limited by mating opportunities and not by the number or rate of production of gametes. The theory then concludes that other differences among males and females follow from this differential investment in gametes. In particular, the evolution of secondary sexual characters in males, is expected to follow either by intrasexual selection, where males compete for access to females, or by intersexual selection where females choose males based on certain characters or quality (Darwin, 1871; Trivers, 1972).

There are other reasons why studies of sexual selection and the evolution of sex roles have focused on differential parental investment. For example, work on mammals reveals extended investment in each offspring exclusively by females during pregnancy and lactation. Even in birds heavy initial investment in eggs limits the ability of females to produce additional clutches rapidly. These kinds of differences in parental investment have been associated with patterns of sexual selection and mating systems. Thus the evolution of sex roles, whether typical or reversed, has usually been seen in the context of sex differences in parental investment (Trivers, 1972).

The simple story of differential parental investment is an incomplete explanation of sex-role evolution. The problem is that although differences in investment in gametes and offspring might be important in determining the reproductive behaviour of each sex and thus the degree to which one sex, often males, will compete for limited fertilization opportunities. This cannot be the full story. Examples of sex role reversal make this clear since what is reversed is not the size and number of gametes produced or the amount each sex invests in fertilized zygotes but the degree to which one sex competes for limited reproductive opportunities through the other sex. This sexual competition drives sexual selection (Clutton-Brock and Vincent, 1991).

The degree to which one sex is limiting is due not only to initial parental investment but also to other factors including the operational sex ratio, costs of defending mates, and costs of parental care. For example, in species with male parental care and females that compete for males, males still might have a higher rate of gamete production (Clutton-Brock, 1991). Investment in offspring is not sufficient to define sex role reversal. Clutton-Brock and Vincent (1991) argue that competitive differences between the sexes arise from sex differences in reproductive rates, measured as the maximum number of independent offspring that parents can produce per unit time, calculated over the period of time when both sexes are reproductively active. These differences in reproductive rate will bias the operational sex ratio (ratio of receptive females to receptive males (Emlen and Oring, 1977). This in turn will govern the comparative intensity of mating competition and thus the pattern of sexual selection. A sex difference in reproductive rate incorporates differences in parental investment but may also include other environmental plus anatomical constraints, so male parental investment is not enough to term a species sex role reversed. Males can be the predominant competitors for mates if their parental care does not depress their reproductive rate below that of females. Males are the predominant care-givers in teleost fishes, but they are also generally the primary competitors for mates, and females are the choosy sex (Breder and Rosen, 1966; Blumer, 1979). For example, the male three spined-stickleback, *Gasterosteus aculeatus*, takes responsibility for care of the eggs but males still compete more intensely with other males for mates than do females and females are the choosy sex (Wootton, 1984). The male stickleback can achieve a higher reproductive rate than the female because he can care for multiple clutches simultaneously (Wootton, 1984), which lead to intense competition among males for mates, and the evolution of secondary sexual characters such as bright red pigmentation in the males.

In this paper I define sex role reversal not as a change in parental investment in zygotes, but as a change in the pattern of competition for limited opportunities for reproduction. In sex role reversed species females compete more intensely than males for access to mates (Vincent *et al.*, 1992). Where males are severely limited in the number of young they can brood in space or time (as in those species where males brood eggs dermally, orally or in pouches) and females have the faster reproductive rate, females would be expected to be the primary competitors for mates. This should result in choosy males, and strong sexual selection on females, which as a result may be larger, more colourful or more ornamented than males. Examples of sex role reversed species include the polyandrous spotted sandpiper (*Actitis macularia*, (Oring and Maxson, 1978; Oring and Lank, 1984)) and two species of pipefishes (*Syngnathus typhle, Nerophis ophidion*), (Figure 1, Berglund *et al.*, 1986a,b, 1989; Berglund, 1991; Berglund and Rosenqvist, 1990; Rosenqvist, 1990). In these species females compete more intensely for mating partners, and females can produce more eggs during the breeding season than males can care for.

Sex role reversal is a question of interest not only to theoreticians. Role reversed species are those in which the direction of sexual selection is opposite to the usual pattern and can thus be extremely valuable in probing generalizations about the evolution of sex differences (Vincent *et al.*, 1992). The family Syngnathidae which includes pipefishes and seahorses, shows specialized male care and has been considered to be sex role reversed (Williams, 1975). Studies of Syngnathids provide an opportunity to explore the causes and consequences of sex role reversal.

NATURAL HISTORY

The family Syngnathidae (which includes pipefishes and seahorses) consists of approximately 175 species. It is cosmopolitan in distribution, most species living in estuarine or coastal marine areas (Dawson, 1985). Many species are very cryptic in shape and appearance. The male assumes a large portion of the parental burdens, providing the offspring with nutrients and oxygen while brooding them on his body or in a brood pouch (Linton and Soloff, 1964; Kronester-Frei, 1975; Haresign and Shumway, 1981; Berglund *et al.*, 1986b).

I will concentrate on one well-studied species of pipefish, *Nerophis ophidion*, common in *Zostera* beds of the Swedish west coast. *N. ophidion* is cryptic and sexually dimorphic. Females are larger than males, have blue sexual colouring and develop a skin fold on their ventral side during the breeding period, which gives them a larger appearance (Figure 2). The skin fold is resorbed after the breeding season. Males lack sexual colour, the skin fold or other features that could be used in attracting females or in competing with other males.

The species has a prolonged courtship during which the female is the active sex, displaying the blue colour and her skin fold for the male. The male follows the female, visually orienting to the blue pattern on the female. Just before copulation the male's nose turns yellow (Fiedler, 1954, pers. obs.). The yellow nose might be an indication for the female that the male is ready to receive the eggs. The female transfers her eggs in pearl-like rows to the male's ventral side with the help of an ovipositor and the male fertilizes them by sinking through his sperm-cloud (Fiedler, 1954). Zygotes are embedded into the epithelial cell-layer on the male's abdomen and nourished during the male's brooding period (Kronester-Frei, 1975; Berglund *et al.*, 1986b). The male is pregnant for about one month, after which independent young emerge.

The female gives all her eggs to one male in one transfer and the male carries eggs from only one female at a time, but he can have more than one brood during a season, and the females may produce more than one clutch of eggs. These fishes show every sign of being sex role reversed: 1) the female has a sexual coloration lacking in males; 2) the female possesses a ventral skin fold; 3) the female is more active during

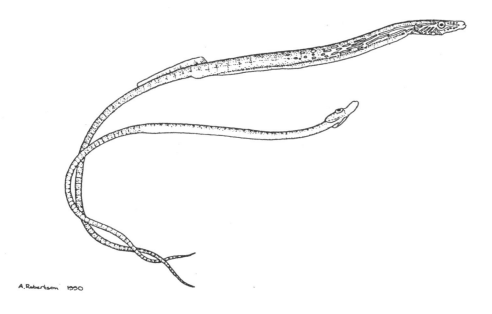

A. Robertson 1990

Figure 2 Female and male of *Nerophis ophidion*. Drawing by Annika Robertson.

courtship. But do females compete among themselves for males? And do they do so because males limit female reproductive rate?

MALES AS A LIMITING RESOURCE FOR FEMALE REPRODUCTIVE SUCCESS

A theory of role reversal based only on differential parental investment (Trivers, 1972) predicts that in a role reversed species the male should invest more in the offspring compared to the female. Pipefish clearly show that such a theory is insufficient. Male parental investment in oxygen and nutrients during his pregnancy cannot explain the reversed sex roles in this species. Males invest a smaller amount of energy in offspring than do females (Berglund *et al.*, 1986b). Field samples showed that there was no correlation between male size and number of offspring the male can carry on his body—a small male can take all eggs from a large female (Berglund *et al.*, 1989). However, by investing in care of the eggs males do potentially decrease their rate of reproduction.

If females can produce more eggs than a male can care for within the time span of one brooding cycle we should expect males to constrain reproductive success of females. In an experiment, the male's pregnancy length was measured as a possible limitation for females. The maximal reproductive output of females when they were not constrained by the availability of males was measured and then compared with male pregnancy length (Berglund *et al.*, 1989). The result was indeed that a female could produce a new clutch of eggs faster than the male could complete a pregnancy. In the experiment, females produced on average 1.8 clutches within the time span of a male's pregnancy (Berglund *et al.*, 1989). In the field the adult sex ratio is equal. This means that females potentially can produce more eggs than naturally available males could handle (Berglund *et al.*, 1986a). A small male can hold all eggs from a large female since the space for attaching embryos is the male's whole ventral body surface (Berglund *et al.*, 1986a). However the male can copulate with only one female as the eggs are loosely attached and by copulating with another female he would probably lose the whole clutch. The embryonic development rate, however, causes the male to brood for a longer time than it takes the female to produce a new clutch. These results show that female reproductive success is limited by males (Berglund *et al.*, 1986a).

Furthermore, males face higher reproductive costs than females in terms of risk of predation (Svensson, 1988). In the experiment pregnant males were preyed upon more frequently than females, but after the reproductive period there was no difference in rate of predation between females and males (Svensson, 1988).

So, potential reproductive rate appears to explain the reversed sex roles in *N. ophidion*: males limit female reproduction through being unable to provide care for eggs as fast as females produce them (Berglund *et al.*, 1989). As theory predicts, the operational sex ratio is female-biased. In the wild there are equal numbers of adult females and males, but the experimental evidence described above shows that the potential reproductive rate of females is much greater than that of males (Berglund *et al.*, 1986a,b, 1989). Given this operational sex ratio, female competition for mates in *N. ophidion* is expected.

FEMALE–FEMALE COMPETITION

Females do compete for access to males, but this competition seems to be more indirect than direct. Females in an experimental group appear to form a dominance hierarchy in which only one female develops the ventral skin fold which attracts males (Rosenqvist, 1990). In normal field conditions females encounter other females and males frequently enough to allow formation of dominance hierarchies.

I found a large variance in skin fold size between females in the field (Rosenqvist, 1990). The female's skin fold is visible only during the reproductive period and is resorbed afterwards. As the dimorphic skin fold seems likely to be a sexually selected character, I was interested in its function. If the skin fold is used in intrasexual competition we would then predict that in a group of females, the variance in skin fold size would be higher, compared to a situation in which there are no other females to compete with. If the skin fold is used to attract males we would also expect it to develop in contact with males. To test these predictions skin fold development was measured in three different experimental treatments: 1) one female together with three other females, 2) one female together with three males and 3) one female completely alone (Rosenqvist, 1990). There was no initial difference in skin fold size of females between the three treatments. After the experiment the skin fold size differed significantly between the three treatments. Females in the female+male treatment on average developed larger skin folds compared to the other two treatments. Females kept with other females had larger variance in skin fold size, however, and this is because in each replicate only one of the four females developed a large skin fold.

In the all-female treatment there was a significant positive correlation between body length and skin fold size after the experiment. Isolated females showed either no change in skin fold size or the size of the skin fold decreased (Rosenqvist, 1990). Field samples showed that the number of eggs in the female i.e. her fecundity, is positively correlated both with body length and size of skin fold (Table 1, Rosenqvist, 1990). This experiment showed that females seem to need a stimulus to develop the skin fold. It is furthermore likely that in a female group one female becomes dominant and depresses the development of skin folds in other females (Rosenqvist, 1990). It is possible that the skin fold functions as a "badge" signalling dominance to other females (Rohwer, 1975) which is disadvantageous for subordinate individuals to display. In addition the skin fold may function as an

Table 1 Correlation analysis of ovary dry weight of *Nerophis ophidion*.

	Correlation coefficient	P	Partial correlation coefficient	P
Body length	0.62	< 0.001	–0.08	n.s.
Blue colour	0.45	< 0.05	–0.107	n.s.
Skin fold	0.43	< 0.05	0.46	< 0.05
Body width	0.70	< 0.001	0.57	< 0.01
		N=30		

indicator of fecundity to males. A male that chooses a female with a large skin fold will gain by getting more eggs than if he chooses a female with a small skin fold.

Although the data on formation of skin folds suggest competition among females, there is no evidence of competition in the behaviour of males to indicate that there is stronger competition between the males than between the females. These experiments and observations suggest the conclusion that females are competing more for partners than males in this species.

MATE CHOICE

Given that females compete for males and males limit female reproductive success in *N. ophidion,* then if either sex exhibits mate choice, we would expect males to be more selective in their choice of mates. Since the reproductive rate of a male is constrained by his brooding capacity, he can produce fewer broods than females can potentially produce. Males should therefore have more to benefit from being choosy than females. The male's reproductive success is highly dependent on his choice of females, mainly because females differ in fecundity (Berglund *et al.,* 1989; Rosenqvist, 1990).

Male choice was tested experimentally in the laboratory. The choice apparatus used one way mirrors designed so that the females could not see the male or each other, but the male could clearly see both females (Figure 3). By using a one way mirror we tried to exclude the possibility of the male's choice being influenced by any female behaviour directed towards him, or the other female. To give the fish the opportunity to acclimatize, they were introduced to the tanks in the evening. Males and females were separated by an opaque partition which was removed in the morning when the observation started. The fish seemed to behave normally in the experimental tanks and did not appear to respond to their own mirror image. Each of three treatments tested the effect of a different sexually dimorphic characteristic of females while holding the others constant: 1) body length, 2) skin fold size and 3) blue colour. The results of this experiment showed that males preferred the longest female, the female with the largest skin fold, and the female with the largest blue colour patch (Figure 4, Berglund *et al.*, 1986a; Rosenqvist, 1990).

Since we found no correlation of female reproductive success with male size, we did not expect to see female choice based on male size. We did test this possibility experimentally, however, using the same design (Figure 3), but giving females a choice between males differing only in size. Females showed no preference for male body size. In half of the replicates the female placed herself in front of the larger male and in the other half in front of the smaller male (Berglund and Rosenqvist, ms). Thus female choice seems to be unimportant in this system relative to male choice.

SEXUAL SELECTION

As would be expected from theory, secondary sexual characters in this sex role reversed species are developed in the female and not in the male (see above). Females face directional selection for increased body size because large size correlates with 1) increased fecundity (natural selection) and 2) increased attractiveness to males, since males prefer females with large body size or with

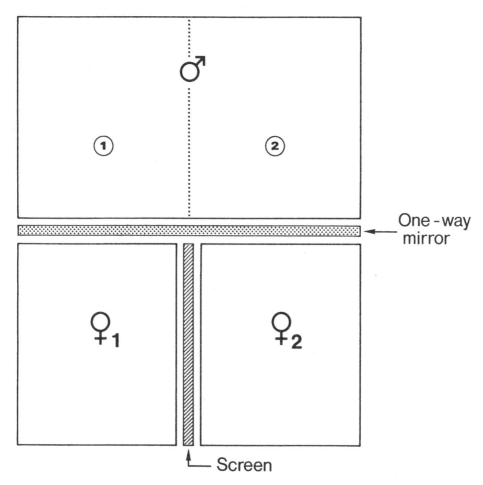

Figure 3 Experimental design of the mate choice experiment. The male's position was determined as being close to female number one if the male was in section 1, and to female number two if the male was in section 2.

characters correlated with body size such as amount of blue coloration and skin fold size (intersexual selection). Males choosing females with these characters will receive more eggs, as these female characters are positively correlated with her fecundity. Increased body size in females may also increase their social rank and thus success in intrasexual selection. There is no strong selection for males to grow larger since there is no correlation between male size and the number of eggs he can carry, or the size of the offspring to which he gives birth. Therefore, a male's reproductive success will depend critically on his choice of females. This creates a strong selective pressure on males to choose more fecund females, as indicated by size, coloration and skin folds. The mating pattern in *Nerophis ophidion* is non-random both due to male mate choice and female–female competition for males. Both natural and sexual selection act on female size and secondary sexual

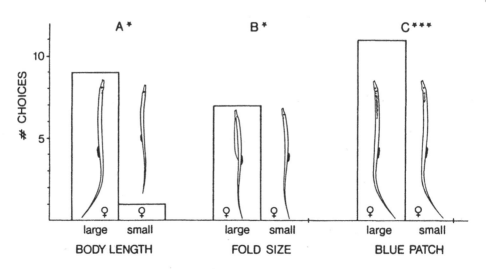

Figure 4 Male mate choice in *Nerophis ophidion* when being allowed to choose between different females. The females differed in: A) length (N=10), B) size of skin fold (N=7) and C) size of blue patch (N=11). *=*P*<0.05, ***=*P*<0.001.

characters, whereas neither strong natural nor strong sexual selection appears to act on male size.

CONCLUSION

As suggested by initial observations, sex roles are reversed in *Nerophis ophidion* even though parental investment offspring was not reversed. Males are not able to brood offspring as fast as the females can produce them. Females compete for access to mates because males limit female reproductive rate. Males are selective in preferring longer females, females with larger blue patches and females possessing larger skin folds. The mating pattern in the sex role reversed pipefish *N. ophidion* is caused by males acting as reproductive bottlenecks, thereby skewing the operational sex ratio so that reproducing females are in excess. Thus, the reversal of the sex roles is strongly linked to male-imposed constraints on female reproductive success.

The family Syngnathidae is an excellent group for studying the evolution of sex roles and sex differences, because males are uniquely specialized for paternal care. Research on *N. ophidion* has provided one example in which male parental care is associated with sex role reversal. This is ultimately the result of morphological and physiological constraints on the number and rate of maturation of offspring by males. It is interesting and important to note that while females are the predominant competitors for mates in *N. ophidion* and some other species, males actually compete most intensely for mates in seahorses and other pipefish species (Vincent, 1990; Vincent *et al.*, 1992). In other words some syngnathids are role reversed while others exhibit normal sex roles despite paternal care of the young. An apparent association between these different patterns of sexual competition and mating systems in particular syngnathid species is providing a rich source for further field and laboratory investigations.

ACKNOWLEDGMENTS

I am thankful to all participants in the workshop on Behavioural Ecology of Fishes for stimulating discussions. I also thank Ingrid Ahnesjö, Anders Berglund, Anne Houde, George Losey, Manfred Milinski, Amanda Vincent and Jonathan Waage for valuable comments on the paper.

References

Bateman, A. J. (1948). Intra-sexual selection in *Drosophila*. *Heredity*, **2**, 349–368.
Baylis, J. R.(1978). Paternal behaviour in fishes: A question of investment, timing or rate? *Nature*, **276**, 738.
Baylis, J. R. (1981). The evolution of parental care in fishes, with reference to Darwin's rule of male sexual selection. *Env. Biol. Fish.*, **6**, 223–251.
Berglund, A. (1991). Egg competition in a sex role reversed pipefish: subdominant females trade reproduction for growth. *Evolution*, **45**, 770–774.
Berglund, A. and Rosenqvist, G. (1990). Male limitation of female reproductive success in a pipefish: effects of body size differences. *Behav. Ecol. Sociobiol.*, **27**, 129–133.
Berglund, A., Rosenqvist, G. and Svensson, I. (1986a). Mate choice, fecundity and sexual dimorphism in two pipefish species (Syngnathidae). *Behav. Ecol. Sociobiol.*, **19**, 301–307.
Berglund, A., Rosenqvist, G. and Svensson, I. (1986b). Reversed sex roles and parental investment in zygotes of two pipefish (Syngnathidae) species. *Mar. Ecol. (Progr. Ser.)*, **29**, 209–215.
Berglund, A., Rosenqvist, G. and Svensson, I. (1989). Reproductive success of females limited by males in two pipefish species. *Am. Nat.*, **133**, 506–516.
Blumer, L. S. (1979). Male parental care in bony fishes. *Q. Rev. Biol.*, **54**, 149–161.
Blumer, L. S. (1982). A bibliography and categorization of bony fishes exhibiting parental care. *Zool. J. Linn. Soc.*, **75/76**, 1–22.
Breder, C. M. and Rosen, D. E. (1966). *Modes of reproduction in fishes*. Garden City: T. F. H. Publication.
Clutton-Brock, T. H. (1991). *The Evolution of Parental Care*. Princeton: Princeton University Press.
Clutton-Brock, T. H. and Vincent, A. C. J. (1991). Sexual selection and the potential reproductive rates of males and females. *Nature*, **351**, 58–60.
Darwin, C. (1871). *The Descent of Man and Selection in Relation to Sex*. London: Murray.
Dawson, C. E. (1985). *Indo-Pacific Pipefishes*. Lawrence: Allan Press Inc.
Emlen, S. T. and Oring, L. W. (1977). Ecology, sexual selection and the evolution of mating systems. *Science*, **197**, 215–223.
Fiedler, K. (1954). Vergleichende Verhaltensstudien and Seenadeln, Schlangennadeln und Seepferdechen (Syngnathidae). *Z. Tierpsychol.*, **11**, 358–416.
Gross, M. R. and Shine, R. (1981). Parental care and mode of fertilization in ectothermic vertebrates. *Evolution*, **35**, 775–793.
Gross, M. R. and Sargent, C. R. (1985). The evolution of male and female parental care in fishes. *Amer. Zool.*, **25**, 807–822.
Haresign, T. W. and Shumway, S. E. (1981). Permeability of the marsupium of the pipefish *Syngnathus fuscus* to 14C-alpha amono isobutyric acid. *Comp. Biochem. Physiol.*, **69A**, 603–604.

Kronester-Frei, A. (1975). Licht- und Eletronenmikroskopische Untersuchungen am Brutepithel des Mannchens von *Nerophis lumbriciformis* (Pennant 1776). Syngnathidae, unter speczieller Berucksichigung der strukturellen veranderung der Eihulle. *Forma et Functio.*, **8**, 419–462.

Linton, J. R. and Soloff, B. L. (1964). The physiology of the brood pouch of the male seahorse *Hippocampus erectus. Bull. Mar. Sci-Gulf Carib.*, **14**, 45–61.

Oring, L. W. and Maxson, S. J. (1978). Instances of simultaneous polyandry by a spotted sandpiper *Actitis macularia. Ibis*, **120**, 349–353.

Oring, L. W. and Lank, D. B. (1984). Breeding area fidelity, natal philopatry, and the social systems of sandpipers. In: *Shorebirds: Breeding Behavior and Populations* (Ed. by J. Burger, and B. L. Olla), pp. 125–146. New York: Plenum Press.

Parker, G. A., Baker, R. R. and Smith, V. G. F. (1972). The origin and evolution of gamete dimorphism and the male–female phenomenon. *J. Theor. Biol.*, **36**, 529–553.

Rohwer, S. 1975. The social significance of avian winter plumage variability. *Evolution,* **29**, 593–610.

Rosenqvist, G. (1990). Male mate choice and female–female competition for mates in the pipefish *Nerophis ophidion. Anim. Behav.*, **39**, 1110–1115.

Svensson, I. (1988). Reproductive costs in two sex-role reversed pipefish species (Syngnathidae). *J. Anim. Ecol.*, **57**, 929–942.

Tresher, R. E. (1984). *Reproduction in Reef Fishes.* Neptune City: T. F. H. Publications.

Trivers, R. L. (1972). Parental investment and sexual selection. In: *Sexual Selection and the Descent of Man, 1871–1971*(Ed. by B. Campbell), pp. 136–179, Chicago: Aldine.

Vincent, A. (1990). Reproductive ecology of seahorses. Ph.D. thesis. University of Cambridge.

Vincent, A., Ahnesjö, I., Berglund, A. and Rosenqvist, G. (1992). Pipefishes and seahorses: are they always sex role reversed? *Trends Ecol. Evol.,* **7**, 237–241.

Williams, G. C.(1975). *Sex and Evolution.* Princeton: Princeton Univ. Press.

Wootton, J. M. (1984). *A Functional Biology of Sticklebacks.* London: Crom Helm.

THE IMPORTANCE OF MALE–MALE COMPETITION AND SEXUALLY SELECTED DIMORPHIC TRAITS FOR MALE REPRODUCTIVE SUCCESS IN SITE-ATTACHED FISHES WITH PATERNAL CARE: THE CASE OF THE FRESHWATER GOBY *PADOGOBIUS MARTENSI*

PATRIZIA TORRICELLI, MARCO LUGLI and LAURA BOBBIO

Dipartimento di Biologia e Fisiologia Generali, Università di Parma, Italy

INTRODUCTION

The presence of sexually selected dimorphic morphological or behavioural traits (i.e. characters evolved by sexual selection that cause differences in individual mating success e.g. Arnold, 1983) is widespread among vertebrates. While members of one sex (typically the male) are often of large body size, colourful and compete strongly for resources or mating opportunities, members of the opposite sex (typically the female) differ in having such traits less pronounced or absent. Darwin (1871) was the first to introduce the concept of sexual selection, as opposed to natural selection, to explain the evolution of many sexually dimorphic traits that seemed to lower chances of survival but, nonetheless, turned out to be necessary for gaining mating advantages. He also distinguished between sexual selection due to competition among members of one sex for mating opportunities (intra-sexual selection) and that due to the choice made by one sex for mates (inter-sexual selection). While the first mechanism of selection is generally accepted as explanation for the evolution of weapons used during fights for the access to mates or resources, the importance of the latter process in shaping sexually dimorphic traits has been controversial. Thus, a considerable amount of empirical and theoretical effort have been devoted to the study of the possible mechanisms underlying female mate choice in sexually dimorphic species characterized by extravagant male sexual displays (e.g. the peacock tail). By focusing mainly on mate choice, much of the attention has been distracted from intrasexual selection as an alternative or concurrent evolutionary force leading to the expression of such traits (but see Zahavi, 1975; Halliday, 1983). For example, it is well known that male traits functioning during male–male agonistic encounters may also be components of the male courtship display (e.g. plumage and songs of songbirds), and that their use may be under the control of both aggressive and sexual motivational systems (e.g. Tinbergen, 1951; Hinde, 1966). Therefore, the same male characteristics may be acted upon by both intra- and inter-sexual selection. In other cases, information relevant for females and competitors may be coded in separate parts of the same complex signal (e.g. songs of birds or advertisement calls of frogs) or may be conveyed by different and yet temporally associated signals. In all these circumstances, the contemporary presence of multiple and multipurpose signals make the assessment of the relative contribution of male–male competition and female choice a difficult task, even by means of controlled laboratory experiments. Such difficulties are particularly striking in field work, where conclusions about sexual selection commonly rely upon correlative analysis among several variables that are often closely dependent on each other. Therefore the importance of mate choice or male–male competition in moulding sexually selected characters remains unclear for most of the species so far investigated.

Sexual Selection, Male Sexual Traits and Paternal Care

In many species of teleosts the male breeding behaviour typically consists of exclusive occupation of a breeding site from where he courts females (Breder and Rosen, 1966). In these cases, a common situation is for the female to leave immediately after oviposition and for the male to stay on the site to acquire new mates and protect eggs from environmental threats. In these systems, the type of paternal benefits may vary considerably according to the species, ranging from simple exclusion of competitors and egg-predators to various forms of care of eggs and fry extended for more-or-less prolonged periods (reviewed in Ridley, 1978). The high incidence of paternal care among fishes is not the only feature that makes them peculiar among the other vertebrate groups; in fact, the fish male, despite his frequent role as unaided egg-keeper, usually remains the more ornamented and competitive sex, while the female retains the traditional role of the more discriminating sex. The piscine pattern of parental care and male mating competition is not easy to explain in terms of Parental Investment Theory (Williams, 1966; Trivers, 1972). As a guarding male may be expected to invest in the care about as much as a female invests in egg production, intra-sexual selection should act on females in a manner not dissimilar from that acting on males of other groups. As an alternative to the parental investment argument, Baylis (1981) pointed to the differential rates of gamete production by males and females as to the most relevant factor for explaining the pattern of sexual selection in fish. More recently, Clutton-Brock and Vincent (1991) have emphasized the importance of potential reproductive rate rather than the rate of gamete production as to the primary variable determining the direction of mating competition and, thus, the direction and intensity of sexual selection. In fact, in many species where the males provide all the parental care they can still achieve faster reproductive rates than females by obtaining and caring for multiple clutches, thus leaving females the limiting sex (see also Rosenqvist, this volume).

This leaves unresolved problems concerning the kind of traits that are most likely to evolve in males and, most important, questions as to which selective forces shape these traits. It is generally held that the pattern of sexual selection is closely associated with the type of mating system (e.g. Alcock, 1984). However, variation in the contribution of intra- and inter-sexual selection is not clearly predictable simply on the basis of mating system (Bradbury and Davies, 1987).

Among site-attached fishes the opportunity for a male to receive and care for eggs is conditional upon the possession of resources, such as a shelter or a territory suitable for oviposition. Therefore, one recurrent prediction is for male–male competition to be a prominent force in the evolution of sexually dimorphic characteristics that confer superior fighting ability for access to such resources. Among these traits, male body size is, undoubtedly, the most widespread. Male body size has been found to be a crucial attribute enhancing the male's fighting ability and strong predictor of dominance in many fishes. In several species even small differences in size may confer substantial advantages in fights; thus, larger males are usually at an advantage over smaller males in obtaining breeding sites (Kodric-Brown, 1978; Thompson, 1986; Bisazza and Marconato, 1988; Petersen, 1988) or spawning surfaces (Lindström, 1988; Magnhagen and Kvarnemo, 1989).

When male parental care is restricted to defence of a territory with oviposition sites, females appear to be most sensitive to characteristics of the territory that enhance eggs survival (e.g. Petersen, 1988; Thompson, 1986; Kodric-Brown, 1983); nevertheless, aspects of the male may be equally relevant for female decision to spawn. For example, in a manipulative field study of the mottled triplefin, *Forsterygion varium*, Thompson (1986) found that both territory quality and male size were influential in the female choice. When the survival of eggs is more critically related to the male's parental abilities, such as sustained fanning or defence of the nest site from competitors or predators, females are expected to make adaptive choices by choosing mates attributes indicating high parental quality (Trivers, 1972). Predation of eggs by conspecifics or heterospecifics is probably one of the greatest sources of egg mortality among territorial species (Dominey and Blumer, 1984). To this context, larger males, due to their higher competitive ability, may be better at defending eggs and, thus, have a greater chance of bringing eggs to hatching (Downhover and Brown, 1980; Bisazza and Marconato, 1988; Côte and Hunte, 1989). Large males may also be better as regards parental duties, since they may suffer less from the care or may provide a more favourable environment for the developing embryos. Experimental evidence that mate body size affects female mating positively was obtained for the river bull-head, *Cottus gobio* (Bisazza and Marconato, 1988).

Unfortunately, precise information about the criteria for female choice is lacking for the majority of the species thus far studied. In fact, body size has been reported to have a strong relationship with other male phenotypic characters, such as courtship intensity or nuptial coloration, that may be important to females. Therefore, the extent to which correlations with mating success reflect causal relationships, and the relative importance of male–male competition and female choice in the evolution of these traits, is difficult to assess without direct and extended behavioural observations or appropriate experimental manipulations (Bradbury and Davies, 1987). Where this has been done (Thresher and Moyer, 1983; Gronell, 1989; Knapp and Warner, 1991), the effect of body size on male mating success is found to be an indirect one, stemming from its association with other male traits that are more directly under the action of female selection.

There is therefore a need for detailed experimental studies on all the relevant aspects of the mating behaviour of a given species if the process of sexual selection is to be fully understood.

The Freshwater Goby Padogobius martensi as a Case Study

Breeding biology

Both sexes are territorial year-round, the core of each territory being a shelter under a stone on the stream bottom. During the breeding season (May to July) ripe females lay their eggs in a single layer on the underside of the male's stone; therefore the occupation and defence of a shelter is a prerequisite for male reproduction in the stream. In *P. martensi* the breeding system is polygamic: males may acquire and tend one or more clutches of eggs, and females mature at least two clutches of eggs in succession during the same breeding season. Paternal care mainly consists of fanning the eggs until hatching (Torricelli *et al.*, 1985) and defending the nest site from intruding competitors. In the laboratory, successful nest take-overs are usually followed by cannibalism of any existing eggs (Parmigiani *et al.*, 1988). Cannibalism has also been documented in the field, although the frequency is unknown (pers. obs.)

The occurrence of chemical communication through emission of a female sexual pheromone is known for this species, as well for other gobies (Colombo *et al.*, 1982). Small quantities of water taken from tanks in which ripe females are housed are effective in evoking male courtship and associated sound production in isolated sexually active males (Lugli, 1989).

Sexual dimorphism

Body size is the most relevant sexually dimorphic trait in this species, males growing to larger sizes than females (Marconato *et al.*, 1989; Lugli *et al.*, 1992). Larger size and higher aggression in males (Gandolfi and Notarbartolo, 1971; Torricelli *et al.*, 1988) both increase male Resource Holding Potential (Parker, 1974).

During mating encounters males are easily recognizable for their brighter colours and nuptial dance (Torricelli *et al.*, 1986). Courting male coloration consists of bluish fins (both paired and unpaired), blue stripe on the first dorsal fin and a dark body (particularly the head). The male's breeding colour pattern is not unique for reproduction in that it closely resembles that described for both sexes during fights (Lugli, 1989). Both intra- and inter-male variability in intensity of breeding livery during mating encounters have been noted in the laboratory; the proximal causes of such changes have not yet been investigated.

The male's nuptial dance consists of a short sequence of brief, stereotyped swimming movements oriented toward the female (the approaching phase) followed by a rapid swim toward the nest (the leading phase). While courting, particularly during the approaching phase, the male emits sounds (Torricelli and Romani, 1986; Lugli, 1989); male vocalizations during courtship and threat are similar in nature, the first being easily distinguishable by their short duration and high stereotype (Torricelli *et al.*, 1990). Variations in rates of emission of courtship sounds have been related to the male's position within his territory, the rate greatly increasing when the male enters his nest (Torricelli *et al.*, 1986).

In contrast to the male, the courted female retains the usual brownish pale cryptic coloration and remains silent all the time, although sound production by females during fights has been documented (unpublished data). She may respond to the male's approach either by resting motionless on the bottom or by moving to the nest in short swimming bouts.

The fresh-water goby *P. martensi* therefore appears to be a suitable subject for examining the relative contribution of several male attributes or quality of defended resource to the male mating success. This species is used here as a model for the study of the role of intra- and inter-sexual selection on the evolution of several sexually selected male traits. To this end, investigations on its breeding system and behaviour have been carried out both in the field and in laboratory.

FIELD STUDIES OF MALE REPRODUCTIVE SUCCESS IN *P. MARTENSI*

Methods

The field study was undertaken during the 1988 and 1989 breeding seasons in a 2.1 kilometer stretch of a hill-stream near Parma (Stirone Stream), in the North-West of Italy. In the course of the study weekly collections of data were made using two sampling methods.

The first method consisted of searching for gobies by simply walking in the stream and lifting all the stones encountered that might have accommodated a sheltered animal. For each netted adult fish (males larger than 50 mm TL; females larger than 45 mm TL) sex and total length were recorded. Stone size was estimated by multiplying the minor and the major axes; water depth and surface current velocity near the stone, as well as the shortest distance from the stone centre to the shore, were also measured. Any eggs present were photographed for counting in laboratory; the number of egg batches was assessed in the field from a unique combination of egg size, colour, transparency and presence of eye pigmentation. In addition, samples of a few eggs were randomly picked from each nest and preserved in the formaldehyde solution (5%) to be examined in laboratory. For a more accurate assessment of egg age, the preserved eggs were compared with eggs collected daily from a batch reared under laboratory conditions. The number of egg batches assessed in the field should be regarded as a minimum estimate of the actual number of clutches obtained since it is likely that in a number of cases we failed to discriminate concurrent or same-day spawns. Therefore the number of eggs per nest, rather than the number of egg batches, was taken as a measure of the male mating success. The two quantities were, nevertheless, highly correlated ($P<0.01$).

The second sampling method consisted of delimiting areas of stream by setting variable size quadrats (from 1 to 6.5 m^2; n = 35) on the stream bottom; the aim here was to investigate the spatial allocation of animals and nest sites, as well as to collect more data on male mating success. All the adults in each quadrat were captured by means of hand-nets. Male and nest site characteristics were measured as above. Further, size and abundance of stones potentially suitable as nests were also evaluated, considering only those stones with a minimum axis greater than 7 cm (earlier field data showed that only 8% of stones used as shelter by adults had a minimum axis equal or less than 7 cm). Based on these measurements, an estimate of the local percentage of stone cover was then calculated as the ratio of the summed sizes of all the stones censused within a quadrat to the area of the quadrat, x100. More detailed information on main features of the study site and on methods of data collection and analysis may be found in Lugli *et al.,* (1992).

Body Size and Competition for Resources

For species that shelter inside holes or under stones and other submerged objects, the stable occupancy of a cavity is generally of overwhelming importance for securing individual protection and reproduction. In these cases, the availability and quality of such resources are two major environmental factors that may affect the spatial distribution of individuals. Under these circumstances two very general predictions can be formulated: firstly, the number of animals settling in a given area will be primarily affected by the local supply of resources; and, secondly, competition will lead to better resources being retained by superior individuals, provided that access to resources is not constrained by factors other than competition. Field data support both predictions, for male and female *P. martensi*. In fact, the density of individuals within quadrats rose significantly with increasing percentage of stone cover ($P<0.05$ for both sexes). Also, the pattern of occupation of stones deviated from randomness in accordance with predictions: large stones, relatively rare on the stream bottom, were used more frequently than expected simply on the basis of distribution of potentially usable stones ($P<0.0001$ for males; $P<0.005$ for females) and larger individuals tended to be found under larger stones ($P<0.001$ for both sexes).

Thus, large stones, or the wide cavity beneath them, appear to be resources of superior quality for which individuals of both sexes are likely to compete. Since body size is known to be a strong predictor of dominance for *P. martensi*, it may be that only bigger individuals are capable of acquiring and retaining such resources. However, competition for shelters with given qualities requires that they are preferred when alternatives are given. Results of choice tests in the laboratory (Lugli *et al.*, 1992) provided evidence of preference for shelters with wider bottom surfaces, by both males and females regardless of body sizes. Benefits from such preference are clear for the reproducing male: shelters with larger inner surfaces increase the potential for polygyny by allowing more concurrent batches of eggs in the same nest (see below). A different explanation, however, is required to explain the preference for larger shelters by females. Additional selective forces are likely to be acting on both sexes. Protection against environmental threats (e.g. predators, current flow) would be a reasonable hypothesis, although at present no supporting evidence exists.

Male Size, Nest Size and Mating Success

Given the advantage conferred by body size in the acquisition and defence of larger shelters it may be worth asking whether male size or resource quality will affect male mating success. Since male size and stone size were to some extent dependent on each other, their separate effect on the probability of finding mated males and on the overall number of eggs in the nest (the two quantities considered here to estimate the male mating success) was examined by means of multivariate statistical tests (Sokal and Rohlf, 1981). Results from log-linear analysis (with male size and stone size as the two variables entered in the model) and correlative methods clearly identified male size as to the main variable affecting male mating success. Large males were more likely to be found with eggs ($P<0.0001$) and cared for higher number of eggs ($P<0.05$), compared to small males (Figure 1a). Stone size, by contrast, appeared unimportant in determining male access to mates, as the probability of male having eggs in his nest was by no means affected by the size of

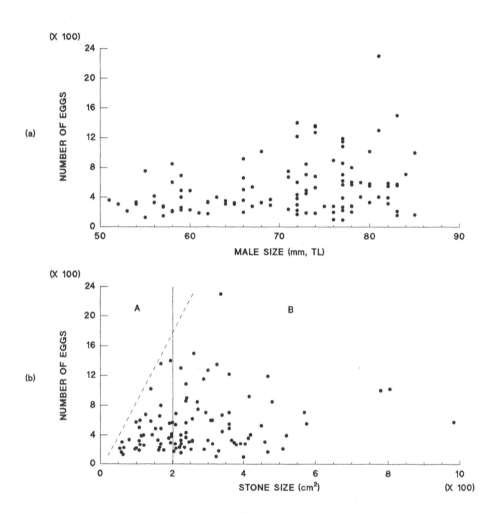

Figure 1 Number of eggs per nest plotted against male body size (a) and stone size (b). Stone size (b) may represent a physical constraint on the male reproductive success, as indicated by the broken line setting an imaginary upper limit to the maximum number of eggs a nest of given size may contain. In order to identify a possible limiting effect of small-sized stones, stones were split in two size categories (A: <200 cm², B: >200 cm²) as illustrated by the solid line, and correlations between number of eggs and, respectively, stone size and male size were calculated separately for the two groups of stones. For the small-sized category of stones (A), number of eggs correlated significantly with stone size (r=0.485, n=38, P<0.005) and only weakly with male size (r=0.284, n=42, P=0.08), while for the large-sized group (B) a significant correlation only exists with male size (for stones size: r=0.151, n=66, ns; for male size: r=0.276, n=66, P<0.05).

the stone ($P=0.7$, ns). Instead, a significant correlation emerged between stone size and egg number ($P<0.05$, and Figure 1b), after having controlled for male size. However, this relationship would be expected if the number of eggs in the nest is constrained by the surface available for additional spawnings; as a consequence, only big stones would be large enough to allow more concurrent spawns. That this was indeed the case is apparent from Figure 1b. Stones smaller than 200 cm^2 appear too small to accommodate large numbers of eggs. Another supporting result is the lack of any association between stone size and egg number when only stones larger than 200 cm^2 are considered. It should be noted that the small proportion of the variance of egg number explained through its correlation with male size strongly calls for other untested variables to be taken into account as possible determinants of the male spawning success (see below).

Male Size, Nest Size and Brood Survival

The superior competitive ability of larger males should make them better parents, as they are better in defending their eggs from nest take-overs and cannibalism. So, when a large male settles and mates on a suitable site it may well have a better chance of successfully defending the site and bringing the eggs to hatching. Supporting evidence is the observation that the proportion of collected nests that were found to exceed the 2/3 of the standard parental cycle (the length of one standard cycle correspond to 18 laboratory days, see methods) was higher when the occupant was a large male (Figure 2). To control for a possible effect of stone quality, log-linear analysis was applied to test for pairwise associations between male size, stone size and the proportion of nests exceeding the 2/3 of the standard parental cycle. Results showed that male size was the only factor affecting the proportion of nests exceeding the 2/3 of the parental cycle ($\chi^2 = 7.01$, df=1, $P<0.008$). Assuming no sampling bias of nests, this would suggest that risk of losing all current offspring is higher for small males and is, to a degree, independent on stone size. Alternatively, small males may cannibalize their own eggs more frequently than large males might do (Dominey and Blumer, 1984). However, the latter hypothesis seems implausible since it is known from laboratory observations that episodes of cannibalism of the entire brood by tending male *P. martensi* mostly occur within a few days of spawning, before the male is being fully involved in the parental activities (see section below and Lugli, unpublished data). This cannot account for the results of Figure 2, where differential loss of broods between large and small males for unknown reasons is more likely in the second half of the parental cycle.

LABORATORY INVESTIGATIONS OF MALE REPRODUCTIVE SUCCESS IN *P. MARTENSI*

Methods

Adult fish were collected in the Stirone Stream just before the onset of the breeding season. Once in laboratory each fish was sexed and measured (to the nearest mm) then, 4 males and 5 females were marked using subcutaneous dye to allow individual identification and placed into a rectangular experimental pool (160 × 80 × 30 cm). The bottom of the pool was a 3 cm layer of sand with scattered stream

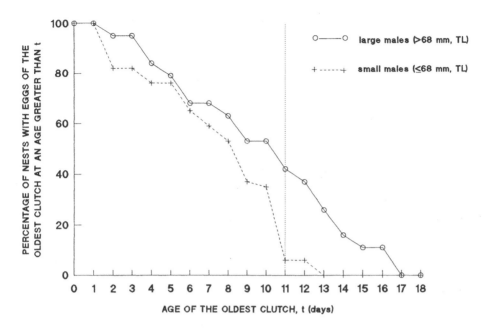

Figure 2 Survivorship curves of eggs from the oldest clutch, cared by large-sized (solid line) and small-sized males (broken line). Age of eggs (in abscissa) has been expressed as standard days, 18 days corresponding to the total duration of parental cycle in laboratory condition (see methods). The vertical dotted line marks the 2/3 of the standard parental cycle.

stones not usable by animals as shelter site; the pool was supplied with a surplus of uniformly spaced artificial shelters of constant size (10 × 9 × 2.5 cm). Two 300 W halogen bulbs illuminated the pool for an automatically regulated photoperiod which approximated the natural light–dark cycle. Preliminary laboratory observations showed, however, that undisturbed gobies are mostly active during the evening crepuscular period and tend to remain inside shelters under condition of full daylight. Therefore, for ease of observation, the artificial photoperiodic regime was adjusted so as to have the lights turned off at 15.00 h and a 150 W red light bulb illuminated the pool from 15.00 h to 18.00 h to simulate roughly the environmental light conditions during sunset. Under red light gobies could be easily observed even at short distance apparently without interfering with the time course of behavioural events. Collection of behavioural data took place only during that period.

Observations started after 3–5 days of acclimatization of fish to laboratory conditions, and continued for at least 4 weeks. On each day, occupied shelters and the identity of the occupant were noted; in addition, occupied shelters were inspected for the presence of new spawns. In order to study male and female breeding behaviour in detail, the activity of fish was investigated by the focal animal technique (Altmann, 1974). To this end, three males and two females (focal animals), arbitrarily selected among those within the pool, were observed for 30 consecutive minutes, about once every two days for the entire duration of the experiment. Behavioural events recorded were: agonistic interactions (chasing, lateral displays, blows; Torricelli *et al.*, 1988), male courtship (approaching-leading movements) and the associated sound production (the description of the electro-acoustic apparatus employed here is reported in Torricelli *et al.*, 1990). For each behavioural event recorded, the time and location of occurrence and the identity of the interacting individuals were reported on a map of the pool. During observations, efforts were also made to record behavioural events of interest that involved individuals other than the focal fish under observation. This allowed the collection of data on courtship events from non focal animals and, in general, provided a more complete description of reproductive and territorial activities of the group as a whole.

A dominance index was obtained for each focal male as proportion of male–male encounters won by the fish within his 30-min observation periods. Dominance indexes for non focal individuals could only be estimated from aggressive events engaged with focal fishes during their observation periods.

The intensity of male courtship was quantified for each male by the number of sounds emitted per minute. Acoustic displays assessed courtship activity better than visual displays, as sounds are easily detectable and relatively easy to count. Moreover, for males inside shelters courtship can only be assessed from the sound emission rate.

In order to study use of space by individuals in the pool, the position of the focal animal was mapped every 30 seconds during the 30-min observation periods. By this method we were able to measure daily the area of activity of the fish and to note possible changes on successive days. Then, a grid of 10 cm square cells was superimposed on the pool map so that the numbers of times the animal was recorded within each cell could be determined. This allowed the intensity of space use by focal animals to be described within selected periods of time.

Each focal animal was observed for a period of 9 hours, overall.

Observations were repeated a second time with new gobies and with a new arrangement of stones and shelters in the pool. In contrast to the previous group of fish, this second group was established in laboratory after the onset of the breeding season. For this reason the first appearance of reproductive activities followed by spawning events occurred in this group before the beginning of observations. Apart from this, methods of observations and data collection were identical to those previously described, with the exception that each focal animal was observed for 7 hours, overall.

Male Size, Territorial Behaviour and Dominance

After the introduction of the fishes into the experimental pool, partitioning of space was initially characterized by animals often changing their area of activity and refuges on successive days. As the time passed, individual area of activity become progressively more stable and defined for most fish of the group. In Figure 3 the mobility of a male within the pool on successive dates is shown as an example. Starting from the tenth day after the establishment of the experimental group, the fish showed persistent occupation of the same area (hereafter referred as habitual area); during this period (Figure 4) the activity of the male centred on shelters inside this area. Shelters, therefore, represent reference points from and to which the territorial male directs his movements.

The time of settlement (Table 1) was different for different males, being largely dependent upon their dominance rank. Indeed, in both experimental groups the time

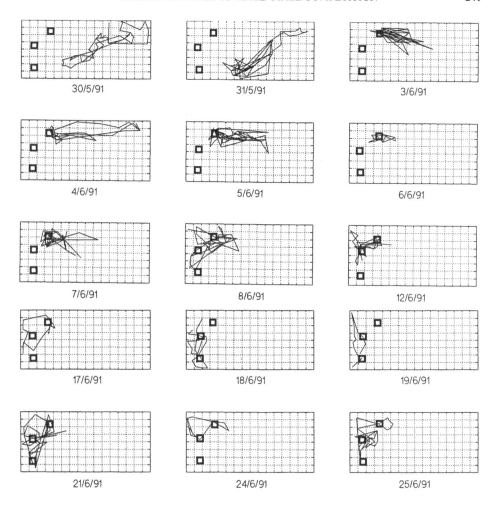

Figure 3 Use of space by 70 mm male in the second experimental pool. The activity of the male in the pool is shown over the entire period of observation. In each diagram lines connect consecutive locations of the fish, over a 30 min period of observation; bold-faced squares represent shelters mostly used by the male.

of settlement of a male was conditional on, and always occurred later than, that of higher ranking males of the group. Thus, the first male to settle was in both cases the one with the highest dominance rank. A second parameter considered here was the size of a males territorial area (Table 1). As for the order of settlement, a strong correspondence emerged in both groups between territory size and dominance rank: dominant males of each group were invariably those with the largest territories. As regards the importance of body size as a correlate of male territoriality, results showed that dominance rank, and the other parameters concerned, were dependent upon the size of the male in the expected manner, i.e. the larger the male, the shorter was its settlement time, the higher was its dominance rank and the wider was the

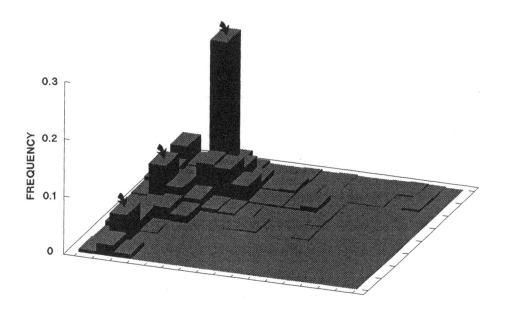

Figure 4 Three-dimensional histogram showing the intensity of space use within the pool by the 70 mm male. The x- and y-axes show the pool length and width, divided into 128 10×10 cm cells; the Z-axis shows relative frequency of locations inside each cell. Arrows indicate cells including shelters. Frequencies were obtained from data summed over the period from the first day of occupancy of the habitual area (June 3, 1991) until the end of observations.

territorial area (Table 1). It should be noted, however, that the largest male in the first group was not the dominant one in that he only retained the second highest dominance rank and settled second.

Compared with males, female territoriality was somewhat less clear. Females were rarely successful in defeating intruding males of even small size, yet they were mostly seen confining their activity within wide but temporally stable areas of the pool. The situation was further complicated by courtship, as ripening females occasionally moved outside their habitual area of activity to approach more distant courting males. A female's territorial behaviour certainly demands more space; this raises questions that go beyond the scope of this paper and so will not be treated further here.

Table 1 Body size (TL, mm) and other individual parameters considered for the 4 males of the two experimental groups. Time of settlement is measured as the number of days elapsed from the establishment of the experimental group until the first date each fish established in his habitual area. The territorial area (cm^2) was calculated here as the fraction of the habitual area in which the fish spent 90% of total time since settlement. Territorial area was not measured for non-focal males in each group. The dominance index of the 72 mm male could not be computed since aggressive events did not occur often enough to make quantification meaningful

	Male size (mm)	Index of dominance	Time of settlement (days)	Territorial area (cm^2)
First experimental group				
	78	0.81	14	2500
	82	0.77	16	?
	68	0.20	18	1500
	63	0.19	–	–
Second experimental group				
	78	1.00	3	2600
	72	?	8	?
	70	0.86	10	2500
	62	0.17	14	1900

Male Dominance, Male Courtship Signals and Female Mating Decisions

In both groups of fishes courting episodes were documented from all the males of the pool, and all females but one spawned at least once in the course of the experiments. This implies that all fishes were sexually active for the duration of the experiment or for some period of it.

Acoustic activity was the most relevant component of male courting behaviour. Sounds were emitted mostly from within shelters (92% in the first pool, 87% in the second pool), thus confirming earlier findings from isolated males (Torricelli *et al.*, 1986). In general, courtship activity of males was confined to their territorial area, and in no instances have males been observed moving through neighbouring male territories while courting females. Females were apparently free to visit several territories in succession without being aggressively interrupted by other males.

Courtship activity in the pools occurred irregularly through observations, as judged from sound production. The intensity of courtship by single males could vary greatly (from a few isolate events to prolonged bouts of sustained activity), even within short periods of time. Short-term variation of male emission rate was readily appreciable whenever a female approached a male courting from the inside of his nest (Figure 5). Usually, as soon as the male establishes visual contact with the incoming female its rate of emission suddenly increases, remaining high until the female leaves or moves out of his sight.

Long-term variation in overall courtship activity was also noticeable. Peaks in the male acoustic activity were observed in association with, as well as one or few days

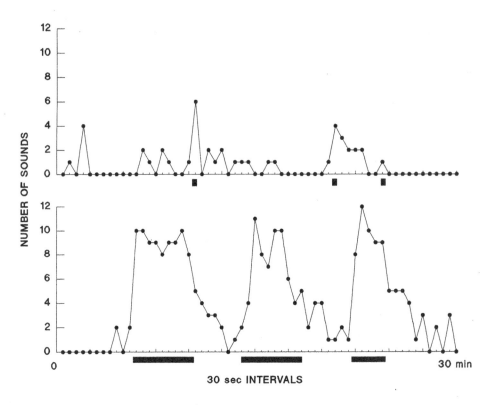

Figure 5   Acoustic behaviour of courting male *P. martensi.* Number of courtship sounds per 30 s
intervals emitted by a male inside his shelter during two different 30 minute focal observations.
Visual perception of nearby females by the courting male and duration of visual contacts are
indicated by the black bars below x-axis.

before, the occurrence of spawning events (Figures 6, 7). Behaviour of females,
although not investigated thoroughly, is unlikely to explain increased levels of male
courtship activity. Increased mobility of ready-to-spawn females just before
spawning was not readily apparent from our observations, although it is documented
for females kept singly in laboratory tanks (Lugli, unpublished data). Moreover,
courtship in proximity to spawning events was often elicited by sub-optimal stimuli,
as courting males were seen responding indiscriminately to ripe females as well as
to non ripe ones or even to males. Such variation is better explained as a consequence
of the releasing effect on courtship caused by the female *P. martensi* sexual
pheromone, to the extent that emission of such sexual chemical stimuli occurs in
concert with gonadal maturation (a well documented fact among gobiids; Colombo
et al., 1982). The releasing effect of female *P. martensi* pheromone on courtship and
acoustic activity has been demonstrated experimentally (Lugli, 1989).

When the courtship activity of each male is thoroughly examined across the
experimental period, noticeable differences emerged among individuals of the same
group (Figures 6, 7), particularly the first.

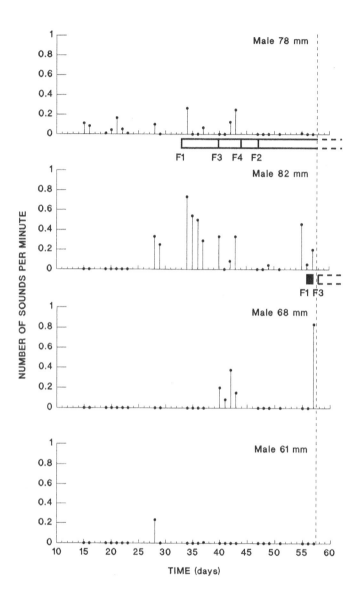

Figure 6 Occurrence and intensity of courtship activity by the four males of the first experimental group across the observations period. The number of sounds per minute (y-axis) is computed over the total daily observation time. Time (x-axis) is expressed as the number of days elapsed since the establishment of the group in the pool. Points are missing where no behavioural data are available. Males are ordered according to their dominance status. For each male horizontal bars below x-axis indicate time of egg-caring, from first spawning event to hatching of the last egg clutch or cannibalism of the entire brood (black squares). Occurrence of each spawning event is indicatead by the symbol F; a vertical line within bar indicates spawnings occurring during the egg-caring period. Body sizes of spawning females are: F1=70 mm, F2=68 mm, F3=66 mm, F4=61 mm. The broken line indicates the end of observations.

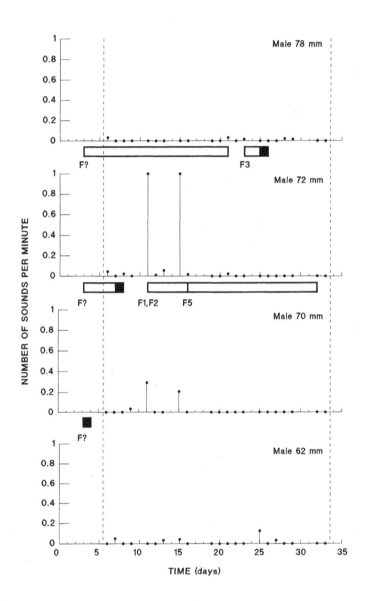

Figure 7 Occurrence and intensity of courtship activity by the four males of the second experimental group across the observations period. The x-axis and y-axis are as in Figure 6 (see caption of Figure 6 for symbols). Body sizes of spawning females are: F1=70 mm, F2=64 mm, F3=61 mm, F4=54 mm, F5=53 mm. Broken lines indicate the beginning and the end of observations.

Table 2   Results of statistical analysis on male courtship activity in both experimental groups. For each group, daily courtship levels of males were converted into ranks; ranks were summed for each male for the period of egg-caring of the dominant male (first pool: from 34th to 57th day; second pool: from 6th to 20th day). Sums of ranks were then compared by means of the Friedman two-tailed test (Siegel, 1956). In the table, the ranked courtship activitiy of males and results from the X^2r test are reported for each group; males are ordered according to their index of dominance

	Male size (mm)	Courtship activity (rank sum)
First experimental group		
	78	28
	82	43
	68	29
	61	20
		$X^2r = 13.70$ $P < 0.01$
Second experimental group		
	78	20
	72	36
	70	22.5
	62	21.5
		$X^2r = 9.87$ $P < 0.02$

Evidences from the first group of fishes suggest that such between-male differences in the courtship activity may partly be attributable to the dominance status of the male. In fact, the dominant male was the only one seen courting from well before the occurrence of the first spawning event in the pool. Further, the two smaller males rarely showed courtship during observations; it was remarkable that in two instances the 68 mm male was seen suffering disruptive attacks by the higher ranking 82 mm male, while courting from the inside of a shelter. A second male characteristic that partially accounted for inter-male courtship differences was parental care. In fact, courtship activity by care-giving males of both groups was reduced or absent during the parental phase of the brood-tending cycle (Figures 6, 7). As a result, courtship levels of males involved in parental care did not match their relative dominance rank, as was indeed the case for non-caring males (for statistical comparisons see Table 2). In sum, dominance and parental care appear to be two interacting male characteristics that explain between-male variability in the occurrence and intensity of courtship during the *P. martensi* breeding period.

What is then the pattern of spawning of females in each experimental group and how do females spawning decisions relate to male characteristics?

In the first group (Figure 6), the male with the highest dominance index (actually the one that had courted for longest), obtained the first available clutch. Thereafter, he continued to be preferred by the three subsequent maturing females despite the significantly higher advertising effort by the next high-ranking male. When the dominant male was presumably strongly involved in egg fanning, and only well after the last clutch was laid in his nest, the second male was successful in obtaining eggs.

The second experimental group (Figure 7) was characterized by three spawning events occurring before the beginning of the observations, when males had not yet settled in territorial areas (Table 1). These spawnings took place under three different shelters in the pool, which is indicative of sub-optimal or random spawning decisions by the three females. Notably, two of these were cannibalized within a few days while the third one was cared until hatching by the largest male of the pool. The three subsequent ovipositions, of which two were concurrent and one following a short time, were all obtained by the 72 mm male. At the time of their occurrence, the 72 mm male was the one performing the most intense courting activity (see Table 2, also for statistical comparisons); this male was also dominant with respect to those currently not involved in parental care. Finally, the largest, dominant male obtained the last spawning shortly after the completion of the first brood cycle.

To summarize, results from both experimental groups clearly indicate a preference by the female *P. martensi* for male dominance, assessed either from the male courtship activity or from the presence of eggs in the nest. The presence of eggs tends to overcome courtship as the most important cue on which females rely for their choices. However, these male characteristics are not the only affecting female spawning decisions. It should be noted that females never spawned in the same nest more than 6 days apart from one another (Figures 6, 7). The age of eggs in the nest therefore appears to be an important aspect of the female mate choice.

A few more words must be devoted commenting that episodes of cannibalism of the entire egg batch occurred in both replicates. Cannibalism of eggs by the parental male (filial cannibalism, Rohwer, 1978) was documented for the 78 mm male (second experimental group) and was strongly suspected for the 82 mm male (first experimental group). Indications as to the nature of the cannibalism are lacking for the two other episodes.

DISCUSSION

Male–Male Competition, Male Body Size and Territoriality

Results from both field and laboratory studies provide evidence for the importance of intermale selection on the evolution of large male body size in *P. martensi*. During breeding, the acquisition of bigger stones with larger spawning surfaces is of paramount importance for the male reproductive success. Since larger stones are, at the same time, limited in supply and preferred by the whole population irrespective of body size, strong competition for them is likely to occur especially among males. Ultimately, only larger males are capable of acquiring and holding such valuable resources. This fact appears of general validity among species that, like *P. martensi*, oviposit eggs in single layer inside enclosures such as shells, holes and stones (e.g. DeMartini, 1988; Lindström, 1988; Magnhagen and Kvarnemo, 1989; but see Hastings, 1988a,b). Defense of resources necessary for reproduction is not the only form of male–male competition selecting for larger male size in *P. martensi*. In the field, spatial allocation of neighbouring males was apparently regulated by body size for the distance between nearest neighbours increased with increasing size of the larger male of the pair (Lugli *et al.,* 1992). Results from laboratory groups of fish have shown that this is indeed to be expected as larger males are able to control larger territorial areas around shelters. However, the importance of male size as trait favoured by intrasexual selection is well beyond that of mere defence of better shelters and wider surrounding area, and extend to embrace the dominance components of male territoriality. Three factors were considered as evidence of size-mediated dominance relationship among individuals in the pools. First, larger males had higher dominance ranks in that they were capable of excluding competitors from their territory more effectively than smaller territorial males. Second, the opportunity for smaller males to settle in a stable territorial area was conditional upon the time of settlement of higher-ranking males. Third, larger, dominant males retained their status even when they moved outside their territorial area, e.g. to invade nearby territories.

Breeding systems characterized by a mix of territoriality and dominance hierarchy have been described for other site-attached fishes showing aggressive partitioning of resources essential for reproduction (e.g. Myrberg, 1972; Itzkowitz, 1978; Kodric-Brown, 1988). Such systems are commonly held to be expression of inter-individual differences in fighting ability, which ultimately lead to unequal shares of resources or mates among males (Magurran, 1986). For example, Cole (1984) described the social system of the marine goby *Coriphopterus nicholsi* in the field as characterized by site constancy with combined territorial and dominance components; moreover, as in *P. martensi*, body size in *C. nicholsi* plays a major role in the expression of both components, in that large-sized males not only defended larger territories but also had higher dominance ranks and mated more compared to medium-sized and small-sized males (Cole, 1982).

The importance of male–male competition for mates in male *P. martensi* is apparent from the contribution of body size on male reproductive success in the field. Males of large size reproduced more times, as they were more able to obtain more spawnings, over the whole breeding season. However, since no direct behavioural observations were carried out in the field, it can not be established whether male size, *per se*, or other male traits were important for females as cues for their choice of mate. On the other hand, correlative analysis between male size and stone size gave no evidence that stone size is used by females as a choice criterion of spawning site; stone size appeared to have an effect only to the degree that it could physically constrain the number of eggs that can be accommodated in the nest. This, of course, does not mean that other nest site features not considered in this study, such as the bottom complexity or nest concealment, are unimportant for female decision to spawn.

Although larger males have been reported to obtain disproportional share of limited resources, including mates, this is not necessarily true for all the species thus far examined (e.g. Schmale, 1981; Kodric-Brown, 1988; Knapp and Warner, 1991; Lugli *et al.,* 1992). In some cases variation in the contribution of male size on male reproductive success has been noted among populations of the same species and tentatively related to population density and intensity of intraspecific competition (Kodric-Brown, 1988; Lugli *et al.,* 1992). In few other cases, changes of the status and, hence, reproductive success, have been reported to vary both inter- and intra-individually throughout the breeding season (Katano, 1990). Further, body size may be a weak predictor of fighting success when, for example, individuals are in poor physical condition due to prolonged and energetically expensive parental care (De

Martini, 1987; Côte and Hunte, 1989; Lugli *et al.*, 1992). It is clear that maintenance of the dominance status, in addition to yielding benefits, bears costs that can easily reverse the balance even for males with high Resource Holding Potential. Therefore, even for those species, like *P. martensi*, where body size is a strong predictor of dominance and fighting ability, it is by no means an infallible one.

Mechanisms of Female Selection and Role of Signals Advertising Dominance

In our well established laboratory groups of individuals of *P. martensi*, larger high-ranking males monopolized all the available spawnings. Since females were probably not hampered by dominant males when visiting the nests of lower-ranking courting males, female preference for male dominance status is likely to occur in *P. martensi*. In the laboratory, courtship interference and spawning disruption perpetrated by dominant males have been documented in this species by Bisazza *et al.* (1989), in the course of male intrusion experiments. It is unlikely, however, that this is the common pattern for male–male competition during breeding, as was indeed suggested. The establishment of dominance relationships between males was, in fact, not allowed in their experiments until the occurrence of spawning. To the extent that the establishment of dominance relationships among individuals is an important aspect of the breeding system of *P. martensi*, a very reasonable hypothesis supported by field data (Lugli *et al.*, 1992), is that female assessment for male dominance should be expected (Borgia, 1979; Alcock, 1984) and, consequently, take over of nests from neighbours uncommon.

Given the role of body size in conferring male dominance and its importance in male mating success, it is worth inquiring about mechanisms of female choice. By definition, direct female mating preference for a specific male trait such as body size can only be established by observing female choice under condition where other correlates of dominance are held constant (Heisler *et al.*, 1987). Such experiments are difficult or even impossible to undertake due to the presence of close associations among several characters that are potentially assessable by females for their mating choices. For example, Bisazza *et al.* (1989) carried out laboratory experiments in which females *P. martensi* were asked to choose between males of different body size, established under identical nests, or under nests differing markedly for size and structural complexity. They found evidence for the occurrence of female choice for nest quality but not for male size. However, due to the short duration of each experiment (see below) and, most important, since they did not control for between-male differences in behaviour (courtship, sound production etc.), their conclusions are open to criticism.

In our laboratory investigation, by using shelters of equal size and uniformly arranged on the bottom, possible confounding effect of nest characteristics on female mate choice were in large part controlled for. Our approach thus provided insights about mechanisms of females mating preferences for male characteristics. Two replicates are admittedly too few for a full comprehension of mechanisms of female mate selection to be achieved; nevertheless, observations across the laboratory breeding period suggested multiple cues for female choice in this species. Among the several correlates of male status (i.e. body size, courtship, territory size and presence of eggs), body size seemed, at best of minor importance as a criterion of choice by females. Rather, the importance of body size for male mating success stems from correlations with other male characters that appeared more reliable indicators of dominance. Based on our laboratory data, the advertisement of dominance through courtship is suggested in *P. martensi* by three facts. First, courting activity of males *P. martensi* is conditional upon their dominance status, in that low-ranking males are rarely seen courting whereas dominant males court more and for longer. Second, courtship by low-ranking males appear to have social costs in that small males attempting to court females may suffer from attacks by high-ranking males. A courting male is certainly more conspicuous and attracts the attention of conspecifics to a higher extent than does a silent, non courting male. Therefore, having a low dominance index may limit the amount of displaying that can be performed without interference from high-ranking males. As courtship apparently can be performed equally well by adults *P. martensi* regardless of their size (Torricelli *et al.*, 1986), it seems plausible that low-ranking males might reduce their investment in signalling to avoid the social costs of advertisement. A third line of evidence is the observation that (in the first group) the male of highest dominance rank starting courting earlier. Schmale (1981) reported that in *Eupomacentrus partitus* females spawned preferentially with males that were the first to initiate courtship bouts. It is possible that the association between dominance rank and the onset of courtship found in *P. martensi* could assist mate choice from visiting females, allowing easier identification of the best male among those courting from their territorial area. After all, data from Gronell (1989) are suggestive of the capacity of female damselfish *Chrysiptera cyanea* to assess the status of several males prior to spawning and to recall early assessments.

Given the purpose of male courtship for the female assessment of male dominance in laboratory *P. martensi*, present evidence, however, does not provide details either on how this is accomplished or on the way by which visual and acoustical displays assist females in their choices. In addition to providing females with information about male characteristics, such as species identity, status, health etc., that are taken into account for their subsequent mating decisions (e.g. Huntingford, 1984), courtship signals are often reported to be redundant and wasteful (Morris, 1957; Zahavi, 1975). Effort in signalling may be viewed as a handicap which may be used by females to assess male quality reliably (Zahavi, 1975, 1977). In fact, a high investment in signal production or elaboration is thought to be a sufficient requirement for honest advertising of male quality, as only fitter males or males in good conditions can afford to expend energy without paying high costs in terms of reduced survival or higher risks of predation (Zahavi, 1975, 1977; Andersson, 1982; Kodrick-Brown and Brown, 1984). Since redundancy and conspicuousness pertain to the courtship of male *P. martensi*, it is also possible that information other than that related to the dominance status, is conveyed by the males display. Redundancy and conspicuousness pertain to courtship of male *P. martensi*, so it is possible that (over and above information on dominance status) their displays provide information on phenotypic or genotypic quality that may be relevant for female choice.

Female preference for male courting rate has been reported in other species of fish with exclusive paternal care (Schmale, 1981; Cole, 1982; Thresher and Moyer, 1983; Gronell, 1989; Knapp and Warner, 1991). Lack of any correlation or even negative correlations between courtship rate and male success have been documented likewise (Thompson, 1986; Hastings, 1988b), suggesting multiple cues for female spawning decisions.

The pattern of spawnings by females *P. martensi* from our laboratory groups provided circumstantial evidence on the importance of recently-laid eggs as the major cue upon which female rely for their assessment of male parental quality. Similarly, in the damselfish *C. cyanea*, the presence of eggs in the nest was of prevailing importance over courtship as indicator of the male status (Gronell, 1989). Preference for eggs by females of this species was interpreted as a copying strategy whose beneficial consequences were in terms of higher egg survival. Indeed, female preference for males with eggs has been widely documented among species with paternal care (e.g. Ridley and Rechten, 1981; Marconato and Bisazza, 1986; Hastings, 1988a,b; Sikkel, 1988; Unger and Sargent, 1988; Gronell, 1989). This may explain why in so many species males without eggs in the nest try to steal eggs from others males' nest or to take over of nests containing eggs (references in Gronell, 1989).

Less clear are the reasons why male should cannibalize their own eggs, as indeed has been widely reported (e.g. Dominey and Blumer, 1984). Parental care is an energetically expensive activity for a male, particularly when several clutches are obtained in succession. The male commitment to care of eggs does not prevent him from behaving egoistically, for example by cannibalizing some eggs for their energetic value. When filial cannibalism is a likely source of egg mortality, a specific prediction from parental investment theory (Trivers, 1972; Rohwer, 1978) is for caring males to cannibalize last received, less valuable, egg clutches preferentially. As a consequence, the female's expected counter strategy is to avoid spawning in nests containing eggs at an advanced stage of development (Rohwer, 1978). Our results for *P. martensi* support this prediction, as females in both pools apparently avoided spawning in nests not containing eggs in the early stages of development. On the other hand, episodes of filial cannibalism of the entire egg clutch have been documented in both experimental pools. Contrary to expectations, however, egg cannibalism always occurred soon after spawning and was accomplished within a few days. Notably, the cannibalistic male retained high levels of courting behaviour throughout this period. Further investigations are required to clarify the causes underlying this phenomenon in *P. martensi*.

To conclude, male traits such as courtship intensity and vigour are variable signals and therefore of high information value, reflecting a male's current social status or physical condition. Our results support the idea that reliability of such signals does not necessarily stem from the high costs associated with their production. In social systems characterized by dominance-like relationships, for a signal to be honest and cheat-proof costs may indeed be quite low, provided that (i) the expression of the signal correctly reveals the current dominance status or competitive ability of the signaller (the risk of being attacked by dominant neighbouring males should be assurance against cheating from low-ranking males), and (ii) the quality of the signaller has previously been assessed by competitors through cost-incurring fights. In such circumstances, the evolution of signals that reliably advertise dominance or competitive ability should be strongly favoured, as these function to decrease the costs associated to maintenance of the status.

Females of species with paternal care, on the other hand, should choose those cues that best reflect the current status of the male and should be sensitive to temporal changes of such signals (Kodrick-Brown, 1990). For example, the presence of recently-spawned eggs in the nest may overcome courtship as the most reliable signals of dominance, as the tendency to court is partially inhibited in males that are involved in parental care. Therefore, females are expected to assess male quality on multiple criteria that maximize the chances of obtaining high quality mates in each phase of the male breeding cycle, provided that enough time is available. Our results are suggesting that optimal female mating decisions may be constrained by the time available for the assessment of the male quality and status within a group of neighbours. As a consequence, sub-optimal mate choice decisions by females may be common during short-term mate choice experiments. In addition, the presence of physical barriers between males or complex unnatural partitioning of the space may alter the usual pattern of interaction among individuals (e.g. McLennan, 1991), for example by preventing or diminishing the occurrence of sexual interference from neighbours through sound interception. Ultimately, this may result in the cheapening of the social costs (either effective or potential) of male sexual displays and lead to erroneous conclusions about prevailing mechanisms of female mate selection of the species.

ACKNOWLEDGMENTS

This research was funded by a grant of Ministero Pubblica Istruzione. We are grateful to Prof. Danilo Mainardi, Director of the School of Ethology of the Ettore Majorana Centre, for the opportunity offered to present this paper during the Workshop "Behavioural Ecology of Fishes". We are also indebted to Dr. Massimo Manghi and Dr. Stefano Orlandini for their invaluable assistance in data processing.

References

Alcock, J. (1984). *Animal Behavior: An Evolutionary Approach.* Sunderland: Sinauer Associates.

Altmann, J. (1974). Observational study of behavior: sampling methods. *Behaviour, 49,* 227–267.

Andersson, M. (1982). Sexual selection, natural selection and quality advertisement. *Biol. J. Linn. Soc., 17,* 375–393.

Arnold, S. J. (1983). Sexual selection: the interface of theory and empiricism. In: *Mate choice* (Ed. by P. Bateson), pp. 67–107. Cambridge: Cambridge Univ. Press.

Baylis, J. R. (1981). The evolution of parental care in fishes, with reference to Darwin's rule of male sexual selection. *Env. Biol. Fish., 6,* 223–251.

Bisazza, A. and Marconato, A. (1988). Female mate choice, male–male competition and parental care in the river bullhead, *Cottus gobio* L. (Pisces, Cottidae). *Anim. Behav., 36,* 1352–1360.

Bisazza, A., Marconato, A. and Marin, G. (1989). Male competition and female choice in *Padogobius martensi* (Pisces, Gobiidae). *Anim. Behav., 38,* 406–413.

Borgia, G. (1979). Sexual selection and the evolution of mating systems. In: *Sexual Selection and Reproductive Competition in Insects* (Ed. by M. S. Blum and N. A. Blum), pp. 19–80. New York: Academic Press.

Bradbury, J. W. and Davies, N. B. (1987). Relative roles of intra- and intersexual selection. In: *Sexual Selection: Testing the Alternatives* (Ed. by J. W. Bradbury and M. B. Andersson), pp. 143–163. New York: Wiley.

Breder, C. M. Jr. and Rosen, D. E. (1966). *Modes of Reproduction in Fishes.* New York: Natural History Press.

Clutton-Brock, T. H. and Vincent, A. C. J. (1991). Sexual selection and the potential reproductive rates of males and females. *Nature, 351,* 58–60.

Cole, K. S. (1982). Male reproductive behaviour and spawning success in a temperate zone goby, *Coryphopterus nicholsi. Can. J. Zool., 60,* 2309–2316.

Cole, K. S. (1984). Social spacing in the temperate marine goby *Coryphopterus nicholsi. Mar. Biol.*, **80**, 307–314.

Colombo, L., Belvedere, P. C., Marconato, A. and Bentivegna, F. (1982). Pheromones in teleost fish. In: *Proc. Int. Symp. Reprod. Physiol. Fish.* (Ed. by C. J. J. Richter and H. J. Th. Goos), pp. 84–94. Wageningen: Pudoc.

Côte, I. M. and Hunte, W. (1989). Male and female mate choice in the redlip blenny: why bigger is better. *Anim. Behav.*, **38**, 78–88.

Darwin, C. (1871). *The Descent of Man and Selection in Relation to Sex.* London: John Murray.

DeMartini, E. E. (1987). Paternal defence, cannibalism and polygamy: factors influencing the reproductive success of painted greenling (Pisces, Hexagrammidae). *Anim. Behav.*, **35**, 1145–1158.

DeMartini, E. E. (1988). Spawning success of the male plainfin midshipman. I. Influences of male body size and area of spawning site. *J. Exp. Mar. Biol. Ecol.*, **121**, 177–192.

Dominey, W. J. and Blumer, L. S. (1984). Cannibalism of early life stages in fishes. In: *Infanticide: Comparative and Evolutionary Perspectives* (Ed. by G. Housfater and S. B. Hardy), pp. 43–64. New York: Aldine.

Downhower, J. F. and Brown, L. (1980). Mate preferences of female mottled sculpins *Cottus bairdi. Anim. Behav.*, **28**, 728–734.

Gandolfi, G. and Notarbartolo, G. (1971). The influence of recent experiences on the conquest of territory in *Padogobius martensi* (Teleostei, Gobiidae). *Atti Accad. naz. Lincei Rc.* (Cl. Sci. fis. mat. nat.) (Serie VIII), **51**, 405–410.

Gronell, A. M. (1989). Visiting behaviour by females of the sexually dichromatic damselfish, *Chrysiptera cyanea* (Teleostei: Pomacentridae): a probable method of assessing male quality. *Ethology*, **81**, 89–122.

Halliday, T. R. (1983). The study of mate choice. In: *Mate Choice* (Ed. by P. Bateson), pp. 3–32. Cambridge: Cambridge Univ. Press.

Halliday, T. R. (1987). Physiological constraints on sexual selection. In: *Sexual Selection: Testing the Alternatives* (Ed. by J. W. Bradbury and M. B. Andersson), pp. 247–264. New York: Wiley.

Hastings, P. A. (1988a). Female choice and male reproductive success in the angel blenny, *Coralliozetus angelica* (Teleostei: Chenopsidae). *Anim. Behav.*, **36**, 115–124.

Hastings, P. A. (1988b). Correlates of male reproductive success in the browncheek blenny, *Achantemblemaria crockeri* (Blennioidea: Chaenopsidae). *Behav. Ecol. Sociobiol.*, **22**, 95–102.

Heisler, L. (group repporteur) (1987). The evolution of mating preferences and sexually selected traits. In: *Sexual Selection: Testing the Alternatives* (Ed. by J. W. Bradbury and M. B.Andersson), pp. 97–118. New York: Wiley.

Hinde, R. A. (1966). *Animal Behaviour, a Synthesis of Ethology and Comparative Psychology.* New York: McGraw-Hill Inc.

Huntingford, F. A.(1984). *The Study of Animal Behaviour.* London: Chapman and Hall.

Itzkowitz, M. (1978). Group organization of a territorial damselfish, *Eupomacentrus planifrons. Behaviour*, **65**, 125–137.

Katano, O. (1990). Dynamic relationships between the dominance of male dark chub, *Zacco temmincki*, and their acquisition of females. *Anim. Behav.*, **40**, 1018–1034.

Knapp, R. A. and Warner, R. R. (1991). Male parental care and female choice in the bicolor damselfish, *Stegastes partitus*: bigger is not always better. *Anim. Behav.*, **41**, 747–756.

Kodric-Brown, A. (1978). Establishment and defence of breeding territories in a pupfish, (Cyprinodontidae: Cyprinodon). *Anim. Behav.*, **26**, 818–834.

Kodric-Brown, A. (1983). Determinants of male reproductive success in pupfish (*Cyprinodon pecosensis*). *Anim. Behav.*, **31**, 128–137.

Kodric-Brown, A. (1988). Effect of population density, size of habitat and ovoposition substrate on the breeding system of pupfish (*Cyprinodon pecosensis*). *Ethology*, **77**, 28–43.

Kodric-Brown, A. 1990. Mechanisms of sexual selection: insights from fishes. *Ann. Zool. Fennici*, **27**, 87–100.

Kodric-Brown, A. and Brown, J. H. (1984). Truth in advertising: the kinds of traits favored by sexual selection. *Am. Nat.*, **124**, 309–323.

Lindström, K. (1988). Male–male competition for nest sites in the sand goby, *Pomatoschistus minutus. Oikos*, **53**, 67–73.

Lugli, M. (1989). Comunicazione acustica nel ghiozzo di fiume Padogobius martensi (Günther, 1861): caratteristiche strutturali, utilizzo e significato dei suoni nel repertorio comportamentale. Ph.D. thesis. University of Parma.

Lugli, M., Torricelli, P. and Bobbio, L. (1992). Breeding ecology and male spawning success in two hill-stream populations of the freshwater goby, *Padogobius martensi. Env. Biol. Fish.*, **35**, 37–48.

Magnhagen, C. and Kvarnemo, L. (1989). Big is better: the importance of size for reproductive success in male *Pomatoschistus minutus* (Pallas) (Pisces, Gobiidae). *J. Fish Biol.*, **35**, 755–763.

Magurran, A. E. (1986). Individual differences in fish behaviour. In: *The Behaviour of Teleost Fishes* (Ed. by T. J. Pitcher), pp. 338–365. London and Sydney: Croom Helm.

Marconato, A. and Bisazza, A. (1986). Males whose nests contain eggs are preferred by female *Cottus gobio* L. (Pisces, Cottidae). *Anim. Behav.*, **34**, 1580–1582.

Marconato, A., Bisazza A. and Marin, G. (1989). Correlates of male reproductive success in *Padogobius martensi* (Gobiidae). *J. Fish Biol.*, **34**, 889–899.

McLennan, D. A. (1991). Integrating phylogeny and experimental ethology: from pattern to process. *Evolution*, **45**, 1773–1789.

Morris, D. (1957). "Typical intensity" and its relation to the problem of ritualization. *Behaviour*, **11**, 1–12.

Myrberg, A. A. Jr. (1972). Social dominance and territoriality in the bicolor damselfish, *Eupomacentrus partitus* (Poey)(Pisces: Pomacentridae), *Behaviour*, **41**, 207–231.

Parker, G. A. (1974). Assessment strategy and the evolution of fighting behaviour. *J. Theor. Biol.*, **47**, 223–243.

Parmigiani, S., Torricelli, P. and Lugli, M. (1988). Intermale aggression in *Padogobius martensi* (Günther) (Pisces: Gobiidae) during the breeding season: effect of size, prior residence, and parental investment. *Monit. Zool. Ital.*, **22**, 161–170.

Petersen, C. V. (1988). Male mating success, sexual size dimorphism, and site fidelity in two species of *Malacoctenus* (Labrisomidae). *Env. Biol. Fish.*, **21**, 173–183.

Ridley, M.(1978). Paternal care. *Anim. Behav.*, **26**, 904–932.

Ridley, M. and Rechten, C. (1981). Female sticklebacks prefer to spawn with males whose nests contain eggs. *Behaviour*, **76**, 152–161.

Rohwer, S. (1978). Parent cannibalism of offspring and egg raiding as a courtship strategy. *Am. Nat.*, **112**, 429–440.

Schmale, M. C. (1981). Sexual selection and reproductive success in males of the bicolor damselfish, *Eupomacentrus partitus* (Pisces: Pomacentridae). *Anim. Behav.*, **29**, 1172–1184.

Siegel, S. (1956). *Non-Parametric Statistics for the Behavioral Sciences*. New York: McGraw-Hill.

Sikkel, P. C. (1988). Factors influencing spawning site choice by female garibaldi, *Hypsypops rubicundus* (Pisces: Pomacentridae). *Copeia*, **3**, 710–718.

Sokal, R. R. and Rohlf, F. J. (1981). *Biometry*. New York: W. H. Freeman and Company.

Thompson, S. (1986). Male spawning success and female choice in the mottled triplefin, *Forsterygion varium* (Pisces: Tripterygiidae). *Anim. Behav.*, **34**, 580–589.

Thresher, R. E. and Moyer, J. T. (1983). Male success, courtship complexity and patterns of sexual selection in three congeneric species of sexually monochromatic and dichromatic damselfishes (Pisces: Pomacentridae). *Anim. Behav.*, **31**, 113–127.

Tinbergen, N. (1951). *The Study of Instinct*. London: Oxford University Press.

Torricelli, P. and Romani, R. (1986). Sound production in the Italian freshwater goby *Padogobius martensi. Copeia*, **1**, 213–216.

Torricelli, P., Lugli, M. and Gandolfi, G. (1985). A quantitative analysis of the fanning activity in the male *Padogobius martensi* (Pisces, Gobiidae). *Behaviour*, **92**, 288–301.

Torricelli, P., Lugli, M. and Gandolfi, G. (1986). A quantitative analysis of the occurrence of visual and acoustic displays during the courtship in the freshwater goby *Padogobius martensi* (Günther 1861) (Pisces, Gobiidae). *Boll. Zool.*, **53**, 85–89.

Torricelli, P., Parmigiani, S., Lugli, M. and Gandolfi, G. (1988). Intermale aggression in *Padogobius martensi* (Günther) (Pisces, Gobiidae): effect of size and prior residence. *Monit. Zool. Ital.*, **22**, 121–131.

Torricelli, P., Lugli, M. and Pavan, G. (1990). Analysis of sounds produced by male *Padogobius martensi* (Pisces, Gobiidae) and factors affecting their structural properties. *Bioacoustics*, **2**, 261–275.

Trivers, R. L. (1972). Parental investment and sexual selection. In: *Sexual Selection and the Descent of Man* (Ed. by B. Campbell), pp. 136–179. Chicago: Aldine.

Unger, L. M. and Sargent, R. C. (1988). Allopaternal care in the fathead minnow, *Pimephales promelas*: females prefer males with eggs. *Behav. Ecol. Sociobiol.*, **23**, 27–32.

Williams, G. C. (1966). *Adaptation and Natural Selection*. Princeton: Princeton Univ. Press.

Zahavi, A. (1975). Mate selection—a selection for a handicap. *J. Theor. Biol.*, **53**, 205–214.

Zahavi, A. (1977). The cost of honesty (Further remarks on the handicap principle). *J. Theor. Biol.*, **67**, 603–605.

MALE COMPETITION, FEMALE MATE CHOICE
AND SEXUAL SIZE DIMORPHISM
IN POECILIID FISHES

ANGELO BISAZZA

Dipartimento di Psicologia Generale, Università di Padova
Piazza Capitaniato 3, 35139 Padova, Italy

INTRODUCTION

A size difference between males and females is a common feature in many animal taxa, although the extent and the direction of such dimorphism is often highly variable. Darwin (1874) suggested that a larger size in males is related to intrasexual competition for access to mates, while a larger size of females is the product of selection for producing more offspring. In the last three decades several other explanatory models have been developed (e.g. Selander, 1966; Gliwics, 1988; Fairbairn, 1990) and the recent development of rigorous methods for comparative analysis has promoted studies in which predictions from different evolutionary models are contrasted. Some work has been done on reptiles, amphibians and arthropoda (e.g. Woolbright, 1983; Carothers, 1984), but most attention has focused on evolution of sexual size dimorphism in birds and mammals, including our own species (Clutton-Brock *et al.,* 1977; Cheverud *et al.,* 1985; Frayer and Wolpoff, 1985; Temeles, 1985; Hoglund, 1989). The terminology itself, reflects this bias. The term "normal" size dimorphism is adopted for species with larger males and the term "reverse" size dimorphism is used for species with larger females, this is in spite of the fact that, except in higher vertebrates, the female is commonly the larger sex (Ghiselin, 1974; Gilbert and Williamson, 1983; Woolbright, 1983).

Poeciliid fishes are characterized by a marked sexual dimorphism in size, males being considerably smaller than females in most species. A smaller size in males is not uncommon among teleost fishes, but dimorphism is often extreme in this family and the males of some livebearers are among the smallest living vertebrates. The degree of dimorphism of the Poeciliidae is variable, ranging from species in which the maximum length of females is more than twice the maximum length of males, to species in which the length in the two sexes is about the same (Jacobs, 1971). Size dimorphism is more evident when we consider body weight; in the sample of eastern mosquitofishes, *Gambusia holbrooki* collected for this study, the ratio of the maximum female to male length is about 1.7 but the largest female outweigh the largest male by 5 times. There have been some discussion of the significance of dimorphism in poeciliids (e.g. Endler, 1983), but before the present study there has been no attempt to test evolutionary hypotheses about its origin. Instead, poeciliid fishes have aroused the interest of evolutionary biologists for another aspect of sexual dimorphism; several species show well developed male secondary sex traits, such as the brilliant and variable coloration of guppies or the elaborate fins of mollies and swordtails, which Darwin (1874) compared in their ornaments to polygamous birds.

In this paper I first introduce the topics of sexual selection in poeciliid fish and summarize the present knowledge in this field. I next examine more in detail the factors affecting reproductive success in one species, *Gambusia holbrooki* and discuss the selective forces acting on body size in both sexes. Current hypotheses about the origin of sexual size dimorphism in poeciliids are then examined and a new model based on the results of the study in *G. holbrooki* is proposed. Finally, the prediction generate by this model are tentatively tested using a comparative approach.

Biology of Poeciliid Fish

The family Poeciliidae (Order Cyprinodontiformes) is composed of nearly 200 species distributed from southern USA to northern Argentina, but mostly inhabiting tropical fresh and brackish waters. Most species are omnivorous, feeding on algae, detritus, small invertebrates and young fish. A few species are herbivorous and one, the pike killifish, *Belonesox belizanus* is strictly piscivorous. All but one species show internal fertilization and give birth to living young. In *Tomeurus gracilis*, probably the most primitive species, there is internal fertilization, but the female has the option of either laying fertilized eggs or retaining the embryos in the ovary (Gordon, 1955). Male poeciliids have a modified anal fin, the gonopodium, which is used as an intromittent organ. In a few species the insertion of the gonopodium into the female's genital pore is probably not necessary and fertilization may occur simply after sperm has been deposited externally or near the genital papilla (Gordon, 1955; Rosen and Tuker, 1961; Constantz, 1989). Female poeciliids store sperm for several months (Dulzetto, 1928; Turner and Snelson, 1984). Matings may thus take place at any time of the reproductive cycle, but because of sperm precedence, copulations occurring after parturition are probably more effective (Bisazza *et al.*, 1989). There are two reproductive modes in females. In most species, a few days after parturition the female fertilizes one batch of eggs. Gestation generally lasts 30 to 40 days, during which period another group of ova accumulates yolk. These species produce large eggs (lecithotrophy) and there is little or no maternal contribution after fertilization has occurred. Some other species show superfetation, namely the capacity of simultaneously carrying broods at different stages of development. The number of simultaneous broods ranges from 2–3 in the genus *Poeciliopsis* to a maximum of 9 in *Heterandria formosa* (Scrimshaw, 1944; Schoenherr, 1977). This reproductive adaptation is typical of the Heterandriini (see Table 1 for a taxonomy of the Poeciliidae) but a few superfetators are present in the other tribes. Superfetation is generally associated with matrotrophy, the passage of nourishment from the mother to the embryos (Reznik and Miles, 1989a).

Table 1 Taxonomic distribution of male mating tactics within the family Poeciliidae
T = copulatory thrusts; C = courtship; TC = both tactics

Tribe TOMEURINI		Tribe GAMBUSIINI	
Genus *Tomeurus*		Genus *Gambusia*	
Tomeurus gracilis	T	Subgenus *Gambusia*	
Tribe POECILIINI		*Gambusia hispagnolae*	T
Genus *Alfaro*		*Gambusia manni*	T C
Alfaro cultratus	T	*Gambusia melapleura*	T
Genus *Poecilia*		*Gambusia wrayi*	T C
Subgenus *Poecilia*		Subgenus *Arthrophallus*	
Poecilia chica	T	*Gambusia aurata*	T C
Poecilia vivipara	T	*Gambusia holbrooki*	T
Poecilia spenops	T	*Gambusia affinis*	T C
Poecilia mexicana	T C	*Gambusia heterochir*	T C
Poecilia latipinna	T C	*Gambusia nobilis*	T
Subgenus *Lebistes*		*Gambusia atrora*	T
Poecilia reticulata	T C	*Gambusia hurtadoi*	T
Poecilia parae	T C	Subgenus *Heterophallina*	
Poecilia picta	T C	*Gambusia vittata*	T
Poecilia branneri	T C	Genus *Belonesox*	
Subgenus *Limia*		*Belonesox belizanus*	T C
Poecilia melanogaster	T C	Genus *Brachyrhaphis*	
Poecilia perugiae	T C	*Brachyrhaphis cascajalensis*	T C
Poecilia nigrofasciata	T C	*Brachyrhaphis parismina*	T
Poecilia vittata	T	*Brachyrhaphis episcopi*	T
Poecilia dominicensis	T	*Brachyrhaphis roseni*	T C
Poecilia zonata	T	Tribe GIRARDININI	
Genus *Xiphophorus*		Genus *Girardinus*	
Platies		*Girardinus metallicus*	T
Xiphophorus gordoni	C	*Girardinus falcatus*	T
Xiphophorus variatus	C	Genus *Quintana*	
Xiphophorus xipidium	T C	*Quintana atrizona*	T
Xiphophorus evelinae	T	Genus *Carlhubbsia*	
Xiphophorus maculatus	T	*Carlhubbsia stuarti*	T
Swortails		*Carlhubbsia kidderi*	T
Xiphophorus milleri	T C	Tribe HETERANDRIINI	

Table 1 (Continued) Taxonomic distribution of male mating tactics within the family Poeciliidae
T = copulatory thrusts; C = courtship; TC = both tactics

Xiphophorus pigmaeus	T	Genus Priapichthys	
Xiphophorus nigrensis	TC	Priapichthys annectens	T
Xiphophorus montezumae	TC	Genus Neoheterandria	
Xiphophorus cortezi	C	Neoheterandria tridentiger	T
Xiphophorus clemenciae	TC	Neoheterandria elegans	T
Xiphophorus alvarezi	TC	Genus Heterandria	
Xiphophorus helleri	TC	Heterandria formosa	T
Xiphophorus signum	TC	Heterandria bimaculata	T
Genus Priapella		Genus Poeciliopsis	
Priapella compressa	T	Poeciliopsis occidentalis	TC
Tribe CNESTERODONTINI		Poeciliopsis monaca	T
Genus Cnesterodon		Poeciliopsis lucida	T
Cnesterodon decemmaculatus	T	Poeciliopsis gracilis	T
Genus Phalloceros		Genus Phallichthys	
Phalloceros caudimaculatus	T	Phallichthys amates	T
Genus Phallottorynus		Tribe XENODEXIINI	
Phallottorynus jucunods		Genus Xenodesia	
Tribe SCOLICHTHYINI		Xenodesia ctenolepis	T
Genus Scholichthys			
Scholichthys greenwayi	T		

MATING BEHAVIOUR AND MECHANISMS OF SEXUAL SELECTION IN POECILIID FISHES

Male Mating Tactics

Male poeciliids are sexually very active and their behaviour is easily observed both in laboratory and in the field. Poeciliids exhibit two mating tactics. The male can court the female to obtain her cooperation during mating, or he can bypass female acceptance and attempt a forced insemination. The action by which males achieve a forced copulation has been given various names such as lunge (Itzkowitz, 1971), rape (Farr, 1980a) and sneak copulation (Endler, 1983); the term gonopodial thrust is the most often used and will be adopted here. In a number of species gonopodial thrust is the only male mating tactic observed, whereas in some others copulation is always preceded by a courtship display. In the remaining poeciliids insemination follows either route. In species such as *Poeciliopsis occidentalis* and *Poecilia reticulata* the same individual can use both tactics depending on circumstances, e.g. its competitive ability or the receptivity of the female (Constantz, 1975; Farr, 1980a). In other species, like *Xiphophorus nigrensis* and *Poecilia latipinna*, there are different types of males that exhibit different behavioural repertoires (Travis and Woodward, 1989; Zimmerer and Kallman, 1989).

Gonopodial thrusts

During a copulatory attempt the male approaches the female from behind, brings the gonopodium forward and tries to insert it into the female's genital pore. In most species a thrust must occur while a female is stationary and the female can avoid copulation by changing her orientation, slapping the male with her tail or moving away from him. In aquaria, female mosquitofish may fold the anal and caudal fins over the genital opening or position their tail close to solid object when harassed by males (personal observation). There are some variations to the basic pattern described here. For example, in *Poecilia reticulata* and *Xiphophorus helleri* the male approaches the female more laterally than in mosquitofish; in *Poecilia melanogaster* copulatory thrusts usually take place while male and female are in rapid motion (personal observation), a variation probably occurring in other species as well (Rosen and Tucker, 1961). Only a small proportion of copulatory thrusts are effective. A recent experiment (Bisazza and Marin, 1991a) showed that in *G. holbrooki* approximately only one in every 30 copulatory thrusts results in a contact between genitalia and analysis of slow motion video recordings suggests that only a small minority of these contacts involves a complete intromission of the gonopodium. In the past, the low efficiency of copulatory thrusts has led some authors to doubt of the effectiveness of these tactics (e.g. Clark *et al.,* 1954). The frequency of male attempts is however very high; in natural populations of *G. holbrooki* there is about one act per male per minute (Martin, 1975; Bisazza and Marin, 1991a) and similar frequencies have been observed in other species (Farr, 1980b; Luyten and Liley, 1985; Travis and Woodward, 1989). Using a conservative estimate of one insemination every 1000 attempts, one male would mate (in the sense of transferring sperm) around one hundred times per season, a figure much greater than in many other vertebrates.

Courtship tactics

During courtship the male generally displays in front of the female with the unpaired fins fully spread. He shows a stereotyped swimming sequence that varies from species to species and that often serves to display conspicuous pigmentation or elaborate fins. The female's acceptance of courtship can be inferred from behaviour that facilitates the occurrence of copulation. This usually consists of staying motionless and tilting the body towards the male, so as to favour contact between genitalia (Liley, 1966; Hughes, 1985a). Females tend to be more responsive to male courtship when virgins, after parturition or after a separation from males (Clark *et al.,* 1954; Liley, 1966; Farr, 1984; Hughes, 1985a).

Taxonomic Distribution of the Two Mating Tactics

In Table 1, I have summarized the present knowledge on the distribution of male mating tactics in the family Poeciliidae. Much of this information has been recently collected by Farr (1989), but some additional data for the genus *Xiphophorus* were found in Heinrich and Schröeder (1988) and information for *Gambusia affinis* and *G. holbrooki* are based on two recent studies (Hughes, 1985a; 1986; Bisazza and

Marin, 1991a; 1991b). The data for a few other species come from my own unpublished aquarium observation. In a few cases in which literature presented conflicting data, I accepted the findings of the most detailed study. In one case my own observations disagreed slightly with the published literature. Farr (1984) reports that in *Poecilia* (*Limia*) *melanogaster* there is virtually no gonopodial thrust, but in the aquarium stock kept in our laboratory this behaviour was fairly frequent, always occurring while both fishes were in fast motion. I have also observed copulatory thrusts in *Belonesox belizanus* (see Farr, 1989), but these were so rare that it is doubtful whether this species should be considered having two mating tactics. The references for many species can be found in Appendix. About half of the species (58%) exhibit only gonopodial thrusts and the majority of the others exhibit both tactics (38%). Only three species (4%) lack gonopodial thrusts and achieve mating only through courtship; with the possible exception of *Belonesox belizanus* all species lacking gonopodial thrusts belong to the genus *Xiphophorus*. Most genera contain both species with courtship and species without. Exceptions are represented by genera of the tribes Cnesterodontini and Girardinini (and three other tribes represented with one single species), in which males perform only gonopodial thrusts. This might be the consequence of the different size of the tribes or of a greater attention received by some taxa. Although there are data for more than one third of Poeciliid species, the above conclusions should be accepted with caution. There are in fact detailed studies for less than one dozen species, all the other information coming from incidental observations. Moreover, in many cases the influence of intersexual selection is excluded on the basis of a lack of a definite courtship display in the male, but as discussed later there are other possibilities for the female to exert mate preferences.

Female Mate Choice

The criteria for mate choice are known for some species. For example in the western mosquitofish, *Gambusia affinis* and in *Xiphophorus nigrensis*, larger males are preferred over small ones (Hughes, 1985a; Ryan and Wagner, 1987). In the guppy, *Poecilia reticulata*, females prefer mates that exhibit more intense coloration, larger tails, higher rates of courtship and a rare phenotype (Farr, 1977, 1980c; Bischoff *et al.*, 1985; Kodric-Brown, 1985). Recent studies have demonstrated geographic variation in female mate preferences in guppies (Houde, 1988a; Stoner and Breden, 1988).

  Courtship behaviour in males and receptivity and sexual preferences in females have generally evolved concurrently but there are exceptions. In *Xiphophorus pygmaeus* males are extremely small, show little secondary sex differentiation and do not court females. Ryan and Wagner (1987) have recently shown that female *X. pygmaeus* prefer to mate with males of the sister species *X. nigrensis*, which are larger, show a courtship display and have a long swordtail. One interpretation proposes that female mating bias may be a primitive condition and that male *X. nigrensis* have evolved a trait that exploited a pre-existing preference (Ryan, 1990). Populations of *X. nigrensis* show a broad range in male body size, which derives from polymorphism at a single locus (the P locus). Small males use gonopodial thrusts almost exclusively; large males have a proportionally larger swordtail and a brilliant coloration and they court females (Rosen, 1960; Ryan and Causey, 1989; Zimmerer and Kallman, 1989). Males *X. pygmaeus* are similar to the small males of *X. nigrensis* in size, morphology and behaviour. Therefore an alternative explanation for mate preferences in *X. pygmaeus* is that male courtship display was present in the common ancestor but that alleles producing larger courting males were secondarily lost in this species. Female preference for a sword-like fin has been reported in another species, the platy *X. maculatus*, in which males lack such elaboration (Basolo, 1989; cited in Ryan, 1990). Interestingly however, the males of this and other platys seem to possess unexpressed genes for the "sword" and the growth of the caudal appendage can be induced by hormone treatment (Zander and Dzwillo, 1969).

The female can exercise some choice of mate even in the absence of courtship behaviour in the male. She may, for example, enhance the likelihood of mating with more competitive individuals by advertising her fertility. This was first suggested for *Poecilia latipinna* (Farr and Travis, 1986), but may occur in other species too (Parzefall, 1973; Bisazza *et al.*, 1989). There are even more subtle ways to be selective. In *G. holbrooki* there is no male pre-copulatory behaviour or direct female cooperation. In situations of competition the dominant male assumes a distinct coloration and a characteristic behaviour aimed at protecting the female from gonopodial thrusts of the other males. Females spend most of their time in the vicinity of the dominant males, in this way enhancing the probability of mating with them (Bisazza and Marin, 1991b; Bisazza, unpublished observations). A similar situation occurs in *Poeciliopsis occidentalis* where females tend to swim away more slowly when approached by a territorial male as opposed to a non-territorial one (Constantz, 1975). I have sometimes observed male mosquitofish to exhibit the behaviour and coloration typical of a guarding individual even in the absence of rivals, especially when very small females were present. By doing so, males may be exploiting female preference for dominant individuals. In the related species *G. affinis*, where males have a courtship display, this is very similar to the display used to threaten other males (Hughes, 1985a). I have observed a similar situation in *G. wray* where males are also brightly coloured, and it is possible that in the genus *Gambusia* courtship display is derived from stereotyped postures indicating dominance (see also Farr, 1989). The use of the same display in both agonistic and reproductive context has been reported also in *Xiphophorus maculatus* (Clark *et al.*, 1954).

Intrasexual Selection

Male–male competition has been reported in many poeciliid fish (reviewed by Farr, 1989) and recent studies have established the influence of fighting ability on male mating success in *Gambusia affinis*, *G. heterochir*, *Poecilia latipinna*, *Xiphophorus helleri* and *X. nigrensis* (Beaugrand *et al.*, 1984; Hughes,1985a; Farr *et al.*, 1986; Yan, 1987; Zimmerer and Kallman, 1989). In *Poeciliopsis occidentalis*, males are known to defend breeding territories (Constantz, 1975), and territoriality has occasionally been reported for other species (Braddock, 1949; George, 1960 cited in Martin, 1975; Zayan, 1975). Size appears the main factor influencing the outcome of fighting, but there are exceptions (Farr, 1980b). Some species, including *Belonesox belizanus*, *Neoheteramdria elegans*, *Scholichthys greenway*, *Poecilia picta* and *P. parae*, exhibit little or no intermale aggression (Farr, 1989; Liley, 1966; Bisazza, unpublished observations). In the guppy, *Poecilia reticulata*, early reports suggested a role of aggression and dominance on male mating success (Gandolfi, 1971; Gorlick, 1976), but all subsequent studies have indicated little intrasexual selection in this species (Farr, 1980c; Houde, 1988b). Data on male competition have come mostly from laboratory studies and many authors suggest that under these conditions the importance of intersexual selection may be in general over-estimated (Luiten and Liley, 1991; Farr, 1989; Travis and Woodward, 1989).

Sexual Selection and Male Ornamentation

Although the majority of poeciliids have little or no secondary sex differentiation, several species are highly dimorphic in colour or fin shape. Current theories of sexual selection highlight the role of female choice in the evolution of male ornaments (Kirkpatrick, 1987; Ryan, 1990; Maynard Smith, 1991). However, recent studies indicate that in many species male secondary sexual traits function in intrasexual competition or serve in both contexts (e.g. Gronell, 1989; Moore, 1990; Searcy and Yasukawa, 1990; Petrie et al., 1991). Among poeciliids the sword of the green swordtail, *Xiphphorus helleri* and the dorsal fin of the sailfin molly, *Poecilia velifera* have a significance in both courtship and male–male competition (Hemens, 1966; Parzefall, 1969; Bildsøe, 1988; Basolo, 1990), whereas sexual dimorphism of guppies, *P. reticulata* is probably related to epigamic selection only (Houde, 1988b).

Evolution of Mating Tactics in Poeciliids

Examining the mating tactics in the tribe Gambusiini, Farr (1989) has suggested that male courtship may have been present in the ancestral species and was subsequently lost in the majority of species of the genus *Gambusia*. On the other hand the observation that gonopodial thrust is apparently the only tactic present in the primitive *Tomeurus gracilis* (Gordon, 1955) and that a behaviour similar to the gonopodial thrust of the poeciliids is present in oviparous cyprinodontiforms (Kulkarni, 1940) suggest that the presence of copulatory thrusts is a primitive condition. As indicated in Table I male mating tactics often vary within a genus, even between sister species (Bisazza and Marin, 1981; Rayan and Wagner, 1987). This suggests that there are few phylogenetic constraints to evolution of male mating behaviour and that mode of fertilization may evolve rapidly in response to environmental demands. Are there ecological conditions that favour the evolution of male courtship and female choice and others that cause copulatory thrusts to become the main method of insemination? Farr (1984, 1989) stressed the importance of selection on females in understanding the determinants of male mating success. He argued that many poeciliids are ecological generalists and live in habitats that are highly variable in space and in time. In these species female choice did not evolve because there are few male traits that would result in increased offspring fitness upon which choice could be based. In contrast, female choice for male indicators of vigour (courtship display, secondary sexual coloration, etc.) is expected to evolve in poeciliids that live in environments that are stable or that change in a predictable way. In support of this view is the observation that in the subgenus *Limia* and in platifish, species with courtship are apparently more ecologically specialized (Rosen, 1960; Farr, 1984). The mode of sexual selection may also be determined by the costs and benefits of different male behavioural options, which in turn may relate to the physical or biotic features of each environment. There is evidence that this occurs at least at the intraspecific level. In Trinidad populations of *Poecilia reticulata*, males from populations that inhabit lowland turbid water show higher frequencies of copulatory thrusts, while males from clear mountain headstreams rely mostly on courtship display, a difference which is probably related to a greater effectiveness of courtship displays in clear water and to a different susceptibility to predation of males employing the two tactics (Luyten and Liley, 1985, 1991; Endler, 1987). Similar factors might affect secondary sexual traits in the genus *Xiphophorus*; species with a swordtail are usually found in clear water with little or no vegetation, while those without tend to inhabits turbid water with dense vegetation, where such male ornamentation would be of limited effectiveness (Rosen, 1960).

BODY SIZE AND REPRODUCTIVE SUCCESS IN THE EASTERN MOSQUITOFISH (*GAMBUSIA HOLBROOKI*)

Size and Female Reproductive Success

Figure 1a shows the relationship between body size and fecundity in a sample of female mosquitofish collected at Lake Fimon (northern Italy) during the reproductive season of 1988. As is usual for fishes, brood size increases significantly with body length. Large females invest more in reproduction per unit of body weight, as evidenced by the relationship between length and gonosomatic index (Figure 1b). The weight of fully developed embryos significantly increase with female length (R = 0.42, n = 43, $P < 0.01$), which suggests that larger females also produce larger offspring.

Size-related variation in reproductive effort is common among fishes and is generally attributed to changes with age in the allocation between growth and reproduction (Roff, 1983; Reiss, 1987). However, it is not clear why larger or older females should produce larger offspring. Some models predict that under specified conditions larger females should produce larger offspring (e.g. Parker and Begon, 1986; Sargent et al., 1987), but in general life history theories suggest that there should be one single optimum for all females in a population (Smith and Fretwell, 1974; Brockelman, 1975). A correlation between the size of the mother and that of eggs or offspring has been observed in some other fishes (Sargent et al., 1987; Marconato and Bisazza, 1988; Docker and Beamish, 1991); among poeciliids, it has been found in species as diverse as *Poeciliopsis occidentalis*, *Poecilia reticulata*, *Belonesox belizanus* and *Heterandria formosa* (Constantz, 1974; Reznic and Endler, 1982; Turner and Snelson, 1984; Travis et al., 1987) and it is probably a general characteristic of this family. In some turtles, variation in offspring size appears to be related to a morphological constraint, the width of pelvic opening (Congdon et al., 1983). Among viviparous fish, the production of larger offspring is unlikely to be constrained in this way, but other factors such as a longer developmental time of larger embryos (Steele, 1977) or a proportionally reduced respiratory rate (Schmidt-Nielsen, 1984), might be important.

Size and Male Reproductive Success

To examine the relationship between body size and male mating success, three male and three female *G. holbrooki* were chosen at random from a sample just

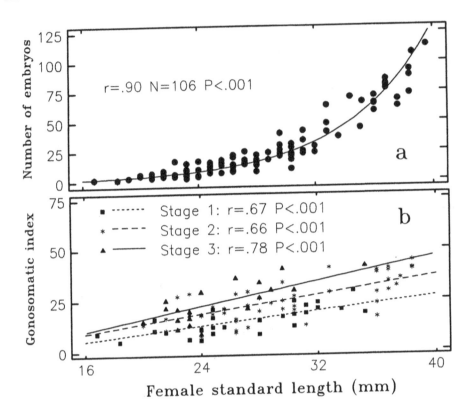

Figure 1 Size and reproductive success in female eastern mosquitofish. a) Fecundity increases with body length. b) Gonosomatic index (gonadal weight/somatic weight × 100) measured at three developmental stages of the embryos (1 = uneyed; 2 = early-eyed; 3 = late-eyed) increases with female length.

netted at Fimon Lake and placed immediately in a large tank simulating their natural environment. After 10 min of acclimatisation, they were observed for an hour. The number of mating attempts culminating with a contact between genitalia was used as an estimate of male mating success. The male were then anaesthetized and measured. The analysis of the 15 replicas indicates that in *G. holbrooki* the number of matings is negatively correlated with male standard length (Figure 2). This result is somewhat puzzling. Since both intra- and inter-sexual selection should favour larger males, a positive correlation between size and mating success is expected. Negative correlations have been reported only for one bird and a few species of diptera (Petrie, 1983; McLachlan and Allen, 1987; Boake, 1989). However, a series of recent experiments hastened to clarify the relationship between male size and male mating success.

Factors affecting the success of gonopodial thrusts

One experiment consisted of observing pairs of fishes of varying size and measuring the success of each gonopodial thrust (Bisazza and Marin, 1991). Slow

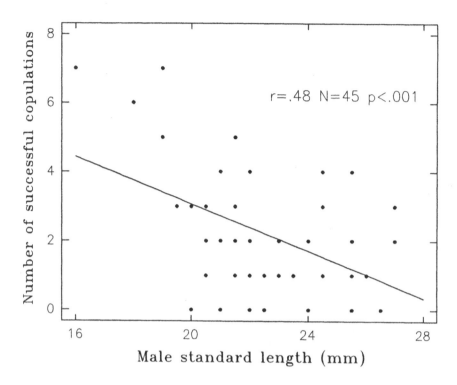

Figure 2   Size and reproductive success in male eastern mosquito fish. The number of copulatory thrusts culminating in a contact between genitalia decrease significantly with increasing male length.

motion videorecording showed that only 97 out of 3196 mating attempts ended with a contact between genitalia. Body size had a dramatic negative effect on the outcome of sexual behaviour, frequency of success decreasing exponentially with length of the male that attempted the copulation. In contrast, success increased with the length of the female, and the best predictor of the outcome of a gonopodial thrust was the ratio of male to female length. When the male was half the length of the female, he had a tenfold greater probability of mating than if he was the same size. The entire sequence of a gonopodial thrust takes place within the blind portion of the visual field of the female and small males are probably less likely to be seen while approaching. Small males may also be better able to manoeuvre while they try to insert the gonopodium. Little data is available on other species to assess the generality of this phenomenon, although Farr *et al.* (1986), testing individual males of the sailfin molly, report that males that mated were, on average, smaller than unsuccessful males.

Influence of male competition

In another experiment, groups of two to five males were observed to be competing for a female in large tanks. Under these conditions a hierarchy was rapidly established and the larger male spent most of his time standing behind the female so to prevent copulatory thrusts from subordinates. The guarding behaviour of the dominant male was very efficient and he accomplished more than ninety percent of all copulatory attempts regardless of the number of competitors (Bisazza and Marin, 1991a).

Female mate choice

In the two previous experiments using *G. holbrooki* courtship display in the male and cooperative behaviour in the female were never observed. However, in the sister species, *G. affinis*, females in a two choice apparatus showed a preference for larger mates after being deprived of contact with males for some time prior to the experiment (Hughes, 1985a). Identical tests conducted on *G. holbrooki* failed to show any preference, either using male-deprived, virgin or post-partum females (Bisazza and Marin, 1991b). When observed in a more natural situation (a large tank with three males of different size), females approached more often the males whose behaviour indicated dominance and they spent longer time in their vicinity (Bisazza and Marin, 1991b; Bisazza, unpublished observations). Thus, although female *G. holbrooki* do not actively cooperate during mating, her behaviour favours their guarding behaviour of the larger, dominant males and indirectly enhances the probability that their progeny is sired by these individuals.

Selection on Male Body Size

The above account indicates that sexual selection acts on male body size in a complex way. Small males are much more efficient at gonopodial thrusts, but when a larger male is present, chances of approaching a female and attempting copulation are considerably reduced. In general, small males should be favoured when the frequency of male competition is low, but it is not easy to predict which phenotype will be more successful in nature. The solution is related to factors such as the sex-ratio, the size of the females, the encounter rate of males and the probability of a given male being the largest in his group. Bisazza and Marin (1991a) ran a computer simulation based on the demographic data of the Fimon lake population. The simulation indicated an overall mating advantage for small males; this conclusion was relatively robust in the face of changes to the model's assumptions. The advantage of small size is predicted to be maximal at the beginning of the reproductive season (when the population density is lowest and sex-ratio is female-biased) and then to decrease in the subsequent month; (due to the increasing influence of male competition). In the last month of the season there should be little size-related variation in reproductive success or even a slight advantage of large males (depending on the assumptions). Even when high levels of intermale competition are maintained for the whole season (as happens in population inhabiting thermal ponds, Zulian, 1990) the predicted optimum for male size is always far below the average size of the females in the population.

This picture may however change considerably if factors other than those considered, in the model cause selection on male body size. Few other selective factors have been examined in this species, but at least two are significantly associated with body size. Large males are more resistant to cold and starvation and are less exposed to winter mortality (Zulian, 1990). Large males take a longer time to reach maturity and so probably suffer a greater pre-reproductive mortality and experience a shorter reproductive life (Zulian, 1990). For both factors, the amount of variation attributable to body size is small and, when introduced in the simulation, they determine little variation in the results (Bisazza and Marin, unpublished data). Another factor is potentially very important. Although the standard length of males usually ranges between 16 and 27 mm, very small adult males (13–14 mm) are sometimes captured with juveniles in dense vegetation (personal observation). These individuals cannot be used in experiments since they are quickly eaten by females, suggesting that the risk of falling into the class of edible individuals may pose a limit to small size in male mosquitofish.

Indirect evidence supports the results of the simulation. In a study of variation of size at maturity in natural populations, the average size of males was found be positively correlate with the density of males in the population. Moreover, in all populations examined, males tended to mature at a minimum size early in the season when male competition is low; the average size at maturity increased later on, parallelling increases in adult density and in the proportion of males in the population (Zulian, 1990)

SEXUAL SIZE DIMORPHISM IN POECILIID FISH

Alternative Models of the Origin of Dimorphism

In this section I examine some of the factors that may account for the reverse sexual size dimorphism of poeciliid fish. Some of the proposed models, most of which are based on just one selective force acting on a sex, can generate testable predictions about variation in the degree of dimorphism within this family

Selection for large size in females

A slightly larger size of females is common in fishes and it has been attributed to the fecundity advantage (Darwin, 1874). Recently this idea has been criticized (Shine, 1988), and in any case it cannot explain why poeciliids are more dimorphic than most fishes. The problem is best viewed into the more general framework of the trade-off between allocation of energy to growth and to reproduction. There is some evidence from intraspecific comparison that variation in factors such as age-specific mortality or food availability can result in selection for different life-history tactics in female poeciliids (Reznik and Briga, 1987). However, data are too scarce to allow a test of this hypothesis at interspecific level. One special case of the fecundity advantage is seen when females incur a high fecundity-independent cost of reproduction; in this case they benefit from producing few, large batches (Bull and Shine, 1979). In livebearing fishes this may occur, for example, if females are particularly vulnerable in the last stages of gestation or during parturition. In poeciliids such as mosquitofishes that have highly cannibalistic tendencies (e.g. Meffe and Crump, 1987) there is probably a selective advantage in producing larger offspring. Indirect evidence of this comes from the observation that females G. holbrooki produce much larger offspring in populations with high density of adults than they do at low adult densities (Merlin, 1989). If small females are physiologically or morphologically constrained in their ability to produce large offspring, this may determine selection for large female size in these species. Ralls (1976) pointed out that reverse sexual size dimorphism may arise because of female competition for mates, food or other resources confer size-related advantages. Female competition for mates is a rare phenomenon and I have no knowledge of its occurrence among poeciliids. In my experience, in most poeciliids there is very little female aggression over food even when this is provided in large pieces. In aquaria, at the time of parturition, large female regularly defend small territories from all conspecifics and I have occasionally observed this occurrence in the field (unpublished observation; see also Winkler, 1979). Females do not usually feed during parturition and therefore territorial behaviour may greatly reduce cannibalistic predation on newborn fry giving a reproductive advantage to larger females. There are no data on other species, but female guppies do not seem to behave in this manner (unpublished observation).

Selection for small size in males

I consider here selection for small size in males in contexts other than during direct sexual encounters. A further model based on the greater efficiency of small males in inseminating females is proposed afterward.

In male poeciliids body size is positively correlated with age at maturity (Farr, 1980b; Farr *et al.*, 1986; Zulian, 1990). There are several potential advantages of early reproduction (Caswell and Hastings, 1980; Singer, 1982; Stearns and Koella, 1986); in poeciliids these may include a shorter generation time, a reduced pre-reproductive mortality and, for species whose individuals breed just one season, a longer reproductive life. If body size is not related to survival or reproduction, selection should in general favour early maturation in males. However, the extent of small male advantage should vary in relation to ecological factors such as age-specific mortality, fluctuations in size of the population, length of reproductive season, etc. Again too few ecological data are available for interspecific comparison to be used to test these predictions.

Some authors have suggested that, in species with a prolonged breeding season, the energetic costs of reproductive behaviour and restrictions on foraging may cause a reduction in male growth (Warner and Harlan, 1982; Endler, 1983; Woolbright, 1983). This may occur in some species of poeciliids at least. In guppies and mosquitofish for example, males are sexually active throughout the day and exhibit about one sexual act per minute (Luyten and Liley, 1985; Endler, 1987; Bisazza, unpublished observations) so that there is probably very little time left for other activities such as feeding.

The above explanations are not necessarily mutually exclusive. For example, early maturing males might have a shorter generation time, enjoy a longer reproductive life, allocate more resources to reproduction and have a greater fertilization success. Similarly, sexual size dimorphism can be seen as the result of the different selective forces acting in the two sexes. In fact, realistic models cannot be based on a single selective force acting on one sex because, if there is genetic correlation between sexes, no sexual size dimorphism is predicted at equilibrium (Lande, 1980). Theories based on one single factor must therefore implicitly assume that there are selective forces keeping the other sex close to the survival optimum.

Other explanations developed for specific taxa may have a validity for poeciliid fishes. Several authors have suggested that sexual size dimorphism may originate from selection on both sexes to reduce competition for food (Selander, 1966; Ligon, 1968; Cammilleri and Shine, 1990). There is no general consensus on this possibility. Dietary specialization, even when demonstrated, may be a consequence of a pre-existing difference in body size rather than its cause. Moreover this hypothesis does not explain why, in many groups, one sex is consistently larger than the other. Some other explanations invoke non-adaptive factors, for example differential responses of male and female to selection of the same intensity (Cheverud *et al.*, 1985). Another possibility is that sexual size dimorphism reflects differential mortality rate in the two sexes (Shine, 1990). Although the female biased sex-ratio observed in many poeciliids is commonly ascribed to a greater male mortality (Snelson and Wetherington, 1980), this factor alone is unlikely to be an explanation for dimorphism in these fishes, since it is clear that in most species sexual size dimorphism is primarily the consequence of males being 'frozen' at the size they have at the time of sexual maturation. The next paragraph will deal with proximate causation of sexual size dimorphism more in detail.

Proximate Determinants of Dimorphism

Before sexual maturity there is little or no difference in growth rate between sexes in poeciliid fishes (Kallman and Schreibman, 1973; Snelson, 1982; Stearns, 1983; Borowsky, 1987; Yan, 1987). Females show a reduction of growth rate after maturation (Krumholz, 1963; Yan, 1987) and in general their relative growth continues to decrease exponentially as they get larger (Kallman and Schreibman, 1973; Snelson, 1982). Male poeciliids virtually stop growing once they attain sexual maturity (Turner, 1941; Schultz, 1961; Krumholtz, 1963; Borowsky, 1973a; Kallman and Schreibman, 1973). In *G. holbrooki* there is a 5% increase in length in the first five weeks after maturation is reached (judged by full development of the gonopodium), but no appreciable growth later on (Zulian, 1990). In *G. heterochir*, *P. latipinna* and *X. maculatus* males probably continue to grow through all life (Kallman and Borkoski, 1978; Snelson, 1982; Yan, 1987; Travis *et al.*, 1989). In all cases post-maturation growth occurs at a very low rate and thus, even in these species, the size of adult males in a population essentially reflects the size at which they become mature. Size at maturity is known to be under genetic control in many species (Kallman and Borkoski, 1978; Kallman, 1983; Reznik and Bryga, 1987; Yan, 1987). To some extent this trait can be influenced by environmental factors such as food availability, salinity, temperature or photoperiod (Stearns and Sage, 1980; Travis *et al.*, 1989; Zulian, 1990). In *Girardinus metallicus*, *Poecilia reticulata*, *Xiphophorus maculatus* and *X. variatus* size at maturity is affected by juvenile agonistic interactions and males raised in groups tend to delay maturation until their size exceeds that of mature males already present (Nagoshi, 1967; Borowsky, 1973b; Sohn, 1977a; Farr, 1980b). Apparently in natural populations, as a consequence of social inhibition of maturation, males grow larger when intrasexual competition is more intense and therefore some authors have regarded this form of plasticity as an adaptation (Sohn, 1977b; Hughes, 1985b). The above account indicates that sexual size dimorphism in poeciliids is in general the consequence of the different pattern of post-maturation growth in the two sexes. Roughly, variation in the degree of sexual size dimorphism may originate in two possible ways: 1) it may be determined mainly by growth and survival rate of females. In other word males and females tend to mature at a similar size but in more dimorphic species females grow faster or live longer. 2) it may be determined mainly by the degree and direction of size differences present at attainment of sexual maturity. Under this hypothesis post-maturation growth of females varies little across species and the degree of sexual size dimorphism is related to early or late maturation of males. To assess the contribution of these two factors I have collected data from the literature on the average length of adult males and females and length at first reproduction in females of 16 species. Data for 10 species were obtained from Reznik and Miles, 1989b,c; populations generally referred as *Gambusia affinis* were pooled together and size at first reproduction in *Belonesox belizanus* was extrapolated from Figure 2 of Turner and Snelson, 1984. Six other species were *Girardinus falcatus* and *G. uninotatus* (Barus and Libosvarsky, 1988), *Poeciliopsis occidentalis* (Constantz, 1984; Schoenherr, 1977), *Quintana atrizona* (Barus *et al.*, 1981), *Heterandria bimaculata* (Miller, 1974), *Xiphophorus variatus* (Borowski, 1978; Borowski and Diffley, 1981). I used average male size as an estimate of size at maturity. This is probably correct for most species, but may have over-estimated size at maturity in those species like *Poecilia latipinna* that grow a little after sexual maturity. Sexual size dimorphism was defined as the ratio of average female length to average male length. When more than one population of the same species was studied I used the unweighed average of the population values. I also calculated the ratio of female to male size at maturity and the percentage increment of female length after maturity. Average growth of females after maturation is not a good predictor of sexual size dimorphism; it explains less than 1% of the variation among species and the correlation is not significant ($R = 0.08$, $P = 0.76$). Instead the ratio of female to male size at maturity is highly correlated with sexual size dimorphism ($R = 0.71$, $P < 0.005$) and explains more than 50% of the variation in dimorphism. Highly dimorphic species (sexual size dimorphism between 1.3 and 1.6) have males that tend to mature earlier than females while species with sexual size dimorphism between 1 and 1.3 show the reverse pattern. As an example in a species with little dimorphism in *X. helleri* size at maturity of males is about 1.5 that of females, while in highly dimorphic species like *G. uninotatus* or *G. marshi* males tend to mature when about 90% of the size at first reproduction in females. The above results suggest that the cause of the wide variation in sexual size dimorphism is not a trivial one and they allow the number of plausible hypotheses to be reduced. However, the observation that sexual size dimorphism is in great part determined by the difference in the time at which the two sexes mature is compatible with two opposite interpretations. A large dimorphism in one species may be the result of either selection for early maturation in males or selection for delayed maturation in females. In my opinion, the existence among poeciliids of very complex mechanisms determining male size (almost unique among fishes), gives more support to the first interpretation and at the same time reduces the likelihood of all non-adaptive hypotheses.

A New Hypothesis for the Origin of Dimorphism in Poeciliids

The discovery in *G. holbrooki* that the efficiency of copulatory thrusts dramatically drops as the length of male increases and the indication that this may be true for other species, suggest that sexual size dimorphism may have arisen in the family Poeciliidae as a consequence of evolution of this peculiar way of insemination. A somewhat similar explanation, a greater acrobatic performance of small individuals, have been suggested as an explanation for reverse sexual size dimorphism in shorebirds and flies (Jehl and Murray, 1986; McLachlan and Allen, 1987)

Since the importance of gonopodial thrusts is not equal for all species and since this male mating tactics is absent in some poeciliids, the intensity of selection for small body size stemming from this cause should be quite variable. Sexual size dimorphism is expected to be maximal in those species in which copulatory thrust is the only way of insemination and reduced or absent when copulations are always preceded by courtship display. When males show both tactics the degree of sexual size dimorphism should be proportional to the relative importance of the two ways of insemination. Additional predictions are related to the action of intrasexual selection. Since large males are generally favoured in competition for mates, in species where this mechanism is effective it will tend to counteract the advantage of small males. However, studies of G. holbrooki indicate that male competition does not neutralise the advantage of small male in copulatory thrusts and, on the basis of these results, male competition would seem to be a less powerful factor than the method of fertilization in determining the degree of dimorphism. Finally, if male ornaments are related to epigamic selection, the degree of secondary sexual differentiation should be inversely correlated with the importance of gonopodial thrusts in a species.

Studies of the species in which male show alternative mating tactics give some sort of support to the above hypothesis. For example in P. latipinna, males that use courtship behaviour as the principal mean of insemination grow larger than the average female length, while males that use copulatory thrusts mature early at a much smaller size; some males use both tactics and are intermediate in size (Travis and Woodward, 1989). An almost identical situation is found in X. nigrensis (Ryan and Causey, 1989; Ryan et al., 1990). However, in both cases it is possible that early maturation has some other advantages and that small, less preferred males do their best by using a mating tactic that does not require female cooperation.

One way of testing the association between sexual selection and sexual size dimorphism is to study the correlation between these variables in natural populations. I have mentioned before the co-variation in G. holbrooki between size at maturity and intensity of male–male competition, both among populations and across breeding season. Male G. holbrooki show only gonopodial thrusts and thus some of the predictions cannot be tested in this species. A more general test of the hypothesis is possible by studying species whose males exhibit both courtship and thrusts. In Poecilia reticulata for example, there is sufficient variation in sexual size dimorphism among populations (Liley and Seghers, 1975; Reznik and Miles, 1988c) and broad hereditary variations in the relative rate of the two mating tactics have been described in males from different populations (Luyten and Liley, 1985 and references therein).

A Test of the Hypothesis Using Interspecific Comparison

An alternative to studying within-species variation is comparison among species. There have been few previous attempts to test evolutionary hypotheses about poeciliids using interspecific comparisons and none of these have considered the possible relationship between sexual selection and sexual size dimorphism. Rosen and Tuker (1961) studied the variation in external male genitalia among 13 genera of poeciliid fish and concluded that males generally have a shorter gonopodium when there is an elaborate precopulatory display and female cooperate during mating. Endler (1983) examined the relationship between colour pattern dimorphism, dichromism, sexual size dimorphism and some life history traits. Sexual dichromism was associated with larger body size and larger brood size; few other significant relationships were found in this study, possibly as a consequence of having used data from aquarist literature. More recently, Reznick and Miles (1989a) have studied the association between life history traits and different reproductive adaptations in female poeciliids. Farr (1989) has summarized existing information about mechanisms of sexual selection and secondary sex differentiation in poeciliids. His analysis demonstrated a positive association between male courtship behaviour and male ornaments.

The data for my analysis (see Appendix) consists primarily of published records of the size of males and females, secondary sexual differentiation and male mating and aggressive behaviour in poeciliid fishes. Some information comes from my own unpublished observations. To determine dimorphism I have utilized records of the standard length of at least ten males and ten females from the same population or captive stock. For most analyses, sexual size dimorphism was calculated as female/male body length ratio. A few data were estimate from figures. When the data from several populations were available, I computed the unweighted average. In some cases standard length was computed as 0.8 total length. For two species (*X. variatus* and *P. occidentalis*) size data for the two sexes were derived from two separate sources and there is no certainty that place and date of collection are the same. The data are rather heterogeneous and include records based on many natural populations and others based on only a few specimens. In many cases the method used for collecting a particular sample is not specified and there is no certainty that it represents a random sample from the population.

The mating tactics used by males were scored in the following way: 1 for species with only gonopodial thrust, 3 for species that use equally both tactics and 5 when copulation occurs only after male courtship and female cooperation; species in which gonopodial thrust is the main way of insemination, but courtship is sometimes observed, were scored as 2; courting species with occasional gonopodial thrusts were scored 4. To score the influence of intrasexual competition I assigned a value of 1 to one species if males are territorial or if there is evidence that male competition influences reproductive success. Species in which male competition was shown to be irrelevant for mating success and species in which there is no manifest male aggression were scored as −1. All the remaining species were scored as 0. A sexual selection index was obtained from algebraical sum of the mating tactic and intrasexual competition scores. In this way the minimum score of 0 is obtained by species which show only gonopodial thrusts and manifest no male competition. Species with aggressive males and courtship tactic received a maximum score of 6. To score secondary sexual traits I assigned a separate score to dimorphism in colour (0 to 3) and fin shape (0 to 3) and then summed the two values. I have integrated data from literature cited in Appendix with informations from Jacobs, 1961 and Farr, 1989, photographs and observation of live specimens.

The analysis of the data evidences that the sexual size dimorphism is inversely correlated with both mating tactic ($R = -0.65$, $P = .001$) and intrasexual competition ($R = -0.46$, $P = 0.01$), indicating that dimorphism tend to be more pronounced when males use only gonopodial thrusts and competition is unimportant. When these variables are used in a multiple regression, the mating tactics of the male is still a good predictor of sexual size dimorphism (t = 3.9, P = 0.001), while the contribution of male competition become marginally significant (t = 1.86, P = 0.07). The sexual selection index, which accounts for both these factors, is the best predictor of sexual size dimorphism (y = –0.055 x + 1.39, r = 0.69, P < 0.001; Figure 3) explaining about fifty percent of the variance in dimorphism.

In many taxa sexual size dimorphism varies with body size and in some cases allometry is responsible for a great part of the interspecific variation. Usually, when males are the larger sex, dimorphism increase with body size and the opposite occur with reverse sexual size dimorphism (see references in Fairbairn, 1990). In the sample analyzed sexual size dimorphism was not significantly correlated with either male or female size (R = –0.16 and R = 0.09 respectively). As a confirmation of a lack of allometric relationship, there was no difference in the results when I used the residuals of the regression between logarithm of male and female length as dependent variable (e.g. the correlation between sexual selection index and dimorphism was R = –0.71, P < 0.001). One possible criticism to the above analyses is that all variables were treated as continuous and fitted in a linear model. It is worth noting however that almost identical conclusions were obtained using either analysis of variance or non-parametric correlation tests.

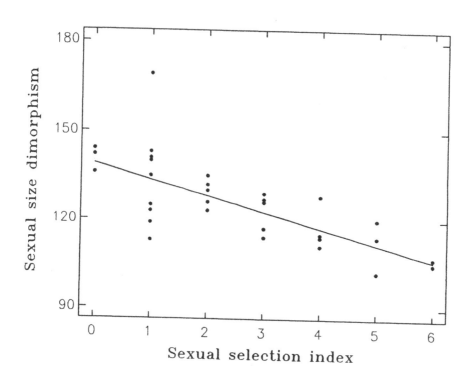

Figure 3 Interspecific variation in sexual size dimorphism. Size dimorphism (female/male length ratio) is accurately predicted by sexual selection index. Species without male competition and with copulatory thrusts as the main male mating tactics show the greatest dimorphism.

Male ornamentation appears in relationship with sexual selection mechanisms. The index of secondary sexual traits is more correlated with mating tactic ($r = 0.57$, $P = 0.001$) than with intrasexual competition ($r = 0.37$, $P = 0.04$) and when both variables are entered in a multiple regression model only the former retain its significant relationship (MT: $t = 3.19$, $P < 0.01$; IC: $t = 1.26$, $P = 0.22$). The sexual selection index explains about 36% of the variance in colour and fin dimorphism ($y = 0.195$ x $+0.55$, $r = 0.60$, $P < 0.001$). Sexual size dimorphism was negatively correlated with secondary sexual trait ($r = 0.52$, $P < 0.01$); in other words species with more ornamented males show the least size difference between sexes. In these analyses, male secondary sexual traits appear more strictly related to intersexual selection than to male aggression but the result may simply reflect lack of consistency in the data on intrasexual competition.

The previous analysis indicates a number of significant relationships between the variables examined. However, closely related species may have inherited the association between two traits from a common ancestor. Treating species as independent points in a statistical analysis may, therefore, greatly overestimate the true number of degrees of freedom. Several different methods have been developed to account for phylogenetic relationship in comparative studies, but they generally require that relationships between species are known without error (Harvey and Pagel, 1991). Recently Grafen (1989) developed a method, the "phylogenetic regression" designed to be used when the knowledge of the phylogeny is incomplete. This method can therefore be applied even to groups where only the taxonomy is known. I have preliminarily applied this method to poeciliid data using the most recent classification (Parenti and Rauchenberger, 1989), which differs only slightly from that proposed by Rosen and Bailey (1963).

The results of phylogenetic regression substantially confirmed previous analyses. The sexual selection index was a significant predictor of sexual size dimorphism ($F(1,13) = 12.0$, $P < 0.01$). Similar results were obtained when I used mating tactic and intrasexual competition as independent variables ($F(2,12) = 5.6$, $P < 0.05$) and when size was controlled for ($F(1,13) = 12.1$, $P < 0.01$). In contrast, the relationship between sexual selection index and secondary sexual traits resulted only marginally significant once phylogeny was accounted for ($F(1,13) = 4.4$, $P = 0.07$).

CONCLUSIONS

This review of the literature confirms that reverse sexual size dimorphism is a general condition in the family Poeciliidae. Great interspecific variability emerges from the data; in some species there is almost no size difference, while in other females are on average fifty percent larger than males of the same population. Determinants of male reproductive success are also extremely variable across the family. Male competition is negligible or absent in many species, but becomes very important in species that are territorial or that possess other mechanism of excluding rival males. Modality of insemination is even more crucial. In many poeciliids males have a precopulatory displays and male mating success is primarily determined by female mate choice; selection for male showiness is expected to occur in these species. Males may also bypass the need of female cooperation and achieve inseminations through gonopodial thrusts. The negative relationship between male size and success of gonopodial thrusts should impose selection for a small size in those species where this is the sole or principal male mating tactics.

Various explanations of the reverse sexual size dimorphism of the Poeciliidae were examined. Although many models are potentially capable of explaining interspecific variation, there are at the present insufficient data to test any of them properly. A new model is proposed which relates sexual dimorphism to insemination through copulatory thrusts; it makes testable predictions about the influence of male mating tactics and male competition on the degree of sexual size dimorphism. Interspecific comparison supports the prediction that species which predominantly use gonopodial thrusts are more dimorphic than those in which male achieve matings mainly by courting the female. The available data only partially support the hypothesis of a role of male competition. The conclusions of Farr (1989) that secondary sexual differentiation is associated with the presence of male courtship and female cooperation receive confirmation from my analysis. Interestingly this association loses significance once the effect of phylogeny has been removed, possibly because most species with showy males belong to one single tribe.

The present analysis must be considered tentative for several reason. The lack of homogeneity of the length data has been outlined before. Data on mating behaviour and sex differentiation suffer from the same limitation. Only a few species have been the subject of detailed studies and information on the action of sexual selection often comes from incidental observations. Data on natural history are clearly scarce for many poeciliid fish and there is a need for more studies especially of those genera that are least known. Some of the assumptions inherent in this model need to be confirmed. One is that the advantage of small males in gonopodial thrust is a widespread phenomenon. Given the great diversity in the size and morphology of the gonopodium and associated structures (Rosen and Gordon, 1953; Rosen and Tucker, 1961) and in consideration of the great variability of body shape in the male compared with the relative uniformity of females in different species, it would not be surprising if the relationship were not general. At last, the attempt to explain the pattern of sexual size dimorphism based on one factor only is probably unrealistic and future research should attempt to develop more complex models.

Notwithstanding the limitations, I believed that the present analysis and discussion have demonstrated new areas of interest and may stimulate a more systematic collection of data. The recently renewed interest in the biology of poeciliids will certainly permit tests of these alternative explanations of sexual size dimorphism in these fish in the near future. A reliable phylogeny of the Poeciliidae is also needed, not only to determine the appropriate degrees of freedom for statistical analysis but also to either better understanding of the evolution of male and female reproductive strategies in this family.

Finally the analysis may benefit from being extended to related taxa. According to the data provided by Jacobs (1971), males are smaller than females in many other viviparous fish belonging to the same order and in the family Jenynsidae, dimorphism appears to be even more pronounced than in Poeciliid fish. Before further conclusions about the significance of sexual size dimorphism can be drawn, it is necessary to know more about the reproductive adaptation and the size dimorphism of the other Cyprinodontiformes.

A. BISAZZA

ACKNOWLEDGMENTS

A large part of this work was done while I was a guest of ABRG in the Department of Zoology of Oxford University. I wish to thank Jeremy John who introduced me to the use of phylogenetic regression and Derek Lambert who gave me the possibility of observing several species of poeciliids. Many participants at the Erice meeting made helpful suggestions. I am indebted to Douglas J. Futuyma, Andrea Marconato and Guglielmo Marin for critically reading the manuscript. I am especially grateful to Felicity Huntingford for taking the time and effort to substantially improve this paper with her careful review.

APPENDIX 1 Summary of the comparative data on mode of sexual selection and intraspecific variation in sexual dimorphism. SLFEM = mean standard length of adult females; SSD = sexual size dimorphism; SSI = sexual selection index; SST = sexual dimorphism in colour and fin shape. Sources: 1) Balsano et al., 1985; 2) Barus and Libosvarsky 1988; 3) Barus et al., 1981; 4) Bisazza and Marin 1991a; 5) Bisazza, unpublished observations; 6) Borowsky 1973a; 7) Borowsky 1978; 8) Borowsky and Diffley 1981; 9) Brett and Grosse 1982; 10) Constantz 1974; 11) Constantz 1975; 12) Endler 1982; 13) Farr 1980a; 14) Farr 1984; 15) Farr 1989; 16)Far et al., 1986; 17) Frank 1964; 18) Gordon and Rosen 1962; 19) Greenfield 1990; 20) Hemens 1966; 21) Houde 1988b; 22) Hughes 1985a; 23) Jacobs 1971; 24) Krumholtz 1948; 25) Krunholtz 1963; 26) Liley 1966; 27) Merlin 1989; 28) Miller 1974; 29) Miller 1975; 30) Miller and Minkley 1963; 31) Miller and Minkley 1970; 32) Milton and Arthington 1983; 33) Parzefall 1969; 34) Parzefall 1979; 35) Peden 1972; 36) Philippi 1909; 37) Reznik and Miles 1989b; 38) Reznik and Miles 1989c; 39) Rosen and Bailey 1959; 40) Rosen and Tucker 1961; 41) Schoenherr 1977; 42) Sohn 1977b; 43) Turner and Snelson 1984.

APPENDIX 1

Genus	Species	SLFEM	SSD	SSI	SST	Source
Alfaro	cultratus	48.0	125	1	0	5, 40
Poecilia	chica	33.4	130	2	2	9, 29
"	sphenops	29.5	123	2	1	18, 33, 34
"	mexicana	44.5	115	4	1	1, 19
"	latipinna	37.9	111	4	5	16, 37
"	reticulata	19.9	126	2	5	13, 21, 38
"	picta	20.9	114	4	4	26, 37
"	melanogaster	39.7	120	5	2	5, 14
Xiphophorus	variatus	28.9	107	6	4	6, 7, 8, 17
"	xiphidium	28.2	114	5	3	5
"	maculatus	28.1	119	1	2	17, 32, 37
"	gordoni	23.6	105	6	2	5, 30
"	helleri	35.6	102	5	5	17, 20, 32, 37
Priapella	compressa	39.9	141	1	0	5
Cnesterodon	decemmaculatus	25.9	144	0	1	5, 36, 40
Phallaceros	caudimaculatus	23.1	113	1	0	12, 36

APPENDIX 1   (Continued)

Genus	Species	SLFEM	SSD	SSI	SST	Source
Scholichthys	greenway	42.3	136	0	0	5
Gambusia	aurata	24.0	117	3	0	23, 31, 35
"	holbrooki	27.0	135	2	1	4, 5, 27
"	affinis	24.4	128	4	1	22, 24
"	manni	25.1	114	3	2	25, 37, 40, 42
"	wrayi	27.6	126	3	3	5
Belonesox	belizanus	88.1	129	3	0	5, 15, 43
Girardinus	falcatus	25.8	123	1	1	2, 23, 40
Quintana	atrizona	25.7	135	1	0	3, 23, 40
Carlhubbsia	stuarti	46.9	119	1	1	15, 39
"	kidderi	31.1	169	1	1	15, 39
Neoheterandria	elegans	20.3	142	0	0	5
Heterandria	formosa	21.8	132	2	0	37, 40
"	bimaculata	54.0	140	1	0	28, 40
Poeciliopsis	occidentalis	29.8	127	3	1	10, 11, 41
Phallichthys	amates	35.3	143	1	1	39, 40

References

Balsano, J. S., Randle E. J., Rasch, M. E. and Monaco, P. J. (1985). Reproductive behavior and the maintenance of all-female *Poecilia*. *Env. Biol. Fish.*, **12**, 251–263.

Barus, V. and Libosvarsky, J. (1988). Observations on *Glaridichthys uninotatus* and *G. falcatus* (Poeciliidae) from Cuba. *Folia Zool.*, **37**, 167–182.

Barus, V., Libosvarsky, J. and Guerra Padron, F. (1981). Observations on *Quintana atrizona* from Cuba, reared in aquaria. *Folia Zool.*, **30**, 203–214.

Basolo, A. L. (1989). Female preference for caudal fin elaboration in the platyfish, *Xiphophorus maculatus*. *Amer. Soc. Ich. Herp.*, San Francisco (Abstract) .

Basolo, A. L. (1990). Female preference for male sword length in the green swordtail, *Xiphophorus helleri* (Pisces: Poeciliidae). *Anim. Behav.*, **40**, 332–338.

Beaugrand, J. P., Caron, J. and Comeau, L. (1984). Social organization of small heterosexual groups of green swordtails (*Xiphophorus helleri*, Pisces, Poeciliidae) under conditions of captivity. *Behaviour*, **91**, 24–60.

Bildsøe, M. (1988). Aggressive, sexual and foraging behaviour in *Poecilia velifera* (Pisces: Poeciliidae) during captivity. *Ethology*, **79**, 1–12.

Bisazza, A. and Marin, G. (1991a). Sexual selection and sexual size dimorphism in the eastern mosquitofish. *Contributi di Psicologia*, **4**, 39–57.

Bisazza, A. and Marin, G. (1991b). Male size and female mate choice in the eastern mosquitofish (*Gambusia holbrooki*: Poeciliidae). *Copeia*, 730–735.

Bisazza, A., Marconato, A. and Marin, G. (1989). Male mate preferences in the mosquitofish *Gambusia holbrooki*. *Ethology*, **83**, 335–343.

Bischoff, R. J., Gould, J. L. and Rubenstein, D. J. (1985). Tail size and female choice in the guppy (*Poecilia reticulata*). *Behav. Ecol. Sociobiol.*, **17**, 253–255.

Boake, C. R. B. (1989). Correlation between courtship success, aggressive success, and body size in a picture-winged fly, *Drosophila silvestris*. *Ethology*, **80**, 318–329.

Borowsky, R. L. (1973a). Relative size and the development of fin coloration in *Xiphophorus variatus*. *Physiol. Zool.*, **46**, 22–28.

Borowsky, R. L. (1973b). Social control of adult size in males *Xiphophorus variatus*. *Nature*, **245**, 332–335.

Borowsky, R. L. (1978). The tailspot polymorphism of *Xiphophorus* (Pisces: Poeciliidae). *Evolution*, **32**, 886–893.

Borowsky, R. L. (1987). Genetic polymorphism in adult male size in *Xiphophorus variatus* (Atheriniformes: Poeciliidae). *Copeia*, **3**, 782–787.

Borowsky, R. and Diffley, J. (1981). Synchronized maturation and breeding in natural populations of *Xiphophorus variatus* (Poeciliidae). *Env. Biol. Fish.*, **6**, 49–58.

Braddock, J. C. (1949). The effect of prior residence upon the dominance in the fish *Platypoecilius maculatus*. *Physiol. Zool.*, **22**, 161–169.

Brett, B. L. H. and Grosse, D. J. (1982). A reproductive pheromone in the Mexican poeciliid fish *Poecilia chica*. *Copeia*, 219–223.

Brockelman, W. J. (1975). Competition, the fitness of offspring, and the optimal clutch size. *Am. Nat.*, **109**, 677–699.

Bull, J. J. and Shine, R. (1979). Iteroparous animals that skip opportunities for reproduction. *Am. Nat.*, **114**, 296–303.

Camilleri, C. and Shine, R. (1990). Sexual dimorphism and dietary divergence: Difference in the trophic morphology between male and female snakes. *Copeia*, 649–658.

Caroters, J. H. (1984). Sexual selection and sexual dimorphism in some herbivorous lizards. *Am. Nat.*, **124**, 244–254.

Caswell, H. and Hastings, A. (1980). Fecundity, developmental time, and population growth rate: an analytical solution. *Theor. Popul. Biol.*, **17**, 71–79.

Cheverud, J. M., Dow, M. M. and Leutenegger, W. (1985). The quantitative assessment of phylogenetic constraints in comparative analyses: Sexual dimorphism in body weight among primates. *Evolution*, **39**, 1335–1351.

Clark, E. L., Aronson R. L. and Gordon M. (1954). Mating behavior patterns in two sympatric species of Xiphophorin fishes: their inheritance and significance in sexual isolation. *Bull. Amer. Mus. Natur. Hist.*, **103**, 135–226.

Clutton-Brock, T. H., Harvey, P. H. and Rudder, B. (1977). Sexual dimorphism, socionomics sex ratio and body weight in primates. *Nature*, **269**, 797–800.

Congdon, J. D., Gibbons, J. W. and Greene, J. L. (1983). Parental investment in the chicken turtle (*Deirochelys reticularia*). *Ecology*, **64**, 419–425.

Constantz, G. D. (1974). Reproductive effort in *Poeciliopsis occidentalis* (Poeciliidae). *Southwest. Natur.*, **19**, 17–42.

Constantz, G. D. (1975). Behavioural ecology of mating in the male Gila topminnow, *Poeciliopsis occidentalis* (Cyprinodontiformes: Poeciliidae). *Ecology*, **56**, 966–973.

Constantz, G. D. (1989). Reproductive biology of Poeciliid fishes In: *Ecology and Evolution of Livebearing Fishes* (Ed. by G. K. Meffe and F. F. Snelson Jr.), pp. 33–50. Englewood Cliffs: Prentice Hall.

Darwin, C. (1874). *The Descent of Man, and Selection in Relation to Sex.* 2nd edn. London: John Murray.

Docker, M. F. and Beamish, W. H. (1991). Growth, fecundity, and egg size of the least brook lamprey, *Lampetra aepyptera*. *Env. Biol. Fish.*, **31**, 219–227.

Dulzetto, F. (1928). Osservazioni sulla vita sessuale di *Gambusia holbrooki*. *Atti R. Acc. Lincei, Rend.*, **8**, 96–101.

Endler, J. A. (1982). Convergent and divergent effects of natural selection on color patterns in two fish faunas. *Evolution*, **36**, 178–188.

Endler, J. A. (1983). Natural and sexual selection on color patterns in poeciliid fishes. *Env. Biol. Fish.*, **9**, 173–190.

Endler, J. A. (1987). Predation, light intensity and courtship behaviour in *Poecilia reticulata* (Pisces: Poeciliidae). *Anim. Behav.*, **35**, 1376–1385.

Fairbairn, D. J. (1990). Factors influencing sexual size dimorphism in temperate waterstriders. *Am. Nat.*, **136**, 61–86.

Farr, J. A. (1977). Male rarity or novelty, female choice behavior, and sexual selection in the guppy, *Poecilia reticulata* Peters (Pisces: Poeciliidae). *Evolution*, **31**, 162–168.

Farr, J. A. (1980a). The effect of sexual experience and female receptivity on courtship-rape decisions in male guppies, *Poecilia reticulata* (Pisces: Poeciliidae). *Anim. Behav.*, **28**, 1195–1201.

Farr, J. A. (1980b). The effect of juvenile social interaction on growth rate, size and age at maturity, and adult social behavior in *Girardinus metallicus* Poey (Pisces: Poeciliidae). *Z. Tierpsychol.*, **52**, 247–268.

Farr, J. A. (1980c). Social behaviour patterns as determinants of reproductive success in the guppy, *Poecilia reticulata* Peters (Pisces: Poeciliidae): an experimental study of the effects of intermale competition, female choice and sexual selection. *Behaviour*, **74**, 38–91.

Farr, J. A.. (1984). Premating behavior in the subgenus *Limia* (Pisces: Poeciliidae): sexual selection and the evolution of courtship. *Z. Tierpsychol.*, **65**, 152–165.

Farr, J. A. (1989). Sexual selection and secondary sexual differentiation in poeciliids: determinants of male mating success and the evolution of female choice. In: *Ecology and Evolution of Livebearing Fishes (Poeciliidae)* (Ed. by G. K. Meffe and F. F. Snelson), pp. 91–123. Englewood Cliffs: Prentice Hall.

Farr, J. A. and Travis, J. (1986). Fertility advertisement by female sailfin mollies, *Poecilia latipinna* (Pisces: Poeciliidae). *Copeia*, 467–472.

Farr, J. A., Travis, J. and Trexler, J. C. (1986). Behavioural allometry and interdemic variation in sexual behaviour of the sailfin molly *Poecilia latipinna* (Pisces: Poeciliidae). *Anim. Behav.*, **34**, 497–509.

Frank, D. (1964). Vergleichende Verhaltensstudien an lebendgebarenden Zahnkarpfen der Gattung *Xiphophorus. Zool. Jb. Physiol.*, **71**, 117–170.

Frayer, D. W. and Wolpoff, M. H. (1985). Sexual dimorphism. *Ann. Rev. Anthropol.*, **14**, 429–473.

Gandolfi, G. (1971). Sexual selection in relation to social status of males in *Poecilia reticulata* (Teleostei: Poeciliidae). *Boll. Zool.*, **38**, 35–48.

George, C. (1960). Behavioral interaction of the pickerel (*Esox niger* LeSueur and *Esox americanus* LeSueur) and the mosquito fish (*Gambusia patruelis* Baird and Girard). Ph.D. thesis. Harvard, Cambridge, USA.

Ghiselin, M. T. (1974). *The Economy of Nature and the Evolution of Sex.* Berkeley: Univ. California Press.

Gilbert, J. J. and Williamson, C. E. (1983). Sexual dimorphism in zooplankton (Copepoda, Cladocera, and Rotifera). *Ann. Rev. Ecol. Syst.*, **14**, 1–33.

Gliwics, J. (1988). Sexual dimorphism in small mustelids: body diameter limitation. *Oikos*, **53**, 411–414.

Gordon, M. (1955). Those puzzling "little toms". *Anim. Kingdom*, **58**, 50–55.

Gordon, M. S. and Rosen, D. E. (1962). A Cavernicolous Form of Poeciliid Fish, *Poecilia sphenops* from Tabasco, Mexico. *Copeia*, 360–368.

Gorlick, D. L. (1976). Dominance hierarchies and factors influencing dominance in the guppy *Poecilia reticulata* (Peters). *Anim. Behav.*, **24**, 336–346.

Grafen, A. (1989). The phylogenetic regression. *Phil. Trans. R. Soc. Lond.*, **B 326**, 119–157.

Greenfield, D. W. (1990). *Poecilia teresae*, a new species of poeciliid fish from Belize, Central America. *Copeia*, 449–454.

Gronell, A. M. (1989). Visiting behaviour by females of the sexually dichromatic damselfish, *Chrysiptera cyanea* (Teleostei: Pomacentridae): a probable method of assessing male quality. *Ethology*, **81**, 89–122.

Harvey, P. H. and Pagel, M. D. (1991). *The Comparative Method in Evolutionary Biology.* Oxford: Oxford University Press.

Heinrich, W. and Schröeder, J. H. (1988). Tentative findings on the phylogenetic relationship within the genus *Xiphophorus* with regard to the frequency distribution of sexual behavior patterns. *Ber. Nat. -Med. Verein Innsbruck*, **73**, 187–197.

Hemens, J. (1966). The ethological significance of the sword-tail in *Xiphophorus helleri* (Haekel). *Behaviour*, **27**, 290–315.

Hoglung, J. (1989). Size and plumage dimorphism in lek-breeding birds: a comparative analysis. *Am. Nat.*, **134**, 72–87.

Houde, A. E. (1988a). Genetic difference in female choice between two guppy populations. *Anim. Behav.*, **36**, 510–516.

Houde, A. E. (1988b). The effects of female choice and male–male competition on the mating success of male guppies. *Anim. Behav.*, **36**, 888–896.

Hughes, A. L. (1985a). Male size, mating success, and mating strategy in the mosquitofish *Gambusia affinis* (Poeciliidae). *Behav. Ecol. Sociobiol.*, **17**, 271–278.

Hughes, A. L. (1985b). Seasonal trends in body size of adult male mosquitofish, *Gambusia affinis*, with evidence for their social control. *Env. Biol. Fish.*, **14**, 251–258.

Hughes, A. L. (1986). Seasonal changes in sexual receptivity of female mosquitofish *Gambusia affinis. Biol. Behav.*, **11**, 3–15.

Itzkowitz, M. (1971). Preliminary study of the social behavior of male *Gambusia affinis* (Baird and Girard) (Pisces: Poeciliidae) in aquaria. *Chesapeake Sci.*, **12**, 219–224.

Jacobs, K. (1971). *Livebearing Aquarium Fishes.* New York: MacMillan.

Jehl, J. R. Jr. and Murray, B. G. Jr. (1986). The evolution of normal and reverse sexual size dimorphism in shorebirds and other birds. *Curr. Ornithol.*, **3**, 1–86.

Kallman, K. D. (1983). The sex determining mechanism of the poeciliid fish, *Xiphophorus montezumae*, and the genetic control of the sexual maturation process and adult size. *Copeia*, 755–769.

Kallman, K. D. and Borkoski, V. (1978). A sex-linked gene controlling the onset of sexual maturity in female and male platyfish (*Xiphophorus maculatus*), fecundity in females and adult size in males. *Genetics*, **89**, 79–119.

Kallman, K. D. and Schreibman, M. P. (1973). A sex-linked gene controlling gonadotrop differentiation and its significance in determining the age of sexual maturation and size of the platyfish, *Xiphophorus maculatus. Gen. Comp. Endocrinol.*, **21**, 287–304.

Kirkpatrick, M. (1987). Sexual selection by female choice in polygynous animals. *Ann. Rev. Ecol. Syst.*, **18**, 43–70.

Kodric-Brown, A. (1985). Female preference and sexual selection for male colouration in the guppy (*Poecilia reticulata*). *Behav. Ecol. Sociobiol.*, **17**, 199–205.

Krumholz, L. A. (1948). Reproduction in the western mosquitofish *Gambusia affinis affinis* (Baird and Girard) and its use in mosquito control. *Ecol. Monogr.*, **18**, 1–43.

Krumholz, L. A. (1963). Relationships between fertility, sex ratio, and exposure to predation in populations of the mosquitofish *Gambusia manni* Hubbs at Bimini, Bahamas. *Int. Rev. Ges. Hydrobiol.*, **48**, 201–256.

Kulkarni, C. V. (1940). On the systematic position, structural modification, bionomics and development of a remarkable new family of cyprinodont fishes from the province of Bombay. *Rec. Indian Mus.*, **42**, 379–423.

Lande, R. (1980). Sexual dimorphism, sexual selection and adaptation in polygenic characters. *Evolution*, **34**, 292–305.

Ligon, J. D. (1968). Sexual differences in foraging behavior in two species of *Dendrocopus* woodpeckers. *Auk*, **85**, 203–215.

Liley, N. R. (1966). Ethological isolating mechanisms in four sympatric species of poeciliid fishes. *Behav. Suppl.*, **13**, 1–197.

Liley, N. R. and Segers, B.H. (1975). Factors affecting the morphology and behaviour of guppies in Trinidad. In: *Function and Evolution in Behaviour* (Ed. by G. B. Baerends, C. Beer and A. Manning), pp. 92–118. Oxford: Oxford University Press.

Luyten, P. H. and Liley, N. R. (1985). Geographic variation in the sexual behaviour of the guppy, *Poecilia reticulata* (Peters). *Behaviour*, **95**, 164–179.

Luyten, P. H. and Liley, N. R. (1991). Sexual selection and competitive mating success of male guppies (*Poecilia reticulata*) from four Trinidad populations. *Behav. Ecol. Sociobiol.*, **28**, 329–336.

Marconato, A. and Bisazza, A. (1988). Mate choice, egg cannibalism and reproductive success in the river bullhead, *Cottus gobio* L. *J. Fish Biol.*, **33**, 905–916.

Martin, R. G. (1975). Sexual and aggressive behavior, density and social structure in a natural population of mosquitofish, *Gambusia affinis holbrooki*. *Copeia*, 445–454.

Maynard Smith, J. (1991). Theories of sexual selection. *Trends Ecol. Evol.*, **6**, 146–151.

McLachlan, A. J. and Allen, D. F. (1987). Male mating success in Diptera: advantages of small size. *Oikos*, **48**, 11–14.

Meffe, G. K. and Crump, M. L. (1987). Possible growth and reproductive benefit of cannibalism in the mosquitofish. *Am. Nat.*, **129**, 203–212.

Merlin, E. (1989). Biologia riproduttiva di *Gambusia holbrooki*: Analisi di sette popolazioni naturali. *Tesi di laurea*. University of Padova, Italy.

Miller, R. R. (1974). Mexican species of the genus *Heterandria*, subgenus Pseudoxiphophorus (Pisces: Poeciliidae). *Trans. San Diego Soc. Natur. Hist.*, **17**, 235–250.

Miller, R. R. (1975). Five new species of Mexican poeciliid fishes of the genera *Poecilia*, *Gambusia*, and *Poeciliopsis*. *Occas. Pap. Mus. Zool. Univ. Michigan*, **672**, 1–44.

Miller, R. R. and Minkley, W. L. (1963). *Xiphophorus gordoni*, a new species of platyfish from Coahuila, Mexico. *Copeia*, 538–546.

Miller, R. R. and Minkley, W. L. (1970). *Gambusia aurata*, a new species of poeciliid fish from northeastern Mexico. *Southwest. Natur.*, **15**, 249–259.

Milton, D. A. and Arthinghton, A. H. (1983). Reproductive biology of *Gambusia affinis holbrooki* Baird and Girard, *Xiphophorus helleri* (Günther) and *X. maculatus* (Heckel) (Pisces; Poeciliidae) in Queensland, Australia. *J. Fish Biol.*, **23**, 23–41.

Moore, A. J. (1990). The evolution of sexual dimorphism by sexual selection: the separate effect of intrasexual selection and intersexual selection. *Evolution*, **44**, 315–331.

Nagoshi, M. (1967). Experiments on the effect of size hierarchy upon the growth of guppy (*Poecilia reticulata*). *J. Fac. Fish. Prefect. Univ. Mie*, **7**, 165–189.

Parenti, L. R. and Rauchenberger, M. (1989). Systematic overview of the Poeciliines. In: *Ecology and Evolution of Livebearing Fishes (Poeciliidae)* (Ed. by G. K. Meffe and F. F. Snelson), pp. 3–12. Englewood Cliffs: Prentice Hall.

Parker, G. A. and Begon, M. (1986). Optimal egg size and clutch size: effects of environment and maternal phenotype. *Am. Nat.*, **128**, 573–592.

Parzefall, J. (1969). Zur vergleichenden Ethologie verschiedener *Mollinesia-* Arten einschlieslich einer Hoehlenform von *M. sphenops*. *Behaviour*, **33**, 1–37.

Parzefall, J. (1973). Attraction and sexual cycle of poeciliids. In: *Genetics and Mutagenesis of Fish* (Ed. by J. H. Schröder), pp. 177–183. Berlin: Springer-Verlag.

Parzefall, J. (1979). Zur Genetik und biologischen Bedeutung des Aggressionsverhaltens von *Poecilia sphenops* (Pisces, Poeciliidae). *Z. Tierpsychol.*, **50**, 399–422.

Peden, A. E. (1972). The function of gonopodial parts and behavioural pattern during copulation by *Gambusia* (Poeciliidae). *Can. J. Zool.*, **50**, 955–968.

Petrie, M. (1983). Female moorhens compete for small fat males. *Science*, **220**, 413–415.

Petrie, M., Halliday, T. and Sanders, C. (1991). Peahens prefer peacocks with elaborate trains. *Anim. Behav.*, **41**, 323–331.

Philippi, E. (1909). Fortpflanzungsgeschichte der viviparen Teleosteer *Glaridichthys januarius* und *G. decem-maculatus* in ihrem Einfluss auf Lebensweise, makroskopische und mikroskopische Anatomie. *Zool. Jahrbücher Abt. f. Anat.*, **27**, 1–94.

Ralls, K. (1976). Mammals in which females are larger than males. *Quart. Rev. Biol.*, **51**, 245–276.

Reiss, M. J. (1987). The intraspecific relationship of parental investment to female body weight. *Funct. Ecol.*, **1**, 105–107.

Reznick, D. N. and Briga, H. (1987). Life history evolution in guppies (*Poecilia reticulata*): 1. Phenotypic and genetic changes in an introduction experiment. *Evolution,* **41**, 1370–1385.

Reznick, D. and Endler, J. A. (1982). The impact of predation on life history evolution in Trinidadian guppies (*Poecilia reticulata*). *Evolution*, **36**, 160–177.

Reznick, D. N. and Miles, D. B. (1989a). Review of life history patterns in poeciliid fishes. In: *Ecology and Evolution of Livebearing Fishes (Poeciliidae)* (Ed. by G. K. Meffe and F. F. Snelson), pp. 125–148. Englewood Cliffs: Prentice Hall.

Reznick, D. N. and Miles, D. B. (1989b). Poeciliids life history patterns. In: *Ecology and Evolution of Livebearing Fishes (Poeciliidae)* (Ed. by G. K. Meffe and F. F. Snelson), pp. 373–377. Englewood Cliffs: Prentice Hall.

Reznick, D. N. and Miles, D. B. (1989c). Intraspecific life history variation. In: *Ecology and Evolution of Livebearing Fishes (Poeciliidae)* (Ed. by G. K. Meffe and F. F. Snelson), pp. 379–381. Englewood Cliffs: Prentice Hall.

Roff, D. A. (1983). An allocation model for growth and reproduction in fish. *Can. J. Fish. Aq. Sci.*, **40**, 1395–1404.

Rosen, D. E. (1960). Middle-American poeciliid fishes of the genus *Xiphophorus*. *Bull. Fla. State Mus. Biol. Sci.*, **5**, 57–242.

Rosen, D. E. and Bailey, R. M. (1959). Middle-American poeciliid fishes of the genera *Carlhubbsia* and *Phallichthys*, with description of two new species. *Zoologica*, **44**, 1–44.

Rosen, D. E. and Bailey, R. M. (1963). The poeciliid fishes (Cyprinodontiformes), their structure, zoogeography and systematics. *Bull. Amer. Mus. Natur. Hist.*, **126**, 1–176.

Rosen, D. E. and Gordon, M. (1953). Functional anatomy and evolution of male genitalia in poeciliid fishes. *Zoologica*, **25**, 1–48.

Rosen, D. E. and Tucker, A. (1961). Evolution of secondary sexual characters and sexual behavior patterns in a family of viviparous fishes (Cyprinodontiformes: Poeciliidae). *Copeia*, 201–212.

Ryan, M. J. (1990). Sexual selection, sensory systems and sensory exploitation. *Oxford Surv. Evol. Biol.*, **7**, 157–195.

Ryan, M. J. and Causey, B. A. (1989). "Alternative mating behavior in the swortails, *Xiphophorus nigrensis* and *Xiphophorus pygmaeus* (Pisces: Poeciliidae). *Behav. Ecol. Sociobiol.*, **24**, 341–348.

Ryan, M. J. and Wagner, W. E. Jr. (1987). Asymmetries in mating preferences between species: female swordtails prefer heterospecific males. *Science*, **236**, 595–597.

Ryan, M. J., Hews, D. K. and Wagner, W. E. Jr. (1990). Sexual selection on alleles that determine body size in the swordtail *Xiphophorus nigrensis*. *Behav. Ecol. Sociobiol.*, **26**, 231–237.

Sargent, R. C., Taylor, P. D. and Gross, M. R. (1987). Parental care and the evolution of egg size in fishes. *Am. Nat.*, **129**, 32–46.

Schmidt-Nielsen, K. (1984). *Scaling: Why is Animal Size So Important?* Cambridge: Cambridge University Press.

Schoenherr, A. A. (1977). Density dependent and density independent regulation of reproduction in the gila topminnow, *Poeciliopsis occidentalis* (Baird and Girard). *Ecology*, **58**, 438–444.

Schultz, R. J. (1961). Reproductive mechanisms of unisexual and bisexual strains of the viviparous fish *Poeciliopsis*. *Evolution*, **15**, 302–325.

Scrimshaw, N. S. (1944). Embryonic growth in the viviparous poeciliid, *Heterandria formosa*. *Biol. Bull.*, **87**, 37–51.

Searcy, W. A. and Yasukawa, K. (1990). Use of the song repertoire in intersexual and intrasexual contexts by male red-winged blackbirds. *Behav. Ecol. Sociobiol.*, **27**, 123–128.

Selander, R. K. (1966). Sexual dimorphism and different niche utilization in birds. *Condor*, **68**, 113–151.

Shine, R. (1988). The evolution of large body size in females: a critique of Darwin's "fecundity advantage" model. *Am. Nat.*, **131**, 124–131.

Shine, R. (1990). Proximate determinants of sexual differences in adult body size. *Am. Nat.*, **135**, 278–283.

Singer, M. C. (1982). Sexual selection for small size in male butterflies. *Am. Nat.*, **119**, 440–443.

Smith, C. C. and Fretwell, S. D. (1974). The optimal balance between size and number of offspring. *Am. Nat.*, **108**, 499–506.

Snelson, F. F. Jr. (1982). Indeterminate growth in males of the sailfin molly, *Poecilia latipinna*. *Copeia*, 296–304.

Snelson, F. F. Jr. and Wetherington, J. D. (1980). Sex ratio in the sailfin molly, *Poecilia latipinna*. *Evolution*, **34**, 308–319.

Sohn, J. J. (1977a). Socially induced inhibition of genetically determined maturation in the platyfish, *Xiphophorus maculatus*. *Science*, **195**, 199–201.

Sohn, J. J. (1977b). The consequences of predation and competition upon the demography of *Gambusia manni* (Pisces: Poeciliidae). *Copeia*, 224–227.

Stearns, S. C. (1983a). The evolution of life-history traits in mosquitofish since their introduction to Hawaii in 1905: rates of evolution, heritabilities, and developmental plasticity. *Amer. Zool.*, **23**, 65–75.

Stearns, S. C. and Koella, J. C. (1986). The evolution of phenotypic plasticity in life history traits: prediction of reaction norms for age and size at maturity. *Evolution*, **40**, 893–913.

Stearns, S. C. and Sage, R. D. (1980). Maladaptation in a marginal population of the mosquitofish *Gambusia affinis*. *Evolution*, **34**, 65–75.

Steele, D. H. (1977). Correlation between egg size and development period. *Am. Nat.*, **111**, 371–372.

Stoner, G. and Breden, F. (1988). Phenotypic differentiation in female preference related to geographic variation in male predation risk in the Trinidad guppy (*Poecilia reticulata*). *Behav. Ecol. Sociobiol.*, **22**, 285–291.

Temeles, E. J. (1985). Sexual size dimorphism of bird-eating hawks: the effect of prey vulnerability. *Am. Nat.*, **125**, 485–499.

Travis, J. and Woodward, B. D. (1989). Social context and courtship flexibility in males sailfin mollies, *Poecilia latipinna* (Pisces: Poeciliidae). *Animal Behaviour*, **38**, 1001–1011.

Travis, J., Farr, J. A., Henrich, S. and Cheong, R. T. (1987). Testing theories of clutch overlap with the reproductive ecology of *Heterandria formosa*. *Ecology*, **68**, 611–623.

Travis, J., Farr, J. A., McManus, M. and Trexler, J. C. (1989). Environmental effects on adult growth in the male sailfin molly (*Poecilia latipinna*, Poeciliidae). *Env. Biol. Fish*, **26**, 119–127.

Turner, C. L. (1941). Morphogenesis of the gonopodium in *Gambusia affinis affinis*. *J. Morphol.*, **69**, 161–185.

Turner, J. S. and Snelson, F. F. Jr. (1984). Population structure, reproduction and laboratory behaviour of the introduced *Belonesox belizanus* (Poeciliidae) in Florida. *Env. Biol. Fish.*, **10**, 89–100.

Warner, R. R. and Harlan, R. K. (1982). Sperm competition and sperm storage as determinants of sexual dimorphism in the dwarf surfperch, *Micrometrus minutus*. *Evolution*, **36**, 44–55.

Winkler, P. (1979). Thermal preference of *Gambusia affinis affinis* as determined under field and laboratory conditions. *Copeia*, 60–64.

Woolbright, L. L. (1983). Sexual selection and size dimorphism in anuran amphibia. *Am. Nat.*, **121**, 110–119.

Yan, H. Y. (1987). Size at maturity in male *Gambusia heterochir*. *J. Fish Biol.*, **30**, 731–741.

Zander, C. D. and Dzwillo, M. (1969). Untersuchungen zur Entwicklung und Vererbung des Caudalfortsatzes der *Xiphophorus*-Arten (Pisces). *Zeit. Wiss. Zool.*, **178**, 276–315.

Zayan, R. C. (1975). Defense du territoire et reconnaissance individuelle chez *Xiphophorus* (Pisces, Poeciliidae). *Behaviour*, **52**, 267–312.

Zimmerer, E. J. and Kallman, K. D. (1989). Genetic basis for alternative reproductive tactics in the pigmy swordtail *Xiphophorus nigrensis*. *Evolution*, **43**, 1298–1307.

Zulian, E. (1990). Selezione e dimensioni nei maschi di *Gambusia holbrooki* (Pisces: Poeciliidae). Analisi dei meccanismi che determinano la lunghezza alla maturita. Ph.D. thesis. University of Padova, Italy.

THE ADVANTAGES OF BEING RED: SEXUAL SELECTION IN THE STICKLEBACK

THEO C. M. BAKKER and MANFRED MILINSKI

Universität Bern, Zoologisches Institut, Abt. Verhaltensökologie, Wohlenstrasse 50a, CH-3032 Hinterkappelen, Switzerland

Male three-spined stickleback develop a conspicuous red breeding coloration. The intensity of red correlates positively with physical condition and decreases by parasitization. In some populations, the male's red coloration functions in male–male competition through intimidation of rival males either directly or via associated characters. A comparison of the function of red in male–male competition among populations and among stickleback species suggests that the signalling of red coloration in dominance contests is a derived function. The primary function probably is to determine female choice. Recent experimental manipulation of red coloration showed that ripe female sticklebacks base their mate choice on the intensity of the male's red coloration. In sequential choice situations, choice is based on the attractiveness of the present and previously encountered males.

HANDICAPS

In most animal species it is the male sex which possesses conspicuous secondary sexual characters. A typical example is the conspicuous nuptial coloration of the three-spined stickleback (*Gasterosteus aculeatus* L.) with the orange/red throat and forebelly and the blue eyes. The female stickleback looks inconspicuous and is, in contrast to the male, adapted to a life as a potential prey of various predators. The pronounced secondary sexual characters of the male must be a true handicap under the risk of predation, which has been experimentally proven (Moodie, 1972). What are then the advantages of such a conspicuous nuptial coloration that the stickleback has developed and maintained against the pressure of natural selection?

SEXUAL SELECTION

Since Darwin (1871) the answer to this is unequivocal: sexual selection. In species in which the males have a higher potential reproductive rate than the females, we can expect strong competition to occur among males for breeding sites and access to females, and females to be selective in their choice of mates (Clutton-Brock and Vincent, 1991). Darwin distinguished these two selection processes for explaining the occurrence of exaggerated secondary sexual traits in males: the sexual ornaments either enhance the male's competitive ability for gaining access to females (intra-sexual selection) or make him more attractive to females (inter-sexual selection). Both selection regimes give more ornamented males a reproductive advantage relative to males with less elaborate traits. There exist examples of male ornaments that have been evolved via either intra- (e.g. horns in fungus beetles, Conner, 1989) or inter-sexual selection (e.g. the peacock's train, Petrie *et al.*, 1991). However, in most cases one cannot exclusively ascribe the conspicuous sexual ornaments of males to one of the two kinds of sexual selection. As a rule intra- and inter-sexual selection act both on the males' sexual ornaments and it is often difficult to separate their respective contributions (Harvey and Bradbury, 1991).

STICKLEBACK REPRODUCTION

The three-spined stickleback is a small fish of the northern hemisphere which reproduces in fresh or brackish water and shows a conspicuous sexual dimorphism in the coloration during the breeding season. In the beginning of the reproductive season the male settles a territory in which he builds a nest. In the nest he tries to collect clutches of several females, thereafter the male cares for eggs and young for about ten days (Wootton, 1976). Three-spined sticklebacks are highly polygynous; a male stickleback may collect as many as 20 clutches of eggs from different females in a single breeding cycle (Kynard, 1978). In addition, he may complete several breeding cycles during a single season (Wootton, 1976).

INTRA-SEXUAL SELECTION: DUMMIES

Darwin's hypothesis of intra-sexual selection did not encounter much opposition of his contemporaries. Research on intra-sexual selection in the stickleback can be traced back to the classical investigations of the late Niko Tinbergen on sign stimuli in the forties (ter Pelkwijk and Tinbergen, 1937; Tinbergen, 1948, 1951). Experiments with dummies of male sticklebacks, which were presented inside the territory of a male, showed the aggression-releasing effect of a red throat. Crude dummies with a red throat elicited more attacks of the territory owner than more realistic but nonred dummies did. The results were that clear-cut, that a quantitative analysis was considered superfluous. Nevertheless, in 1949 such an analysis was made by Collias while being a guest in Tinbergen's laboratory in Leiden, and the results were only recently published (Baerends, 1985; Collias, 1990). His data affirmed the statements of Tinbergen. Yet this was not that self-evident, because several other researchers failed to find evidence for the aggression-releasing effect of a red throat (reviewed in Rowland and Sevenster, 1985). Rowland for instance found in 1982, that when he simultaneously presented a dummy with and one without a red throat and belly, the grey dummy was significantly more often the target of aggression of the territory owner.

How to explain this discrepancy? Tinbergen and co-workers made their dummies out of wax and the belly was painted red by applying shellac mixed with a red powder (P. Sevenster, pers. comm.). A direct comparison of the remainder of probably the only dummy left from that time (present at the zoological laboratory in Leiden) and a Rowland dummy revealed a striking difference between the red colour of Rowland's plastic dummies and Tinbergen's red shellac. Rowland's red colour appears stronger, mainly due to a difference in contrast with the body colour, in hue (Rowland's red is less orange than Tinbergen's colour) and to some extent in colour intensity. Although the colours on the remainder of the Tinbergen dummy may have faded somewhat, in any case they have originally contrasted less strongly with the body colour than in Rowland's dummies (P. Sevenster, pers. comm.; G.P. Baerends, pers. comm.). The strong red of the Rowland dummy may rather have an intimidating effect on the territory owner than an aggression-releasing effect and his results can be explained from this perspective (Rowland, 1982, in press).

INTRA-SEXUAL SELECTION: DOMINANCE

It is obvious that an aggression-releasing effect of a red throat cannot account for the evolution of the red throat and must be a side-effect; it may be rather disadvantageous to its bearer to provoke aggression of rivals. Yet male–male competition may have played a role in the evolution of the male's red breeding coloration, namely through its intimidating effect on rivals. An advantage of a red throat in intra-sexual selection is indicated by its association with the probability to win in male–male competition when new territories are being settled (Bakker and Sevenster, 1983; Bakker, 1986, in press). In the laboratory we can simulate such a situation by introducing two reproductive, isolated males into a tank which is unfamiliar to both and just large enough for the settlement of one territory. After a short and intense fight one of the males will usually dominate the other. The dominant male will start nest building, while the inferior male will remain quiet at the water surface or hidden between plants and will be attacked by the dominant male upon movement.

When dominance tests are carried out with inexperienced males that are introduced simultaneously, experiential and prior residence effects on the outcome of the dyadic combat are ruled out. Under these conditions, dominance ability was positively correlated with the degree of red breeding coloration (Bakker and Sevenster, 1983) suggesting that red, or a trait that correlates with red, intimidates the opponent. The red coloration was quantified on a four-point scale increasing from hardly any red to much red. An intimidating effect of red breeding coloration is only conceivable if this signal is uncheatable and somehow related to fighting ability. The positive correlation between the intensity of red and a male's physical condition make these likely (Milinski and Bakker, 1990; Figure 1).

A behaviour-genetic analysis of variation in dominance ability based on two-way selection experiments showed that this variation could partly be attributed to genotypic variation (Bakker, 1985, 1986, in press). Within each generation of the high and low dominance lines all possible pairwise combinations of males were tested. The criterion of selection was based on the number of tests that each male had won. In each generation, the response to two-way selection for dominance, i.e. for low and high dominance, was determined from the outcome of dominance tests between males of the high and low lines. Selection for low and high dominance ability produced significant divergence between the two lines by the third generation. Dominance tests between males of either selection line and males of an unselected control line in the second generation, suggested that the divergence between the high and low lines was due to a decrease in the dominance ability of low line males rather

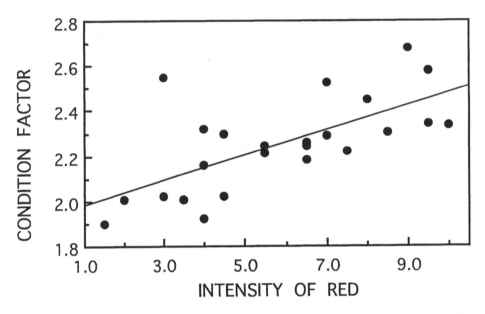

Figure 1 The correlation between the intensity of red breeding coloration and the condition factor. $r^2 = 0.44$, $F = 17.27$, $d.f. = 1,22$, $P < 0.0004$ (after Milinski and Bakker, 1990).

than an increase in that of the high line males. This asymmetry in the response to two-way selection suggests that in nature high dominance abilities had been favoured.

As a rule, artificial selection does not only change the trait chosen as the criterion of selection. An array of other traits that are genetically correlated with the selected trait will also be affected by the applied selection regime. An analysis of these correlated responses of selection for dominance ability revealed an interesting change in the degree of red coloration. The degree of red coloration of males of the low dominance line was on average less than that of high line males or control line males (Bakker, 1986; Figure 2). The latter two did not differ significantly as to their red coloration. Notice also the difference in the degree of red coloration of all males, irrespective of their origin, after presentation of a rival male (Figure 2A) and after presentation of a ripe female (Figure 2B); both stimuli were confined in a glass tube and offered inside the territory. Females made males flush most strongly. We will come back to this point later on.

Three-spined sticklebacks exhibit two contrasting life-history patterns. Anadromous populations overwinter in estuarine or coastal marine habitats and migrate to fresh or brackish water in the spring to reproduce, while freshwater populations reside permanently in rivers, streams, or lakes. The only population that has been studied in which coloration played a prominent role in determining dominance relationships was a freshwater population (Bakker and Sevenster, 1983; Bakker, 1986). Studies in which the degree of red coloration was not correlated with dominance happened all to have used males from anadromous populations (FitzGerald and Kedney, 1987; Rowland, 1989).

To investigate whether this discrepancy could really be ascribed to a different function of red in freshwater and anadromous populations, we compared dominance

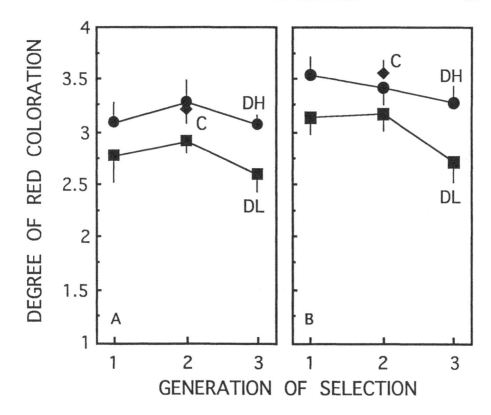

Figure 2 The average (+ or − SE) degree of red breeding coloration of males in three successive generations selected for low (DL line, squares) and high (DH line, circles) levels of dominance ability, and in the second generation of the control (C, diamonds) line; (A) after presentation of a male in a tube, and (B) after presentation of a ripe female in a tube (after Bakker, 1986).

abilities between laboratory-bred males from wild-caught parents of a Dutch anadromous and a Dutch freshwater population. Laboratory-reared anadromous males tended to be slightly, but significantly brighter than laboratory-reared freshwater males, suggesting a genetic basis to the difference (Bakker, in press). Based on our earlier established positive correlation between the degree of red and dominance in the freshwater population, we expected greater dominance of the anadromous males. In dominance tests between males of the two populations, their dominance ability was, however, less than that of their freshwater counterparts. Isolated males from the freshwater population displayed greater dominance ability under various circumstances (Bakker, in press).

Within each population dominance ability was measured by making all pairwise comparisons of relative dominance among a group of individually isolated males. The males can then be arranged in a linear order of dominance based on the probability of winning the dominance contests. There existed a significant positive correlation between the degree of red coloration and dominance ability in the freshwater population ($r_s = 0.72$, $N = 10$, $P < 0.02$) but not in the anadromous population (r_s = 0.20, N = 10, P > 0.25) (Bakker, in press). The correlation in the freshwater population was not attributable to particular males. Body size and aggressiveness were both uncorrelated with dominance ability in these populations (Bakker, in press).

Comparisons of anadromous and freshwater populations have the potential to provide insight into the directions of evolutionary change, because freshwater populations are thought to have been derived from the marine form (Bell and Foster, in press). In addition, phylogenetic studies suggest that the evolution of colour patterns in the stickleback family, the Gasterosteidae, is more strongly correlated with inter-sexual selection than with intra-sexual selection (McLennan *et al.*, 1988; McLennan, 1991). It is thus possible that in anadromous populations, nuptial coloration functions primarily to determine patterns of mate choice (inter-sexual selection) and that the signal function of this coloration in dominance contests is a derived condition.

FEMALE PREFERENCE FOR RED: CIRCUMSTANTIAL EVIDENCE

Considering that male three-spined sticklebacks have very conspicuous breeding colorations among European freshwater fish species (Darwin himself mentioned the stickleback as an example) and the fact that they have been used as experimental animals in ethological studies for more than 55 years, makes it rather surprising that, apart from circumstantial evidence which has been disregarded for a long time, it was only recently proven that the male's red breeding coloration is of crucial importance in female choice.

Ter Pelkwijk and Tinbergen (1937) noticed already that there was little response of ripe females to male dummies that lacked a red throat and belly. The importance of a red throat in inter-sexual selection is also suggested by the fact that males flush their red colours more strongly when given a sight of a ripe female than when given a sight of a rival male, as was indicated before in the results of the dominance selection lines (Figure 2). In an early experiment on sexual selection, Semler (1971) showed that females from a population polymorphic for male breeding coloration preferred red males over non-red males. Female preference for artificially coloured (with nail-polish or lipstick) non-red males over non-manipulated controls indicated that the females were responding to the red coloration alone. However, Semler's experimental design did not exclude effects either of red on male–male competition or of the paint itself. We applied Semler's recipe for a different purpose, but the paint came off during the experiments and the fish seemed to be adversely affected (Bakker and Sevenster, 1983). An interesting example of the various circumstantial evidence for the functioning of the male's red breeding coloration in female mating choice was given by Cronly-Dillon and Sharma (1968). Outside the reproductive season males and females have a similar visual sensitivity for red (Figure 3A). However, during the reproductive season the females' visual sensitivity for red increases and reaches a higher level than that of males (Figure 3B). These diverse, but consistent indications point to the role of inter-sexual selection in the evolution of the red coloration of male sticklebacks. This role is probably even greater than that of intra-sexual selection.

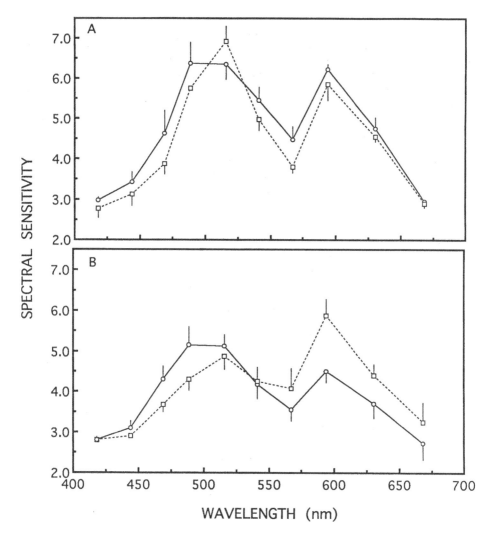

Figure 3   Spectral sensitivity (optomotor response + or − SD) of female (squares, dotted line) and male (circles) sticklebacks outside (A) and during (B) the breeding season (after Cronly-Dillon and Sharma, 1968).

FEMALE PREFERENCE FOR RED: GREEN LIGHT EXPERIMENTS

Recent experiments gave conclusive evidence that female sticklebacks base their mate choice almost exclusively on the intensity of the male's red breeding coloration (Milinski and Bakker, 1990). Unlike correlative studies on female preference for exaggerated male traits (which form the great majority of studies on sexual selection; exceptions e.g. Andersson, 1982; Møller, 1988; Hill, 1991), we ruled out the influence of confounding variables by experimental manipulation of the preferred male trait.

We did our experiments with wild-caught fish. The males showed great variation as to the intensity of their red coloration which was positively correlated with variation in the condition factor. To investigate whether females base their mate choice on the intensity of the males' breeding coloration the tanks of individual males were arranged according to increasing intensity of coloration, and neighbouring males were defined as pairs for presentation to ripe females. In a separate tank positioned centrally in front of each pair of neighbouring tanks a cell containing a single gravid female was placed and her choosing between the two males was registrated. Females were previously selected for their readiness to spawn, i.e. to adopt and maintain the head-up courtship posture while pointing towards one of the two males. In order to be able to exclude the possibility that female choice would be based on some character correlated with the intensity of red, such as the intensity of courtship, the well-known zigzag dance of the stickleback, we carried out the choice experiments under two different lighting conditions. Under the first condition, choice experiments were carried out in normal white light. The other condition was such that females were almost unable to judge differences in the intensity of red coloration. This was achieved by green illumination.

In order to test whether the courtship behaviour of the males differs between light conditions, we confronted each male singly with the cell containing a ripe female under both green and white light. The number of zigzags of the male's courtship display, which is regarded as a reliable measure of male courtship intensity (e.g. Sevenster, 1961), was counted during a standard period of time. The males' courtship intensity did not differ significantly in the two light situations and differences between pair members did not change significantly either between the lighting conditions.

Although the experiment was designed so that the males of each pair were very similar in the intensity of red coloration (see above), under white light there was a significant preference by the females for the brighter male in each pair, this preference being intensified as the difference in coloration increased (Figure 4A). Under green light the trend in favour of the brighter males was not significant any more (Figure 4B). The slope of the regression under white light was significantly greater than that under green light. This inevitably leads to the conclusion that the females based their choice primarily on differences in male red coloration. This makes functional sense because colour intensity correlated significantly with condition (for definition see Milinski and Bakker, 1990) whereas courtship intensity did not correlate significantly with condition. The conclusion that female choice was based mainly on colour cues was affirmed by partial correlations under white light.

FEMALE CHOICE AND PARASITES

Among the good-genes models of the evolution of exaggerated characters by female choice, the Hamilton and Zuk host-parasite model from 1982 attracted considerable attention, because it offers a plausible explanation for a sustained heritability of fitness. Hamilton and Zuk interpret many of the male's exaggerated sexual ornaments as signals indicating parasite resistance. According to their model, these signals are "revealing handicaps", that is, they give females a chance to detect whether a potential mate is parasitized (for instance because he is less brightly coloured) or not.

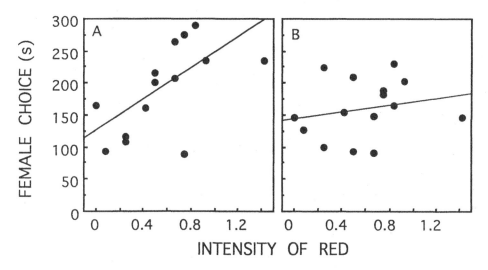

Figure 4   Average active female choice for the brighter male as a function of the difference in the intensity of red coloration of pairs of reproductive males under white (A) and green (B) light (after Milinski and Bakker, 1990).

The intensity of red breeding coloration in sticklebacks is a revealing handicap, because it correlates positively with a male's physical condition. It is also a revealing handicap in the sense of Hamilton and Zuk (1982) as suggested by the following experiment. In order to investigate whether parasites influence both the males' red breeding coloration and the result of active female choice, we infested the brighter male of each pair from the previous mentioned experiment with the ciliate *Ichthyophthirius multifiliis*, a serious and widespread fish disease known as 'white-spot'. Our fish were only mildly infected and recovered within a week. To prevent reinfection, the tanks of all fish (including control fish) were treated with an anti-white spot solution. The fish continued to court stimulus females. Four of the six students who had scored the males' red coloration before parasitization were asked to repeat their estimation of pair differences. They were not informed about the prior parasitization of half of the males. Again we submitted the remaining pairs of males to female choice experiments with new ripe females under white and green light.

Parasitization caused a significant decrease of the intensity of the males' red coloration. Parasitization also caused a significant decrease of the difference in condition factor and this was probably the cause of the reduction in colour intensity. The females significantly decreased their prior preference for the formerly brighter males under white light, but under green light the males' parasitization had no significant effect on female choice. This implies that the females detected the prior parasitization of the males by their decreased intensity of red breeding coloration, which is a necessary condition for coloration to be judged as a revealing handicap.

MECHANISM OF SEQUENTIAL FEMALE CHOICE

The empirical evidence for the evolution of male sexual ornaments by female choice is bound to simultaneous choice situations in which a direct comparison facilitates the preference for the more ornamented male. In nature, simultaneous choice is restricted either to situations in which males are displaying in a common area, a lek, or to cases in which females use acoustical male ornaments for choosing among spread out males. In most of the remaining cases the female must visit males sequentially and compare the present male with what she has stored in her memory of those met previously. This appears to be a more difficult task than that of simultaneous choice. It is, however, a prerequisite for the application of models of sexual selection to most territorial species.

Male three-spined sticklebacks claim a territory of up to several m^2 at the start of the breeding season. The distance between the males and also vegetation often forces a female to compare potential mates sequentially under natural conditions if she is able to do so. Therefore the three-spined stickleback seems to be an ideal species to study sequential female choice.

Male three-spined sticklebacks with varying degrees of breeding coloration were caught from a Swiss freshwater population and placed individually into tanks. After several weeks most males had a complete nest built in a Petri dish provided close to the backwall and courted vigorously. Out of nine males which could easily be categorized as either dull, medium or bright according to the intensity of their red breeding coloration, we used five different combinations of three males, one of each colour type, for sequential presentation to ripe females. Females caught from the same population were used in experiments when they were ready to spawn as judged from their colour and the extension of their bellies and from the opening of their cloacae.

A cell containing a single gravid female was placed in front of a male's tank and the duration of her head-up display, which correlates positively with her probability to spawn with that male (McLennan and McPhail, 1990), was measured during a standard period of time. Half of 28 females were presented to the males in the order: dull, medium, dull, pause of 30 min, bright, medium, bright. The other half saw the males in the reverse order (Figure 5). Each female was used only once.

Each female saw each male twice and the position of each type of male in the order of presentation was the same averaged over all females. Therefore we can compare the duration of head-up display each type of male received per presentation to a female with that of the other two. The duration of the females' head-up posture increased with the intensity of the males' brightness (Bakker and Milinski, 1991; Figure 6). This result demonstrates the females' ability to prefer the brighter male also when they see only one male at a time. Thus female sticklebacks are able to exert sequential mate choice. The result does not prove, however, that the females have actually compared the different males with each other, but can be the consequence of comparing each male with an internal order of standards. Since the probability to spawn with a given male depends on the female's duration of head-up posture (McLennan and McPhail, 1990) she would not need to have an additional rule for continuing to sample further males.

The probability to spawn with a given male may not only depend on its rank in the female's internal standards but could also be influenced by the brightness of the previous male. The female may rate a male higher when the previous male was duller

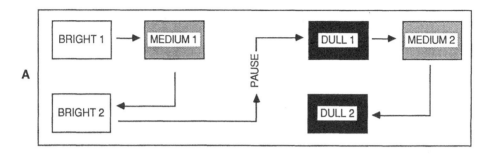

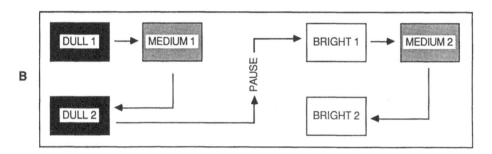

Figure 5 Order of presentation of males of 3 different colour types, i.e. dull, medium, bright, to (**A**) females starting with a bright male, (**B**) females starting with a dull male.

and vice versa. Hence we would expect that the duration of the head-up posture is longer during the second presentation than during the first one with a given colour type when the male met before the second presentation is duller. Vice versa we would expect that this duration is shorter when the male in between is brighter. In our experimental design this means that for females starting with a dull male we expect for the duration of the head-up posture the following inequalities: dull 1 > dull 2, medium 1 > medium 2, bright 2 > bright 1. Similarly, for females starting with a bright male we expect: bright 2 > bright 1, medium 2 > medium 1, dull 1 > dull 2. Each of the observed six inequalities (three for females starting with a dull male, and three for females starting with a bright male) was tested with a 1-tailed Wilcoxon matched-pairs signed-ranks test with $N = 14$ and the six probabilities were combined with a Fisher combination test; the probability of that inequality which was in variance with our hypothesis was entered as $1-P$. A potential habituation effect by which the second presentation of each color type is expected to score lower than the first one is compensated for by this procedure, i.e. three of the predicted inequalities would be in the same direction and the other three would be in opposite direction of a potential habituation effect. The combined probability supports our hypothesis of a previous male effect at the 1% level of significance (Bakker and Milinski, 1991).

Our finding of the females' ability for exerting sequential mate choice and of the 'previous male effect' concerns the mechanism of female mate choice, the existence of which is a necessary prerequisite of most models of sexual selection. It broadens

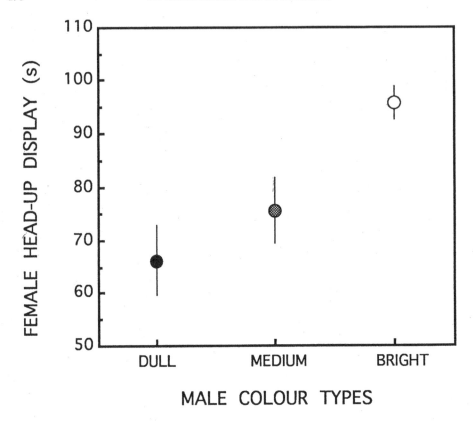

Figure 6 Average (± SE) duration of the female head-up display directed at dull, medium, or bright males. $P < 0.0001$ after Page test for ordered alternatives (after Bakker and Milinski, 1991).

the application of these models to species in which females can choose a potential mate only by visiting a number of males sequentially.

ACKNOWLEDGMENTS

We thank Carin Magnhagen and Gunilla Rosenqvist for helpful comments and the Swiss National Science Foundation for financial support.

References

Andersson, M. (1982). Female choice selects for extreme tail length in a widowbird. *Nature*, **299**, 818–820.

Baerends, G. P. (1985). Do dummy experiments with sticklebacks support the IRM-concept? *Behaviour*, **93**, 258–277.

Bakker, Th. C. M. (1985). Two-way selection for aggression in juvenile, female and male sticklebacks (*Gasterosteus aculeatus* L.), with some notes on hormonal factors. *Behaviour*, **93**, 69–81.

Bakker, Th. C. M. (1986). Aggressiveness in sticklebacks (*Gasterosteus aculeatus* L.): a behaviour-genetic study. *Behaviour*, **98**, 1–144.

Bakker, Th. C. M. (In press). Evolution of aggressive behavior in threespine stickleback. In: *The Evolutionary Biology of the Threespine Stickleback* (Ed. by M. A. Bell and S. A. Foster). Oxford: Oxford University Press.

Bakker, Th. C. M. and Milinski, M. (1991). Sequential female choice and the previous male effect in sticklebacks. *Behav. Ecol. Sociobiol.*, **29**, 205–210.

Bakker, Th. C. M. and Sevenster, P. (1983). Determinants of dominance in male sticklebacks (*Gasterosteus aculeatus* L.). *Behaviour*, **86**, 55–71.

Bell, M. A. and Foster, S. A. (Eds). (In press). *The Evolutionary Biology of the Threespine Stickleback*. Oxford: Oxford University Press.

Clutton-Brock, T. H. and Vincent, A. C. J. (1991). Sexual selection and the potential reproductive rates of males and females. *Nature*, **351**, 58–60.

Collias, N. E. (1990). Statistical evidence for aggressive response to red by male three-spined sticklebacks. *Anim. Behav.*, **39**, 401–403.

Conner, J. (1989). Density-dependent sexual selection in the fungus beetle, *Bolitotherus cornutus*. *Evolution*, **42**, 1378–1386.

Cronly-Dillon, J: and Sharma, S. C. (1968). Effect of season and sex on the photopic spectral sensitivity of the three-spined stickleback. *J. Exp. Biol.*, **49**, 679–687.

Darwin, C. (1871). *The Descent of Man, and Selection in Relation to Sex*. London: John Murray.

FitzGerald, G. J. and Kedney, G. I. (1987). Aggression, fighting, and territoriality in sticklebacks: three different phenomena? *Biol. Behav.*, **12**, 186–195.

Hamilton, W. D. and Zuk, M. (1982). Heritable true fitness and bright birds: a role for parasites? *Science*, **218**, 384–387.

Harvey, P. H. and Bradbury, J. W. (1991). Sexual selection. In: *Behavioural Ecology: An Evolutionary Approach*, 3rd edn (Ed. by J. R. Krebs and N. B. Davies), pp. 203–233. Oxford: Blackwell Scientific Publications.

Hill, G. E. (1991). Plumage coloration is a sexually selected indicator of male quality. *Nature*, **350**, 228–229.

Kynard, B. E. (1978). Breeding behavior of a lacustrine population of threespine sticklebacks (*Gasterosteus aculeatus* L.). *Behaviour*, **67**, 178–207.

Maynard Smith, J. (1985). Mini review: sexual selection, handicaps and true fitness. *J. Theor. Biol.*, **115**, 1–8.

McLennan, D. A. (1991). Integrating phylogeny and experimental ethology: from pattern to process. *Evolution*, **45**, 1773–1789.

McLennan, D. A., Brooks, D. R. and McPhail, J. D. (1988). The benefits of communication between comparative ethology and phylogenetic systematics: a case study using gasterosteid fishes. *Can. J. Zool.*, **66**, 2177–2190.

McLennan, D. A. and McPhail, J. D. (1990). Experimental investigations of the evolutionary significance of sexually dimorphic nuptial colouration in *Gasterosteus aculeatus* (L.): the relationship between male colour and female behaviour. *Can. J. Zool.*, **68**, 482–492.

Milinski, M. and Bakker, Th. C. M. (1990). Female sticklebacks use male coloration in mate choice and hence avoid parasitized males. *Nature*, **344**, 330–333.

Moodie, G. E. E. (1972). Predation, natural selection and adaptation in an unusual threespine stickleback. *Heredity*, **28**, 155–167.

Møller, A. P. (1988). Female choice selects for male sexual tail ornaments in the monogamous swallow. *Nature*, **332**, 640–642.

Petrie, M., Halliday, T. and Sanders, C. (1991). Peahens prefer peacocks with elaborate trains. *Anim. Behav.*, **41**, 323–331.

Rowland, W. J. (1982). The effects of male nuptial coloration on stickleback aggression: a reexamination. *Behaviour*, **80**, 118–126.

Rowland, W. J. (1989). The effects of body-size, aggression and nuptial coloration on competition for territories in male threespine sticklebacks, *Gasterosteus aculeatus. Anim. Behav.*, **37**, 282–289.

Rowland, W. J. (In press). Proximal determinants of stickleback behavior: an evolutionary perspective. In: *The Evolutionary Biology of the Threespine Stickleback* (Ed. by M. A. Bell and S. A. Foster). Oxford: Oxford University Press.

Rowland, W. J. and Sevenster, P. (1985). Sign stimuli in the threespine stickleback (*Gasterosteus aculeatus*): a re-examination and extension of some classic experiments. *Behaviour*, **93**, 241–257.

Semler, D. E. (1971). Some aspects of adaptation in polymorphism for breeding colours in the threespine stickleback (*Gasterosteus aculeatus*). *J. Zool. Lond.*, **165**, 291–302.

Sevenster, P. (1961). A causal analysis of a displacement activity (fanning in *Gasterosteus aculeatus* L.). *Behaviour Suppl.*, **9**, 1–170.

Ter Pelkwijk, J. J. and Tinbergen, N. (1937). Eine reizbiologische Analyse einiger Verhaltensweisen von *Gasterosteus aculeatus* L. *Z. Tierpsychol.*, **1**, 193–200.

Tinbergen, N. (1948). Social releasers and the experimental method required for their study. *Wilson Bull.*, **60**, 6–52.

Tinbergen, N. (1951). *The Study of Instinct*. Oxford: Clarendon Press.

Wootton, R.J. (1976). *The Biology of the Sticklebacks*. London: Academic Press.

BEHAVIOURAL ORGANISATION AND THE EVOLUTION OF BEHAVIOURAL STRATEGIES

GERARD P. BAERENDS

Department of Behavioural Biology, University of Groningen, The Netherlands

BEHAVIOURAL ECOLOGY AND THE STUDY OF CAUSATION

Behavioural ecologists are concerned with the evolution of behavioural strategies under the influence of natural selection. Of necessity, this must have involved modification of existing strategies or the development of new ones. In other words the evolution of behavioural strategies implies changes in the form of elementary activities and, qualitively and quantitatively, in the sequences in which they occur. This holds for activities with a direct instrumental function and, perhaps even more, for communicative behaviour.

Thanks to pioneer studies by Whitman, Heinroth and Lorenz, we know that the concept of homology, as developed in comparative anatomy with respect to morphological structures, can also be successfully applied to behavioural elements. This is due to the fact that both are built up by means of developmental programmes, which are in a way comparable to the software used in computers. These programmes, and the principles upon which they are structured, are at the basis of homologies. The behavioural algorithms are studied with the methods of causal ethological analysis. In essence this is a black box approach, in which the sensory input into the intact animal and its behavioural output are quantitatively determined and their relationships statistically traced. The results generate hypotheses about the way the information is processed in the black box, and thus about the software structures underlying the organisation of the behaviour of an animal. These deductions have to be experimentally verified. Thus, evolution of new behaviour patterns proceeds through modification of existing software. Insight into how such changes could have resulted from alterations in selection pressures can be gained from comparative ethological research. The Cichlid family, with its great diversity of species and strategies, offers excellent opportunities for studying this problem. In this paper I shall report about work carried out in my research team at the University of Groningen on the causation and evolutionary derivation of the displays subserving the reproduction in these fish.

REPRODUCTION STRATEGIES AND COMMUNICATION IN CICHLIDS

Two major reproductive strategies can be distinguished in the Cichlidae: substrate spawning, which is thought to be the original strategy of the family, and oral incubation, taken to have been derived from it (Baerends and Baerends-Van Roon,

1950; Barlow, 1991). In substrate spawners the eggs are attached to a solid substrate (rock or plant); together the partners care for eggs and young. In most mouthbreeders, only the female takes care of the brood. After visiting a male on a communal spawning ground (arena or lek) to get her clutch fertilized, she leaves with the eggs in her mouth. The male remains on the lek and continues attracting females. Consequently the success of male mouthbreeders mainly depends on the *number* of females they can attract, whereas the females—because of the parental investment they make—will mainly be interested in assessing the *quality* of the males. In contrast, in the substrate spawners both sexes will be interested in the vigour of the partner. Moreover, in substrate spawners male and female both need to select a partner with which they can co-operate over a longer period, i.e. one with a matching 'personality'.

In both cases communication between the fish is an essential element of their reproductive behaviour. Although a role for chemical and acoustical communication should not be excluded, visual signals or displays are of major importance. This paper asks how the repertoire of display activities in the cichlids might have evolved and how the adaptive radiation over the different strategies could be explained.

THE CONFLICT HYPOTHESIS

So far the most elaborate postulate we possess with respect to the evolution of visual displays is the "conflict hypothesis" developed by Niko Tinbergen and several of his pupils (Tinbergen, 1952; Morris, 1955; Moynihan, 1955). It is based on the idea that many displays—and in particular those with agonistic functions—have evolved as a consequence of the hierarchical structure of the behavioural organisation. In this structure motivational systems subserving different functions and activated by internal and external stimulation are conceived as mutually inhibitory and thus competing for temporal control. Simultaneous activation of incompatible systems, in particular those controlling attack and escape, would lead to the following phenomena (see Figure 1):

a. Incomplete performance of activities controlled by these systems.

b. Ambivalence i.e. combination of (incomplete) elements of interacting systems. This may involve either the basic form of the activities or their orientation components, in which case the term 're-direction' is used.

c. Displacement i.e. the occurrence of an activity controlled by another system than those primarily activated in the internal conflict, which are thought to be mutually inhibiting one another.

Often, more than one of these principles appear to be involved in the production of a particular behaviour pattern (Baerends, 1975). Since activities generated in this way break the thread of a prevailing functional chain, Baerends (1970) has proposed to indicate them descriptively as "interruptive activities", a term which does not anticipate how they are caused or what function they might serve. According to Tinbergen (1952) these "interruptive activities" would have been a source for the evolution of communicative activities, because they tend to occur when opponents meet. In that situation functionally antagonistic systems are activated and selection pressure for communication, preventing damage from fighting, must exist. It was

DERIVED ACTIVITIES

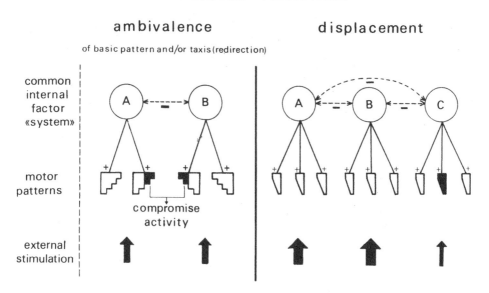

Figure 1 Schematic representation of the concepts of ambivalence and displacement. The incompatible systems A and B are simultaneously activated. In case of ambivalence, parts of model action patterns of A and of B may become combined in a "compromise activity", and/or the resulting activity may become oriented along the resultant of the orientations prevailing in each of the systems (redirection). In case of displacement the mutual interaction of A and B is conceived to weaken their inhibition on other systems. Which of these comes to expression depends upon the strength of the internal factor(s) feeding it and the presence of external triggers.

considered beneficial for both parties to exchange reliable information about the strength of these tendencies. However, with the rise of sociobiology, and especially with the application of game theory to functional thinking in ethology, the latter conclusion was seriously criticized (Maynard Smith, 1974). Moreover, attempts to test the ideas on the organisation of behaviour experimentally suffered from serious imperfections. By the end of the sixties, therefore, the "conflict hypothesis" had fallen into discredit (Baerends, 1975). Given the merits of the observations and ideas which had led to it, as well as absence of alternative constructs, we felt that a rehabilitation should be attempted, taking the criticism to heart and applying modern methods for data recording and processing. The displays of cichlid fish and of gulls were chosen as subjects; here I shall only deal with the results obtained in the former group. For both mouthbreeders and substrate spawners reproductive behaviour starts with the establishment of territories and pair formation begins with the intrusion of a fish into the territory of a conspecific. A causal ethological analysis of the procedures during such encounters was therefore attempted, using mouthbreeders as subjects because they proved easier to experiment with.

CAUSAL ANALYSES OF MOUTHBREEDER DISPLAYS—IDENTIFICATION OF BEHAVIOURAL SYSTEMS

The basis of our approach was laid by Nance Vodegel (1978a), who made a profound study of the causation of the behaviour patterns and the behavioural sequences with which male *Pseudotropheus zebra* respond to the entrance of a conspecific (male or female) in its territory. The various behaviour patterns that comprise this response (about 25 in all) could all be elicited by dummies as well as by real fish. Vodegel classified each of these activities into one of two classes: "dummy-associated" and "dummy-free" behaviour. Dummy-associated behaviour included the overtly aggressive Biting (Bi) and Snapping (Sn), as well as a number of stereotyped displays. Here we shall mainly deal with 5 of them, viz: Butting (Bu), Frontal Display (FD), Lateral Display (LD), Tailbeating (TB) and Vibrating (Vb). Statistical analysis of detailed records of the sequential relationships among all activities observed in the fish (in isolated pairs and in mixed communities) revealed a strong temporal association between Bi, Sn, Bu, FD, LD and TB, which all serve to repel intruders. In contrast, Vb was found to be associated with a different group of activities, used in attracting females. Vb is directed away from the intruder and towards the center of the territory where the eggs would be laid. "Dummy-free" behaviour includes neutral activities not oriented to the intruder, such as swimming, feeding and resting, and fear-indication ('fearful') activities, such as hiding in a shelter at the bottom of the territory and fleeing.

Although all these activities can follow each other in almost any order, transitions are far from random. Stochastically speaking a tendency for the occurrence of particular sequences can be recognized. (Figure 2, top). To study how factors internal and external to the fish determine the structure of these sequences, Vodegel quantified the responses of individual territorial males (8 cm in length) to dummies of different sizes. A small (S; 5.5 cm in length), a medium (M; 6.8 cm in length) and a large (L; 11 cm in length) dummy were presented to the test fish for 2 min, at intervals that were sufficient to exclude after-effects.

During dummy presentation the incidence of dummy-associated behaviour and of fearful activities increased in relation to dummy size (Figure 3). This suggests that the dummy stimulates two opposite tendencies in the test fish: one to approach the dummy and another to flee for it. As further illustrated in Figure 4, increase in dummy size induced changes in the relative frequency of dummy-associated behaviour patterns. Bodily contact with the intruder (represented by Bu) was replaced by distance-keeping displays (represented by FD + LD). The stronger a fish's fear responses, the more its displays shifted from FD to LD or Vb. In Figure 4 this is shown by the difference in the scores of FD between the individuals 5 and 6 when the large dummy was presented: fish 5, who showed a steeper rise in hiding as dummy size increased, responded mainly with LD to the large dummy, whereas fish 6 also showed FD.

These results led Vodegel to postulate that at each moment of the encounter the type of activity performed by the test fish is determined by the ratio between the tendencies to attack and to escape, i.e. the relative aggressiveness (RA). This hypothesis was inspired by a study of Neal Miller (1944) in which he measured in rats (by means of the force exerted on a restraining harness) the changes in strength of the tendencies to approach and to avoid with the distance from a spot at which both food and an electric shock were presented. Both tendencies gradually declined with

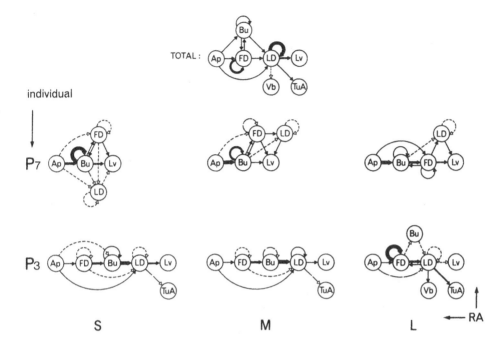

Figure 2 Pathway diagrams indicating the incidence of the transitions observed between different dummy-associated activities. The top diagram represents an average for all tests of all fishes, the second and third row refers to tests of two individual fishes (P7 and P3) with the small, the medium and the large dummy. The RA of the tests decreases from left to right. The RA of P7 is higher than that of P3; notice the difference in the activity most frequently following Ap. (Ap = approach, TuA = turning around, Lv = leaving the dummy). (After Vodegel, 1978a,b).

increasing distance from that spot. However, the gradient of avoidance was steeper than that of approach, so that they intersected (i.e. were in balance) at a particular distance from that spot. Upto that distance the rats approached and beyond it they withdrew. Results such as those pictured in Figure 4 showed that in Cichlids too the tendency to flee decreased more steeply with distance from a stimulus (the dummy) than the tendency to attack. The position of the intersection point depends on the ratio between these tendencies. Figure 5 shows that the different displays tend to occur within a particular range of distances from the intruder, this distance increasing in the order FD, LD, TB, Vb. In addition, orientation of these displays (i.e. the angle away from between the dummy and fish) increases in the same order.

Vodegel tested her hypothesis by investigating whether the values of all parameters measured in the tests were consistent with the regulatory process conceived. Furthermore, to compare the nature of the responses to dummies of different sizes, she calculated RA-values for each of the 2 min tests on the basis of the incidence of Bu, the time that elapsed between first approach and Bu, and the orientation of the fish at particular moments during the test. RA decreased with dummy size, and this decrease was associated with a rise of the incidence of FD and/ or LD. Differences in RA-values went together with consistent changes in the order

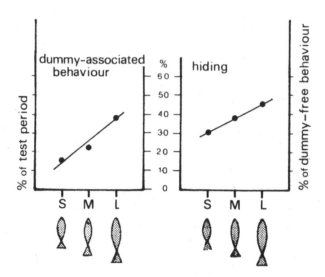

Figure 3 Change with increase of dummy size in the percentage of time of a test period spent on dummy-associated behaviour, and in the percentage of dummy-free behaviour spent hiding. (After Vodegel, 1978a,b).

in which the activities followed one another (Figure 2; compare P3 and P7). The larger the dummy, the lower the level of the RA of a test and the more activities were replaced by others of lower RA. For instance, at high RA values Bu would immediately follow approach, whereas at a lower RA-level Bu tended to be preceded by display.

Although the average RA-level of a test turned out to be a useful measure, RA is not constant throughout a test, as evidenced by the change over time in the kind of activities performed. Because the external situation is constant, this change must be due to motivational changes in the fish and/or changes in its position with respect to the dummy. At the beginning of a bout of dummy-associated behaviour, RA tended to increase with time, but later a decrease set in. Further, the bout length of an activity was found to be proportional to the RA-value of the immediately preceding act. Bout lengths of activities with a lower RA tended to be longer and those of a higher RA activities shorter. These changes seem to be a consequence partly of the performance of motor activity and partly of habituation to the dummy. Separate tests showed that the tendency to attack habituates at a faster rate than the tendency to escape.

On the basis of the results, Vodegel constructed a mathematical model describing the processes underlying the response of male *Pseudotropheus zebra* to intruders. The results of simulations using this model closely resembled the behaviour observed during real confrontations. Consequently we may conclude that the communicative activities and the ways in which they are used are indeed brought about through interaction of at least two conflicting agonistic tendencies.

Vodegel's work made it quite clear that it is essential to treat individuals separately. The fishes varied considerably in their average ratio of attack and escape, so a lumping of data from different fish would have blurred the picture. On the basis

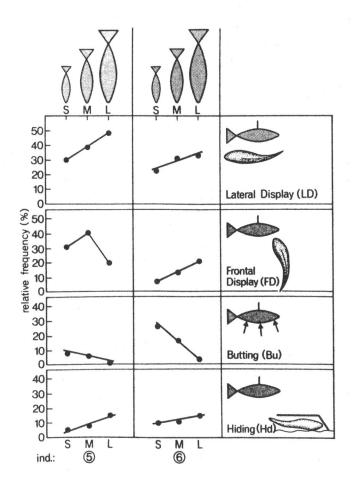

Figure 4 Change with increase of dummy size in the percentage taken up by butting and two display activities of the total time spent on dummy-associated behaviour, and (bottom part) on hiding as a percentage of dummy-free behaviour. (After Vodegel, 1978a,b).

of the RA-values of all tests with the same fish, an average RA-value for the individual could be computed. When all test fish were ranked in an order of decreasing individual RA-value (and so in a range from bold to shy), the activities performed showed a similar shift to that found when all tests of the same fish were ranked in an order of decreasing RA-value. The lower the RA-value of a fish (thus the shyer it was), the more time it spent in display as opposed to overt aggression.

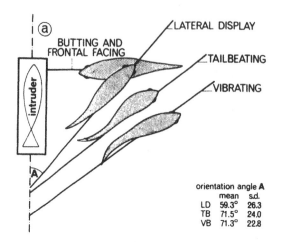

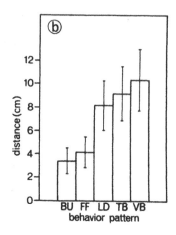

Figure 5 Five behaviour patterns performed by a resident fish towards an intruder (a), with the angle of orientation and the distances at which they are performed (b). (After Carlstead, 1983a).

CAUSAL ANALYSES OF MOUTHBREEDER DISPLAY—EXPERIMENTAL MANIPULATION OF BEHAVIOURAL SYSTEMS

To explore how the interaction between the antagonistic tendencies takes place, the effect of each must be considered separately. Following up Vodegel's study, Carlstead (1983a) attempted this by manipulating the escape factor. Using mouthbreeders of the closely related genus *Haplochromis*, she administered a variety of frightening stimuli such as a bright light or the sudden appearance of the experimenter. The test fish (14 in all) were kept singly in tanks divided into three compartments that offered opportunities for a broad spectrum of behaviour patterns. In the central compartment a standard intruder (a territorially coloured male, restrained in a glass tube) could be presented. In one adjoining compartment, a territorial neighbour was kept behind a transparent screen (except during tests) to ensure that the test fish kept its territorial motivation. The other adjoining compartment, provided the test fish with the possibility of interacting visually with a community of conspecifics. Two experiments were conducted. The first one was designed to assess changes in behaviour of the test fish in the presence of a standard intruder in response to exposure to bright light. In this experiment a 20 min period of exposure to the intruder was immediately followed by an equally long period with extra light. To control for possible changes due to time effects only, a parallel set of two consecutive 20 min tests was run in which the stimulus situation was kept unchanged. In both cases the experimenter was only visible to the fish at the beginning of the first 20 min period, when fixing the restrained intruder. The second experiment was designed to determine the effect of the restrained intruder on the tendency of the test fish to attack. To measure this, 10 juvenile *Tilapia mariae* were present as constant targets for attack throughout 3 different 20 min phases of this experiment, i.e. before the intruder was present, while it was present and after it had been removed. No extra light was given in these tests. The proceedings of all tests were video-taped and these records were analyzed, taking account of a great many behavioural variables and of the locations of the tank (test area, community area, shelter) visited by the test fish. To cope with the inter-individual variation in responsiveness, Carlstead developed a method that enabled her to assess the effects of the fright stimuli on each of 18 variable characteristics of the response. Besides approach behaviour these variables induced attentiveness to various constant stimuli in the environment, time spent in different areas of the test area, and absolute and relative amount of stereotyped behaviour. The data collected in each 20 min presentation of the intruder were subjected to principle component analysis. 77.4% of the total inter-individual variance was explained by 4 components, which could be designated as: 1) Interest in the Intruder, 2) Avoidance of the Intruder, 3) Shelter Attraction, and 4) Persistence (= duration and intensity) of the response. Each fish could therefore be characterized relative to the others on each of these 4 properties of the response. Various statistical methods were applied to search for associations between the scores on each component in successive periods and between individual scores on the 4 components made under different circumstances. On the basis of the relationships found and following the rule of parsimonious reasoning, a 4-factor model was constructed of the processes by which disturbing stimulation influences the response of the resident to a territorial intruder (Figure 6). The structure of the model is mainly based upon the following arguments:

1. The changes in persistence to the intruder and to the little fish, as well as in orientation directed at the intruder, when the bright light was turned on were associated. However, these did not co-vary with the persistence of response to

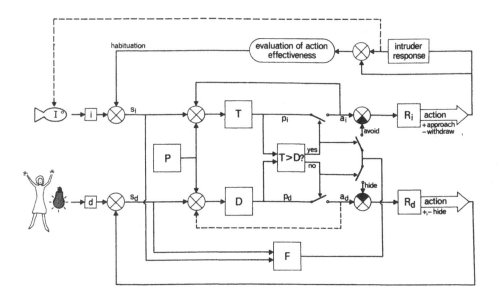

Figure 6 Carlstead's (1983a) model of the behavioural organization underlying responses of a residential cichlid fish to an intruder in its territory and to two types of frightening stimuli. Explanation in text.

disturbing stimulation, which instead was associated with changes in the Shelter Attraction component, which increased with the bright light. Persistent responding to the light or to the appearance of the experimenter was negatively related to persistent responding to the intruder, and changes in the frequency of hiding were inversely related to changes in the frequency of attacks at the little fish. Since the *degree* of these changes was not correlated, it was concluded that different internal factors provided for their control. Consequently the model is primarily based on two motivational units. One of them (T) is thought to be selectively responsive to intruders into the territory, and the other (D) to potentially dangerous disturbances. The evidence suggests that response to either class of stimuli excludes response to the other. Thus T and D are considered to represent competing systems for the control of selective attention. At any moment the system with the highest level is thought to dominate, i.e. to determine the response.

2. Turning on the bright light had no consistent effect on Interest in the Intruder. However, it caused a rise in the Avoidance scores, reflected in hesitancy to approach, staying away from the intruder and an increase in displays oriented away from it (in particular vibrating; Vb). In addition, the extra light, increased the scores for Shelter Attraction. Appearance of the experimenter and the presence of the restrained intruder had a similar effect. Individual scores for Avoidance in the second period of the first experiment (with as well as without the bright light) were positively related to scores on the Shelter Attraction component in the preceding period. A third motivational factor must therefore be involved, influencing both Shelter Attraction and Avoidance of the Intruder. This factor can be taken to represent individual variation in response to any external disturbing stimulation and was consequently called "fear" (F). Suggestions about the way in which F affects the behaviour have to take into account the lack of correlation between Shelter Attraction and Avoidance of the Intruder. In the model this is solved by allowing the input of F to be channelled to either the T or the D System, depending on which of these competing systems is dominating. Integration of the outputs of these two systems determines the qualitative and quantitative aspects of the behaviour of the fish. If T is dominant, F reduces direct contacts with the intruder and increases displays, particularly those directed away from the intruder (e.g. Vb). If D is dominant it controls the persistence with which the fish stays in the shelter, but the *frequency* of hiding and other escape behaviours is increased by raising the level of F.

3. The Persistence of the responses to all novel stimuli (the intruder, the juvenile fish, the bright light or the experimenter) co-varies within individuals. This can be accounted for by postulating one internal factor that varies considerably between individuals (unit P in Figure 6). T and D receive similar output from P, which affects the duration and level (persistence) of both systems. Since orientation towards the intruder increased persistent responding to it as well as to the little fish, a positive feedback from the factors determining direct orientation to T has been included in the model. Negative feedback loops reflect habituation to the unresponsive restrained intruder and to longer lasting disturbances. We may conclude that Carlstead's model gives further substance to the basic idea of the "conflict hypothesis". In addition it shows how two important individual characteristics of a fish, viz. persistence and fear, can be conceived as affecting ongoing behaviour.

ECOLOGY AND THE ORGANISATION OF MOUTHBREEDER DISPLAYS

Subsequently Carlstead (1983b) has seen whether the mechanism represented by the model can also explain behavioural differences between related species. To this end she compared males of 3 *Haplochromis* species (*angustifrons, elegans* and *squamipinnis*) in their responses to a restrained rival and a free swimming female. During all tests the experimenter, sitting in front of the tank in full view of the test fish, featured as a constant disturbing stimulus situation.

Various inter-specific differences of a quantitative nature were found. For instance, *H. angustifrons* males spent less time with the intruder than did males of the other two species. They left their shelter to approach the intruder less frequently (and on average after a longer latency) and they also spent more time watching the observer. These observations indicate that in this species the activation of system D tends to exceed that of T. In *H. elegans* the number of butts per butting bout increased during test sessions. This was attributed to the positive feedback from orientation towards the intruder and suggests that in *elegans* T is pitched at a higher level than D. In *H. squamipinnis* hiding bouts lasted longer on average than in the other two species. Neither the number of butts per butting bout nor the duration of hiding bouts changed during an intruder session; in this species only these two variables were found correlated. Thus *squamipinnis* appears to have the highest response persistence (factor P) of the 3 species.

Examinations of the response sequences led to similar conclusions. For instance, in *H. angustifrons* "turning around" (TuA)—(a behaviour suitable as a reference point because it indicates that the tendencies to approach and to leave the intruder are in balance) tends to be followed by tailbeating (TB), an act associated with interest in the shelter. This is consistent with a relatively high activity in the D system in this species. In *elegans*, TuA was followed by vibration more often than expected, but in *squamipinnis* the subsequent act was usually lateral display (LD). In the model both acts result from interaction between T and F, depending on the relative strength of these 2 influences. Carlstead suggests that in *squamipinnis* the shift towards LD is caused by the greater strength of factor P in that species. This interpretation is supported by the finding that in this species the incidence of LD was positively correlated with the (P-controlled) duration of hiding.

The same inter-specific differences were found in the responses towards females, but in this case with respect to the spawning act, during which the tendencies to approach and to withdraw can be considered to be in balance too. In *angustifrons* spawning was mainly connected with TB, in elegans with Vb and in *squamipinnis* most strongly with LD.

These inter-specific differences in the relative impact of particular components of their behavioural organisation can be understood functionally in terms of differences in the ecological niches that the fish occupy. All three species live in Lake George (Uganda). *H. angustifrons* inhabits inshore regions and feeds on benthos in flocculent mud, where visibility is poor and alertness to disturbances is required. Here a high responsiveness to various disturbances (D) and absence of long lasting displays (LD) seems adaptive. *H. elegans* is restricted to areas along the shores, a variegated environment with good visibility and plenty of shelter. Insect larvae are the main food source. Intraspecific competition may well be severe in this species and this accounts for the high responsiveness of system T. In *H. squamipinnis*, a predator on fish, the emphasis on persistence (P) may have been selected to ensure consistent pursuit of selected prey-fish within a shoal.

THE MOTIVATIONAL BASIS OF ADAPTIVE MATE CHOICE

The extent to which these inter-specific differences are actually used by the fish in the recognition of conspecifics has not yet been investigated experimentally. The tests needed for such a study are difficult to design, the more so because sensory modalities other than vision must be taken care of.

Mouthbreeders

Most mouthbreeders are sexually dimorphic. Resident males on the lek show a species-characteristic nuptial coloration, in contrast to the rather unspecific dull coloration of the females. Only the males behave aggressively and perform the displays described above; females remain passive or flee. Observations of male/female interactions suggest that a male accepts a female as long as she parries rather than reciprocates his initial attacks, which only a conspecific female ready to spawn, will do. In response to this passivity, the male increasingly shifts to displays oriented away from her and directed towards the spawning site (TuA, Vb). A ripe female responds to this by gradually following the soliciting male over greater distances, until she finally stays with him and joins him in cleaning the spawning site. This latter phase does not last long. Soon another activity, "Skimming" over the substrate begins to dominate; thus finally changes into real spawning. Thus, a male obtains a ripe conspecific female because only that class of females is able to respond correctly to the questions asked with his displays. It seems likely that the females can assess the quality of males from details in their display behaviour.

Substrate Spawners

In most substrate spawning cichlids, sexual dimorphism in colour and behaviour is slight or even absent. In such fish too territorial behaviour (which can occur in both sexes) forms the first phase of pair formation though there are indications that in some species pairs can also be formed in shoals. Female territories tend to be strictly defined, whereas the males claim extensive home ranges, encompassing several female residences. Attempts by a male to enter a female's territory are usually met with agonistic displays, homologous to those described for male mouthbreeders, but now performed similarly by both sexes. In addition to parrying the aggressive approaches of the male, the female probably gives female-typical signals. The agonistic behaviour may either escalate to a fight (ending by flight of one of the fishes) or it may subside—often quite quickly. In the latter case, a mate of the appropriate species and sex has been selected but, in contrast to the mouthbreeders, spawning is not yet imminent. For days rather than hours (as in the mouthbreeders) the partners repeatedly and often simultaneously execute various forms of display that are no longer directed towards the other fish but towards the neutral substrate. These displays strongly resemble the beginning of digging or cleaning the substrate. However, these "mouthing" acts occur far more often than is needed for this purpose and are mostly too incomplete to be effective. As to their causation, a study of the pre-spawning behaviour of *Aequidens portalegrensis* (Baerends, 1983) has shown that at the time this "interruptive behaviour" is shown, the tendencies to attack and to escape are still activated, though their level has decreased during the period of agonistic display, whilst the tendency to spawn is rising. Thus the "mouthing" displays can be as seen as originating from a disinhibition of feeding activities, due to the interactions between the three motivational systems (displacement). Body quivering displays can be recognized as incomplete digging, while some of the nest-digging activities can be seen as re-directed attacks.

Thus in substrate spawners and in mouthbreeders the agonistic displays that follow territory intrusion allow fish to select mates of the right species and sex. The subsequent digging and cleaning phase, though found in both strategies, is more elaborate and longer in the substrate spawners. In *Ae. portalegrensis* Baerends (1984) found evidence that the performance of these displays contributes to the synchronisation of the endocrine states of the partners. In contrast, in mouthbreeders the first phase of the encounter effectively ensures this synchronisation. This raises the question of whether the early weak bonding of substrate spawner partners has another function, namely an assessment of personal qualities of a potential mate. Evidence for this is the fact (well known to aquarists) that in order to obtain good breeding pairs it is best to allow the fish to make their own choice from a group, rather than arbitrarily putting male and female together. In *Tilapia zillii* 60% of naturally formed pairs bred successfully, whereas all forced pairs failed at an early stage (Hulscher-Emeis, 1991, pers. comm.). Further the members of such an incompatible combination frequently form successful pairs with other individuals at a later date. In substrate spawners, partners often stay together for a long time, and produce many broods. After having been separated for periods of several months, members of such successful pairs rapidly rejoined one another when brought together again within a group of conspecifics (Baerends, unpubl. data).

Failing pairs are usually characterised by relatively high levels of aggression between the partners. If spawning takes place in spite of this, the eggs or young larvae are eaten, generally by the male first. In successful pairs the female tends to dominate the situation near the brood (even if the male regularly takes part in fanning), whereas the male is vigilant and dominates in the boundary area. In this context, aggressiveness in the male is advantageous to the female, but it must not be so high that she cannot control him near the brood. As we have seen in Vodegel's work, the relative aggressiveness of individuals varies within the population. We have the impression that the RA-ranges of both sexes largely overlap, with males having somewhat higher values than females. However, in successful pairs the RA of the female is always lower than that of the male (the same holds for size, which is strongly correlated with aggressiveness). Therefore, information about the relative aggressiveness of a potential partner is likely to be important in pair formation. The proximate causation of the displays allows the fish to obtain this information. In Carlstead's model the balance between boldness and shyness (and thus RA) reflects the state of unit F and the weighing of T against D, which can potentially be deduced from the behaviour of fish of either sex.

EXCHANGE OF INFORMATION DURING TERRITORIAL FIGHTS

The early suggestion of game theorists (e.g. Maynard Smith, 1974) that it would not be functional for an animal to be honest about the level of its tendencies to attack and to flee during a fight does not apply to displays shown by cichlid fish in response to territorial intruders, because of the multiple function of such displays. These serve not only to repel rivals but also to attract suitable partners, so their evolution must have been influenced by different selection pressures. This probably applies to many other kinds of animal and particularly to monogamous species with biparental care. The early models of game theorists did not provide for this, though more recent models do predict an exchange of information about intentions in a territorial context (Maynard Smith and Riechert, 1984). Interacting cichlids show not only a range of species-typical movements but also a variety of species-typical colour patterns of body and fins. Most species can show one pattern of melanistic transverse bars and another of longitudinal bands. The conspicuousness of both can be varied by changing the intensity of the black areas and by altering the (lighter) background. It has been suggested repeatedly in the literature that these different patterns indicate the extent to which a behavioural system for aggression and one for escape, respectively, are activated. As the interpretations of different authors were not consistent, Hulscher-Emeis (1991) undertook, with *Tilapia zillii* as subject, an intensive study of the occurrence of the different melanistic patterns throughout the life of these fish and in a wide range of circumstances. She came to the conclusion that the tendency to show transverse bars is controlled by motivation to stay put in a restricted area near to the bottom, whereas the tendency to show longitudinal bands is associated with moving about higher up in the water. In both cases, the melanistic markings only appear if the fish is in some way thwarted in its ongoing behaviour. The author suggests that hormonal factors typical of stressful situations cause the blackening of which ever marking pattern is appropriate for the fish's current situation. The vertical bars become more intense as the tendency of the fish to stay put increases, i.e. the more valuable its ties with its present location. Fish with free-swimming fry show a more intensive blackening than fish with less developed young (wrigglers or eggs) or fish occupying a territory without a brood. Longitudinal bands are maximally black in parents that have left the territory with their fry.

Although she has not tested this experimentally, Hulscher-Emeis is of the opinion that the marking patterns are used as a threat display. This is why they were erroneously considered to be controlled by a tendency to attack. A vigorously attacking male, however, does not show them until the end of a fight, when it is forced into the defensive. The evidence strongly suggests that the variable transverse and the longitudinal bars have primarily evolved for the function of camouflage and have secondarily evolved to subserve a threat function. In this latter process they remained under control of higher order control systems determining an appetite for either settling in a restricted locality or moving about.

CONCLUSION

The displays of *Cichlidae*, which serve in pair formation and defence against rivals, have probably evolved from "interruptive behaviour" caused by interaction of incompatible systems in the behavioural organisation of these fish, in particular those underlying the tendencies to attack, to escape and to spawn. Although the displays have become ritualized, the causal ties with these systems have not been lost, their existence being evident in many aspects of the behavioural interactions of these fish. The spectrum of interruptive activities developed in the family ranges from activities mainly caused by ambivalence to those mainly caused by the displacement phenomenon. Homologues of all types can be found in each species. But in mouthbreeders ambivalent displays predominate, while in substrate spawners displays originating from displacement acts are more common. This difference is likely to be related to the biparental care of the offspring practised by most substrate spawners, which requires an assessment of the "personality" of the mate. The variable marking patterns exhibited during encounters originate from cryptic patterns. The type of pattern is controlled by factors determining whether a fish has an appetite for staying in the territory or for moving about. Their intensity is related to current stress factors impinging upon the fish.

References

Baerends, G. P. (1970). A model of the functional organization of incubation behaviour. In: *The Herring Gull and its Egg: Part I* (Ed. by G. P. Baerends and R. H. Drent). *Behaviour* Suppl., **17**, 265–292.

Baerends, G. P. (1975). An evaluation of the conflict hypothesis as an explanatory principle for the evolution of displays. In: *Function and Evolution of Behaviour* (Ed. by G. Baerends, C. Beer and A. Manning), pp. 187–227. Oxford: Clarendon Press.

Baerends, G. P. (1984). The organization of the pre-spawning behaviour in the cichlid fish *Aequidens portalegrensis* (Hensel). *Neth. J. Zool.*, **34**, 233–366.

Baerends, G. P. (1986). On causation and function of the pre-spawning behaviour of cichlid fish. *J. Fish. Biol.*, **29**, Suppl. A, 107–122.

Baerends, G. P. and Baerends-Van Roon, J. M. (1950). An introduction to the study of the ethology of cichlid fishes. *Behaviour, Suppl.*, **1**, 1–242.

Barlow, G. W. 1991. Mating systems among cichlid fishes. In: *Cichlid Fishes: Behaviour, Ecology and Evolution* (Ed. by M. H. A. Keenleyside), pp. 173–190. London: Chapman and Hall.

Carlstead, M. K. (1983a). The behavioural organization of responses to territorial intruders and frightening stimuli in cichlid fish (*Haplochromis* spp.). *Behaviour*, **83**, 18–68.

Carlstead, M. K. (1983b). Influences of motivations on display divergences in three cichlid fish species (*Haplochromis*). *Behaviour*, **83**, 205–228.

Hulscher-Emeis, T. M. (1991). Causal ethological analysis of the display of variable colour patterns in *Tilapia zillii*. Thesis Univ. of Groningen, the Netherlands. pp. 1–131.

Maynard Smith, J. (1974). The theory of games and the evolution of animal conflicts. *J. Theor. Biol.*, **47**, 209–221.

Maynard Smith, J. and Riechert, S. E. (1984). A conflicting tendency model of spider agonistic behaviour: Hybrid–pure population line comparisons. *Anim. Behav.*, **32**, 564–578.

Miller, N. E. (1944). Experimental studies of conflict behaviour. In: *Personality and Behaviour Disorders* (Ed. by J. McV. Hunt), pp. 431–465. New York: Ronald Press.

Morris, D. (1956). The function and causation of courtship ceremonies. In: *L'instinct dans le Comportement des Animaux et de L'homme*, pp. 262–284. Paris: Masson.

Moynihan, M. (1955). Some aspects of reproductive behaviour in the black-headed gull (*Larus r. ridibundus* L.) and related species. *Behaviour, Suppl.*, **4**, 1–201.

Tinbergen, N. (1952). Derived activities: their causation, biological significance, origin and emancipation during evolution. *Quart. Rev. Biol.*, **27**, 1–32.

Vodegel, N. (1978a). A study of the underlying motivation of some communicative behaviours of *Pseudotropheus zebra* (Pisces: Cichlidae): a mathematical model. *Proc. Kon. Ned. Akad. Wet. Ser.*, **C 81**, 211–240.

Vodegel, N. (1978b). A causal analysis of the behaviour of *Pseudotropheus zebra*. Thesis, Univ. of Groningen, The Netherlands, pp. 1–223.

SPECIES LIST

INDEX

Printed and bound by CPI Group (UK) Ltd, Croydon, CR0 4YY

23/10/2024

01778230-0010